Methods in Molecular Biology™

Series Editor
John M. Walker
School of Life Sciences
University of Hertfordshire
Hatfield, Hertfordshire, AL10 9AB, UK

For further volumes:
http://www.springer.com/series/7651

Plant Lipid Signaling Protocols

Edited by

Teun Munnik

Swammerdam Institute for Life Sciences, Section Plant Physiology, University of Amsterdam, Amsterdam, The Netherlands

Ingo Heilmann

Department of Cellular Biochemistry, Institute for Biochemistry and Biotechnology, Martin-Luther-University Halle-Wittenberg, Halle, Germany

Editors
Teun Munnik
Swammerdam Institute for Life Sciences
Section Plant Physiology
University of Amsterdam
Amsterdam, The Netherlands

Ingo Heilmann
Department of Cellular Biochemistry
Institute for Biochemistry and Biotechnology
Martin-Luther-University Halle-Wittenberg
Halle, Germany

ISSN 1064-3745 ISSN 1940-6029 (electronic)
ISBN 978-1-4939-5927-3 ISBN 978-1-62703-401-2 (eBook)
DOI 10.1007/978-1-62703-401-2
Springer New York Heidelberg Dordrecht London

Printed on acid-free paper

Humana Press is a brand of Springer
Springer is part of Springer Science+Business Media (www.springer.com)

Preface

Eukaryotic cells are surrounded by membranes consisting of various lipids, including sterols, sphingolipids, glycolipids, and phospholipids. Besides structural functions, membranes also contain lipids with regulatory and signaling roles. Such lipids include polyphosphoinositides, the low-abundant derivatives of phosphatidylinositol, phosphatidic acid, certain sphingolipids, *N*-acyl ethanolamines, and oxylipins, oxidized derivatives of C18-fatty acids; all these are emerging as important players regulating various physiological processes in plants.

As we are beginning to understand the complexity of lipid signaling and its roles in plant biology, there is an increasing interest in their analysis. However, due to the low abundance and transient nature of some of these hydrophobic compounds, this is not always easy. The difficulty in analyzing signaling lipids arises from some unique analytical challenges. For instance, the structural similarity of polyphosphoinositide species and of derived soluble inositol polyphosphates must be taken into account when using "specific" in vivo markers or when assaying relevant enzyme activities. Furthermore, plants exhibit greater metabolic complexity in comparison to animals or yeast, and the detection of low levels of signaling molecules requires specialized technology. Together, these considerations have created a need for novel means to identify and quantify signaling lipids and have raised new questions regarding the biochemical properties of enzymes involved in their conversion.

This book is dedicated to the various experimental approaches by which plant signaling lipids can be studied. Experts in the field have contributed their knowledge and extensive experience to provide a collection of 27 chapters on the analysis of plant signaling lipids, including detailed protocols to detect various relevant compounds by targeted or nontargeted approaches; to assay relevant enzyme activities in biological material or using recombinant enzymes; to test for specific binding of signaling lipids to protein partners; or to visualize signaling lipids or lipid-derived signals in living plant cells. On the way, a broad spectrum of methods is detailed, such as radiolabeling, diverse chromatographic methods (TLC, HPLC, GC), mass spectrometry, nuclear magnetic resonance spectroscopy, and confocal imaging. By looking left and right at signaling events related to lipids, key protocols for the analysis of other important players are also covered, such as diacylglycerol, galactolipids, and Ca^{2+}. All methods described have been developed or optimized in particular for the use in plants and are described including proper controls and notes for meaningful interpretation to ensure that tools and assays are used correctly and to their full potential. Tools presented often provide alternative options for relevant analyses, either requiring extensive instrumentation or being achievable with the most basic of laboratory equipment, thus helping interested researchers to select methods suitable for their questions and equipment available.

Research groups around the world are pursuing an increasing number of questions about functions of plant lipid signals, including adaptation to environmental stresses, guard cell functioning, vesicle trafficking, or control in development, such as cell polarity and pattern formation. Paying tribute to its emerging complexity, we are convinced that plant lipid signaling is best approached by multiple parallel methods, combining key biochemical tests and cell biological analyses described in this book. It is therefore timely and important to present plant researchers with this full set of tools needed to elucidate the particular roles of signaling lipids in plants. We hope that this collection of current protocols will aid plant researchers in their endeavor to elucidate the roles of lipid signals, and we are eager to see these methods put to use towards exciting new discoveries.

Amsterdam, The Netherlands ***Teun Munnik***
Halle, Germany ***Ingo Heilmann***

Contents

Contributors

STEVEN A. ARISZ • *Swammerdam Institute for Life Sciences, Section Plant Physiology, University of Amsterdam, Amsterdam, The Netherlands*

WENDY F. BOSS • *Department of Plant Biology, North Carolina State University, Raleigh, NC, USA*

IRENA BRGLEZ • *Department of Plant Biology, North Carolina State University, Raleigh, NC, USA*

KENT D. CHAPMAN • *Department of Biological Sciences, Center for Plant Lipid Research, University of North Texas, Denton, TX, USA*

FRANÇOISE COCHET • *Physiologie Cellulaire et Moléculaire des Plantes, Université Pierre et Marie Curie, EAC 7180 CNRS, Paris, France*

SYLVIE COLLIN • *Physiologie Cellulaire et Moléculaire des Plantes, Université Pierre et Marie Curie, EAC 7180 CNRS, Paris, France*

CATHERINE DIECK • *Department of Genetics, North Carolina State University, Raleigh, NC, USA*

ISABEL DOMBRINK • *Institute of Molecular Physiology and Biotechnology of Plants (IMBIO), University of Bonn, Bonn, Germany*

JANET L. DONAHUE • *Virginia Tech, Blacksburg, VA, USA*

PETER DÖRMANN • *Institute of Molecular Physiology and Biotechnology of Plants (IMBIO), University of Bonn, Bonn, Germany*

MUSTAFA ERCETIN • *Virginia Tech, Blacksburg, VA, USA*

LIONEL FAURE • *Department of Biological Sciences, Center for Plant Lipid Research, University of North Texas, Denton, TX, USA*

AGNES FEKETE • *Pharmaceutical Biology, Julius-von-Sachs-Institute for Biosciences, University of Wuerzburg, Wuerzburg, Germany*

GLENDA E. GILLASPY • *Virginia Tech, Blacksburg, VA, USA*

SIMON GILROY • *Department of Botany, University of Wisconsin, Madison, WI, USA*

ZACHARY T. GRABER • *Chemistry Department, Kent State University, Kent, OH, USA*

INGO HEILMANN • *Department of Cellular Biochemistry, Institute for Biochemistry and Biotechnology, Martin-Luther-University Halle-Wittenberg, Halle, Germany*

MAREIKE HEILMANN • *Department of Cellular Biochemistry, Institute for Biochemistry and Biotechnology, Martin-Luther-University Halle-Wittenberg, Halle, Germany*

MICHAEL HEINZE • *Lab of Molecular Cell Biology, Department of Pharmaceutical Biology, Institute of Pharmacy, Martin-Luther-University Halle-Wittenberg, Halle, Germany*

S.M. TERESA HERNÁNDEZ-SOTOMAYOR • *Unidad de Bioquímica y Biología Molecular de Plantas, Centro de Investigación Científica de Yucatán (CICY), Merida, Yucatan, Mexico*

GEORG HÖLZL • *Institute of Molecular Physiology and Biotechnology of Plants (IMBIO), University of Bonn, Bonn, Germany*

YANG JU IM • *Department of Plant Biology, North Carolina State University, Raleigh, NC, USA*
MAGDALENA M. JULKOWSKA • *Swammerdam Institute for Life Sciences, Section Plant Physiology, University of Amsterdam, Amsterdam, The Netherlands*
SANG-CHUL KIM • *Donald Danforth Plant Science Center, St Louis, MO, USA; Department of Biology, University of Missouri at St Louis, St Louis, MO, USA*
EDGAR E. KOOIJMAN • *Department of Biological Sciences, Kent State University, Kent, OH, USA*
MARKUS KRISCHKE • *Pharmaceutical Biology, Julius-von-Sachs-Institute for Biosciences, University of Wuerzburg, Wuerzburg, Germany*
ANA M. LAXALT • *IIB-CONICET-UNMdP, Universidad Nacional de Mar del Plata, Mar del Plata, Argentina*
GERALD H. LUSHINGTON • *K-INBRE Bioinformatics Core Facility, University of Kansas, Lawrence, KS, USA*
SHANTAN REDDY MAREPALLY • *K-INBRE Bioinformatics Core Facility, University of Kansas, Lawrence, KS, USA*
JENNIFER E. MARKHAM • *Department of Biochemistry, University of Nebraska-Lincoln, Lincoln, NE, USA*
JAN MARTINEC • *Institute of Experimental Botany, Academy of Sciences of the Czech Republic, Prague, Czech Republic*
FIONN MCLOUGHLIN • *Swammerdam Institute for Life Sciences, Section Plant Physiology, University of Amsterdam, Amsterdam, The Netherlands*
MARTIN J. MUELLER • *Pharmaceutical Biology, Julius-von-Sachs-Institute for Biosciences, University of Wuerzburg, Wuerzburg, Germany*
TEUN MUNNIK • *Swammerdam Institute for Life Sciences, Section Plant Physiology, University of Amsterdam, Amsterdam, The Netherlands*
J. ARMANDO MUÑOZ-SANCHEZ • *Unidad de Bioquímica y Biología Molecular de Plantas, Centro de Investigación Científica de Yucatán (CICY), Merida, Yucatan, Mexico*
YUKI NAKAMURA • *Institute of Plant and Microbial Biology, Academia Sinica, Taipei, Taiwan, R.O.C*
DAYA SAGAR NUNE • *K-INBRE Bioinformatics Core Facility, University of Kansas, Lawrence, KS, USA*
KIRK L. PAPPAN • *Metabolon, Inc., Durham, NC, USA*
PŘEMYSL PEJCHAR • *Institute of Experimental Botany, Academy of Sciences of the CzechRepublic, Prague, Czech Republic*
IMARA Y. PERERA • *Department of Plant Biology, North Carolina State University, Raleigh, NC, USA*
JOHANNA M. RANKENBERG • *Department of Biological Sciences, Kent State University, Kent OH, USA*
WERNER ROOS • *Lab of Molecular Cell Biology, Department of Pharmaceutical Biology, Institute of Pharmacy, Martin-Luther-University Halle-Wittenberg, Halle, Germany*
MARY R. ROTH • *Division of Biology, Kansas Lipidomics Research Center, Kansas State University, Manhattan, KS, USA*
GÜNTHER F.E. SCHERER • *Section of Applied Molecular Physiology, Institute of Floriculture and Wood Science, Leibniz University of Hannover, Hannover, Germany*

SUNITHA SHIVA • *Division of Biology, Kansas Lipidomics Research Center, Kansas State University, Manhattan, KS, USA*
NADJA STINGL • *Pharmaceutical Biology, Julius-von-Sachs-Institute for Biosciences, University of Wuerzburg, Wuerzburg, Germany*
SARAH J. SWANSON • *Department of Botany, University of Wisconsin, Madison, WI, USA*
CHRISTA TESTERINK • *Swammerdam Institute for Life Sciences, Section Plant Physiology, University of Amsterdam, Amsterdam, The Netherlands*
JOOP E.M. VERMEER • *Department of Plant Molecular Biology, University of Lausanne, Lausanne, Switzerland*
MAHESH VISVANATHAN • *K-INBRE Bioinformatics Core Facility, University of Kansas, Lawrence, KS, USA*
KATHARINA VOM DORP • *Institute of Molecular Physiology and Biotechnology of Plants (IMBIO), University of Bonn, Bonn, Germany*
HIEU SY VU • *Division of Biology, Kansas Lipidomics Research Center, Kansas State University, Manhattan, KS, USA*
XUEMIN WANG • *Department of Biology, Donald Danforth Plant Science Center, University of Missouri, St Louis, MO, USA*
RUTH WELTI • *Division of Biology, Kansas Lipidomics Research Center, Kansas State University, Manhattan, KS, USA*
VERA WEWER • *Institute of Molecular Physiology and Biotechnology of Plants (IMBIO), University of Bonn, Bonn, Germany*
MAGDALENA WIERZCHOWIECKA • *Department of Molecular and Cellular Biology, Institute of Molecular Biology and Biotechnology, Adam Mickiewicz University, Poznan, Poland*
XAVIER ZARZA • *Swammerdam Institute for Life Sciences, Section Plant Physiology, University of Amsterdam, Amsterdam, The Netherlands*
ZHENGUO ZHOU • *K-INBRE Bioinformatics Core Facility, University of Kansas, Lawrence, KS, USA*

Part I

Lipid Analyses

Chapter 1

Analyzing Plant Signaling Phospholipids Through $^{32}P_i$-Labeling and TLC

Teun Munnik and Xavier Zarza

Abstract

Lipidomic analyses through LC-, GC-, and ESI-MS/MS can detect numerous lipid species based on headgroup and fatty acid compositions but usually miss the minor phospholipids involved in cell signaling because of their low chemical abundancy. Due to their high turnover, these signaling lipids are, however, readily picked up by labeling plant material with ^{32}P-orthophosphate and subsequent analysis of the lipid extracts by thin layer chromatography. Here, protocols are described for suspension-cultured tobacco BY-2 cells, young *Arabidopsis* seedlings, *Vicia faba* roots, and *Arabidopsis* leaf disks, which can easily be modified for other plant species and tissues.

Key words Polyphosphoinositides, Phosphatidic acid, Diacylglycerol pyrophosphate, Thin layer chromatography, Isotopic radiolabeling

1 Introduction

The concentration of signaling lipids, such as polyphosphoinositides (PPI), phosphatidic acid (PA), and diacylglycerolpyrophosphate (DGPP), is typically too low to be detected by modern, lipidomic-type analyses such as GC- and LC-MS (1–9). PA is sometimes detectable, as it is also the precursor of phospholipids and galactolipids (1, 2). In addition, it can be massively produced upon PLD hydrolysis of structural phospholipids as an artifact of preparation and, with biological relevance, in response to stresses, such as wounding and freezing (10–12). The PA pool, however, that would be generated upon PPI hydrolysis by phospholipase C (PLC) and phosphorylation of the resulting diacylglycerol (DAG), is much smaller and easily overlooked (2, 13, 14).

A main difference between structural and signaling lipids is their rate of synthesis and breakdown (i.e., turnover), which in signaling lipids is much higher (6, 13, 15). The benefit of this is that these lipids are therefore faster labeled by inorganic, radioactive-labeled phosphate ($^{32}P_i$ or $^{33}P_i$). This type of label is easily

Teun Munnik and Ingo Heilmann (eds.), *Plant Lipid Signaling Protocols*, Methods in Molecular Biology, vol. 1009,
DOI 10.1007/978-1-62703-401-2_1, © Springer Science+Business Media, LLC 2013

taken up by various plant tissues and is rapidly incorporated into the ATP pools, which are required to phosphorylate the precursors of signaling lipids. Structural lipids also incorporate $^{32}P_i$ but this occurs through a relatively slower pathway involving their de novo biosynthesis (6, 13, 15). Here, a quick and easy protocol is described to analyze signaling lipids like PA, DGPP, and PPIs in different plant tissues, i.e., seedlings, leaves, roots, and suspension-cultured cells.

2 Materials

2.1 Plant Media

1. Medium for cell suspensions: For 1 L, 4.4 g Murashige–Skoog (MS) salts with vitamins, 30 g sucrose, 100 μL BAP (6-benzylaminopurine, 10 mM), 100 μl NAA (naphthalene-acetic acid, 54 mM in 70 % EtOH). Adjust pH to 5.8 with KOH. Autoclave 20 min at 120 °C.
2. Medium for *Arabidopsis* seedlings: 0.5× MS medium, supplemented 1 % (w/v) sucrose and 1 % (w/v) agar.
3. Medium for *Vicia* seedlings: 2.72 mM $CaCl_2$, 1.95 mM $MgSO_4$, 2.20 mM KH_2PO_4, 1.26 mM Na_2HPO_4, and 0.08 mM ferric citrate.
4. Phosphate-free labeling medium for *Vicia*: 2.72 mM $CaCl_2$, 1.95 mM $MgSO_4$, 0.08 mM ferric citrate, 10 mM HEPES, pH 6.5.
5. Fertilized pot soil to grow mature plants.

2.2 Plant Material and Cultivation

1. Suspension-cultured plant cells (tobacco BY-2; 4–5-days old (*see* **Note 1**)).
2. Rotary shaker, 125 rpm, 24 °C, in the dark.
3. Seedlings (*Arabidopsis thaliana*, Col-0); ~5-days old.
4. Seedlings *Vicia* (*Vicia sativa* spp. *nigra*); 2–3-days old.
5. Growth chamber, 21 °C, 16 h light, 8 h dark.
6. Mature *Arabidopsis* plants (~3 weeks) for leaf disks.
7. Leaf disk cutter: cork borer (0.5 cm diameter).

2.3 Labeling Components

1. ^{32}P-inorganic phosphate ($^{32}P_i$; carrier-free, 10 μCi/μL).
2. 2 mL "Safe-lock" Eppendorf polypropylene reaction tubes.
3. Three sterile 15 mL tubes.
4. 5 μm, 0.45 μm, and 0.22 μm filters for 10 mL syringes.
5. Labeling buffer for seedlings and leaf disks: 2.5 mM MES-KOH (pH 5.8), 1 mM KCl.
6. Labeling buffer for cell suspensions: CFM (cell-free medium, 0.22 μm filtered).

7. Fume hood.
8. Safety glasses.
9. Safety screen, 1 cm Perspex.
10. Perspex tube holders.
11. Gloves.

2.4 Lipid Extraction Components (see Note 2)

1. Perchloric acid (PCA): 50 % and 5 % (w/v) in H_2O.
2. CMH: chloroform/methanol/37 % HCl (50:100:1, by vol.).
3. Chloroform.
4. 0.9 % (w/v) NaCl.
5. TUP (Theoretical upper phase) (TUP): chloroform/methanol 1 M HCl (3:48:47, by vol.).
6. Reaction tube microcentrifuge, 13,000 × *g*.
7. Vortex shaker.
8. Glass Pasteur pipets.
9. *Iso*-propanol.
10. Vacuum centrifuge with cold-trap.
11. Nitrogen (gas).

2.5 Thin Layer Chromatograph

1. Thin layer chromatograph (TLC) plates: Merck silica 60 TLC plates, 20 × 20 × 0.2 cm.
2. Impregnation solution: 1 % (w/v) potassium oxalate, 2 mM EDTA in MeOH/H_2O (2:3, by vol.).
3. Oven at 115 °C.
4. Graduated cylinders (100 mL; 25 mL; 5 mL).
5. Alkaline TLC solvent: $CHCl_3$/MeOH/25 % NH_4OH/H_2O (90:70:4:16, by vol.).
6. Acidic TLC solvent: $CHCl_3$/CH_3COCH_3/MeOH/HAc/H_2O (40:15:14:13:7.5, by vol.).
7. Filter paper (21 × 21 cm).
8. Two glass rods (~1 cm thick, 2–3 cm long).
9. Clear plastic wrap.
10. TLC Tank.
11. Hair dryer.
12. Autoradiography film (Kodak, X-Omat S).
13. PhosphorImager and screen.
14. Light cassette.

3 Methods

Described below are the procedures to label and identify signaling lipids in: (1) suspension-cultured plant cells, (2) seedlings, (3) roots, and (4) leaf disks. Once labeling is stopped through PCA, a similar lipid extraction procedure and TLC analysis is performed in all cases. As examples, we used tobacco BY-2 cells, young *Arabidopsis* seedlings, roots of *Vicia* and *Arabidopsis* leaf disks of mature *Arabidopsis* plants, but obviously, other species and materials can be analyzed in a similar fashion.

3.1 Lipid Labeling and Extraction of Cell Suspensions

Typically, tobacco BY-2 cells are used but we have also good experience with other cell suspensions, including tomato, potato, *Arabidopsis*, *Medicago*, coffee, or unicellular algae like *Chlamydomonas* (4, 16–19). The protocol also works for pollen tubes and microspores (20) (Parzer and Munnik, unpublished) and *Phytophthora* zoospores (21).

1. Cell suspensions are grown on a rotary shaker (125 rpm) at 24 °C in the dark and subcultured weekly.
2. Pour ~10 mL of 4–5-day-old cell suspensions into a sterile 15 mL test tube.
3. Let the big cell clumps sink for 30 s and pipet 1,200 μL of cell suspension (for 12 samples) from the top into a 2 mL reaction tube (*see* **Note 3**). Use residual cell suspension to prepare *Cell-Free Medium (CFM) if required.
4. Preparation of CFM: Let the cells sink to the bottom of the tube. Pipet off the liquid and filter through 5 μm, 0.45 μm and 0.22 μm syringe filters, respectively. This CFM is used for cell treatments as 1:1 dilutions.
5. Add 60 μCi $[^{32}P]PO_4^{3-}$ (~5 μCi/sample) in the fume hood of an isotope lab.
6. Aliquot 12 samples of 80 μL into 2 mL "Safe Lock" reaction tubes in a Perspex rack using large orifice tips (*see* **Note 4**). Label for ~3 h; close tubes to prevent water loss and keep the rack in the fume hood to reduce your exposure to radiation.
7. Apply stresses and controls by adding 1:1 volumes (i.e., 80 μL) in CFM* at $t = 0$ with 15 s intervals (*see* **Note 5**); Incubate at RT.
8. Stop reactions by adding 20 μL (=1/10 vol.) of 50 % PCA (5 % final concentration) with 15 s intervals; vortex immediately for 10 s.
9. Shake for 5 min.
10. Spin 10 s (*see* **Note 6**).

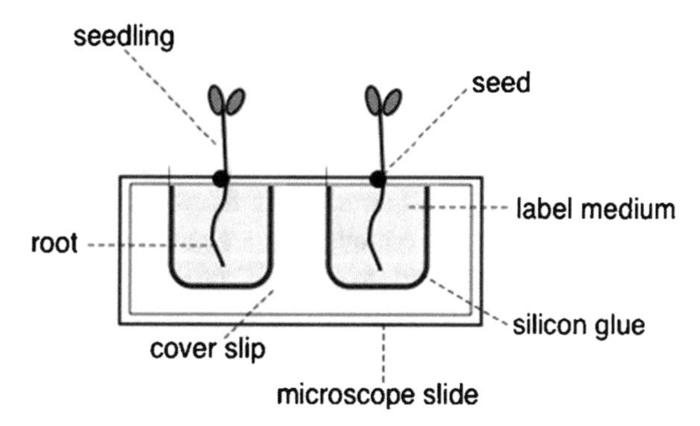

Fig. 1 Schematic drawing of a Fåhreus slide modified for $^{32}P_i$-labeling of *V. sativa* seedlings. The slide contains two chambers that were created with silicon glue between a coverslip and a microscope slide. The silicon glue pastes the coverslip to the slide and serves as spacer as well. Adopted from (15)

3.4 Lipid Labeling and Extraction of Arabidopsis Leaf Disks

The protocol described below is for *Arabidopsis* leaves but can be used for any other plant species, e.g., *Craterostigma plantagineum* (22), sorghum (23) and rice (24), or tissues, e.g., flower petals of carnation or petunia (25, 26).

1. Punch leaf disks (0.5 cm in diameter) from 2–3-weeks-old *Arabidopsis* plants.
2. Transfer disks to 2 mL Safe Lock reaction tubes containing 190 μL labeling buffer. Use 1 leaf disk per tube and use a yellow tip, with a drop of buffer, for the transfer and guidance into the tube. Make sure all leaf disk orientations are the same (e.g., right sight up).
3. Add 10 μL $[^{32}P]PO_4^{3-}$ (5–10 μCi/sample), diluted in buffer to each sample.
4. Incubate with radiolabel over night. Leave the lights of the fume hood on to enable photosynthesis and close the tubes to prevent water loss (O_2 will be produced!).
5. Next day, treatments can be applied. Apply (e.g., salt stress) in rel. large volumes (1:1) so that one does not have to mix rigorously (e.g., 200 μL); pipet next to the disks in the liquid. For controls, use buffer alone. Start treatments with 15 s intervals.
6. Stop reactions by adding 1/10 vol. of 50 % perchloric acid (PCA ~5 % final) with 15 s intervals (*see* **Note 14**).
7. Shake samples for ~15 min (until the whole green tissue turns brownish).
8. Continue as in Subheading 3.2, step 8.

3.5 Thin Layer Chromatography

1. For good separation of the PPIs, K-oxalate-impregnated plates are required (*see* **Note 15**). Impregnate at least 2 days in advance by dipping the TLC plates for 10 s in 1 % (w/v) potassium oxalate, 2 mM EDTA in MeOH/H_2O (2:3, by vol.).

2. Dry and store at 115 °C (*see* **Note 16**).
3. Before loading, mark the TLC plate with a soft pencil where the samples are to be spotted: draw a line, 2 cm from the bottom. Mark the samples with a small dot, 0.5–1.5 cm apart. The solvent front tends to "smile" so distribute the sample equally and stay at least 2 cm away from the sides of the TLC plate (20 × 20 cm).
4. Return the TLC plate in the oven to keep it heat-activated (*see* **Note 16**).
5. Prepare the TLC solvent. For alkaline TLC this is: 90 mL $CHCl_3$, 70 mL MeOH and 20 mL ammonia (25 %, w/v)/H_2O (4:16, by vol.). Mix well and keep 80 mL separate (to run the TLC).
6. Use the remaining solvent to "wash" the tank and to flush the filter paper (~21 × 21 cm) placed at the back of the tank that helps to keep the solvent atmosphere in the chamber saturated.
7. Flush three times and, after the last wash, immediately add two small glass rods that will keep the TLC plate above the solvent when pre-equilibrating. Close the tank quickly with the glass lid and put a weight on top as the solvent pressure my lift the lid and move it to leak. The tank is ready now and can be used three to four runs within a week.
8. Get your lipid samples out of the −20 °C and spin them for 2–4 min in a microcentrifuge to let the samples quickly adjust to room temperature (*see* **Note 17**).
9. Meanwhile, take the TLC plate from the oven and put it with the silica-side onto a sheet of plastic wrap, just above the 2 cm line marked with a pencil, and wrap it. The plastic protects the rest of the plate when loading.
10. Place the TLC plate horizontally on the table, behind a 1-cm thick Perspex screen. Put another plate (glass or perspex) on top, but keep the 2-cm line clear to load.
11. Place a hair dryer ~30 cm away from the TLC plate, so that a gentle stream of air is streaming over the plate. This speeds up the evaporation of the chloroform and reduces the size of the loading spots.
12. Place the ^{32}P-lipid samples in a Perspex block behind the TLC.
13. Use a reaction tube filled with chloroform to wash your pipet tip in between the samples.
14. Equilibrate the tip of 20 μL pipet with chloroform by pipetting it up-and down for a number of times (*see* **Note 18**).
15. Load the pipet tip with 20 μL lipid extract and spot gently onto the TLC without damaging the silica-coating (*see* **Note 19**).
16. Rinse the tip with chloroform to spot the next sample.

17. Repeat until all samples are loaded.
18. After the last sample, wait approx. 1 min to dry the last spot.
19. Remove the plastic wrap and place the TLC plate in the tank on the glass rods so that the plate rests above the alkaline solvent. Quickly replace the lid to limit the loss of solvent vapor.
20. Pre-equilibrate for 30 min.
21. Start the TLC by placing the plate into the solvent. Let the plate stand almost vertically and quickly close the tank again to prevent the loss of solvent vapor.
22. Run the TLC for 1.5 h (*see* **Note 20**).
23. Take the plate out and let it dry in the fume hood for 1 h to eliminate solvent fumes which will affect the PhosphorImager screen or autoradiography film (*see* **Note 21**).
24. Wrap the plate in plastic wrap and expose a PhosphorImager screen for ~1 h.
25. For higher resolution images, expose the TLC plate to autoradiography film for ~1 h and over night for a short and long exposure, respectively.
26. A typical alkaline TLC pattern of salt- and heat-stressed *Arabidopsis* seedlings is shown in Fig. 2.

4 Notes

1. The protocol works for any cell culture, including *Arabidopsis*, coffee, tomato and *Medicago*, green algae like *Chlamydomonas*, and pollen tubes and microspores (4, 16–20).
2. Use cold solvents. They pipet better (less gas expansion within the pipet) and decrease chemical breakdown.
3. As most cell suspensions tend to clump a bit, the use of large orifice tips is preferred. Alternatively, simply cut off the pipet tips to increase the opening. Note that the latter option is less accurate.
4. The use of Eppendorf Safe Lock reaction tubes is preferred as normal tubes may open due to the gas pressure of the solvent.
5. The CFM is a conditioned medium to which the cells are not, or hardly, responding. This is in huge contrast to using plane water or a buffer, which activates lipid signaling instantly, likely via osmotic stress.
6. A short spin in the microcentrifuge is sufficient to remove the radioactive liquid from the lid to prevent contamination when opening.

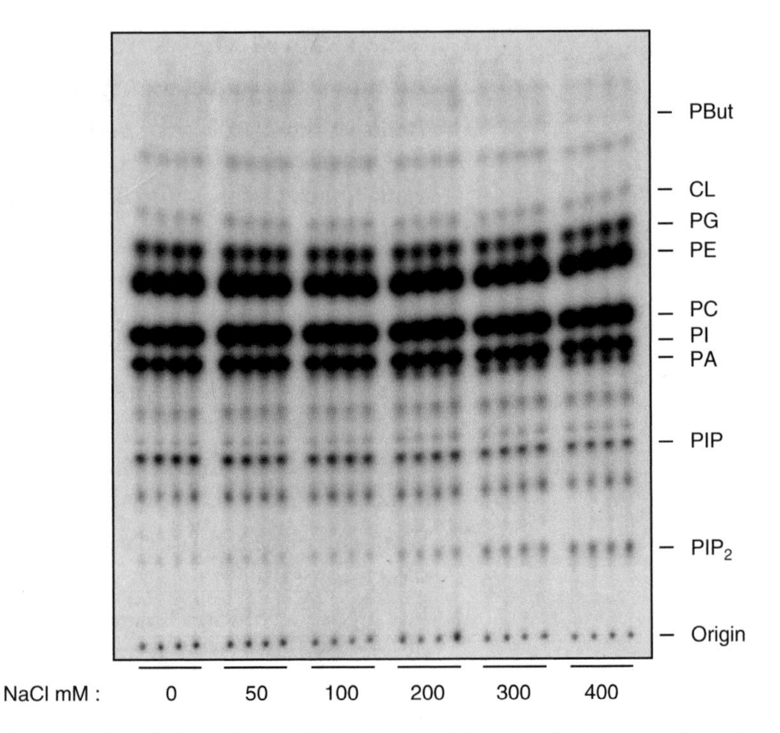

Fig. 2 Alkaline-TLC profile of $^{32}P_i$-labeled *Arabidopsis* wild-type seedlings in response to 30 min of increasing concentrations of salt. Five-day-old seedlings of *A. thaliana* (*Col*-0) were prelabeled overnight with $^{32}P_i$ and then treated with NaCl (0, 50, 100, 200, 300 or 400 mM) for 30 min. Lipids were subsequently extracted, separated by Alkaline TLC and visualized by autoradiography. Treatments were performed in quadruplicates and in the presence of 0.5 % 1-butanol to visualize PLD activity, producing phosphatidylbutanol (PBut) (2, 4, 6, 14, 26, 27). Each sample contained three seedlings

7. It is very important that the CMH completely dissolves the aqueous fraction. By using a 3.75 vol. ratio, this will always occur. The MeOH enables to mix the water and chloroform phases. By increasing either chloroform or water fraction, the solution will split into two phases; an aqueous upper phase and an organic lower phase. This effect is used later (Subheading 3.1, step 16) to separate the lipids from other cellular contents, such as proteins, sugars and DNA.
8. Use the two phases in the Pasteur pipet to separate the lipid fraction from the upper aqueous fraction and transfer it to a clean 2 mL Safe-lock Eppendorf tube, already containing 750 μL TUP.
9. It is very important to remove all of the aqueous-upper phase. Water will cause chemical hydrolysis of the lipid extract. First

remove the upper phase as much as possible. Then aspire all extract and use the organic phase to "wash" the wall of the tube two to three times. Every time, aspire the complete extract again and let the phases separate (Note that it is important to keep enough "balloon air"). Use the two-phase separation in the Pasteur pipet to remove aqueous fraction. Due to a faster evaporation of the chloroform than the water, two phases may reappear and this cause chemical hydrolysis (breakdown) of the lipids.

10. The iso-propanol keeps the solution in one phase, keeping the water dissolved into the chloroform.
11. First transfer the seedlings to a drop of buffer (2.5 mM Mes-KOH, pH 5.8, 1 mM KCl) in an empty petri dish. This discharges the static electricity of the roots and allows a "clean" transfer of all three seedlings in one step into the reaction tube (do not touch the wall!). Use a yellow 20–200 μL tip for the transfers, not tweezers (damage). Alternatively, use a 24 wells microtiter plate containing 400 μL of buffer with 2 or 3 seedlings/well.
12. Add the ^{32}P-label diluted, i.e., as 10 μL in buffer to lower the variability of labeling. Usually, the label is 10 μCi/μL, carrier free.
13. The extract now splits into two phases with an "aqueous" upper phase (free label) and an organic lower phase (lipids). At the interphase, solid material (seedling remainings, protein precipitates) settles.
14. We use an Eppendorf multistepper/repeater pipet for this.
15. Potassium oxalate chelates cations like Ca^{2+} and Mg^{2+}, which cause PPIs to "smear".
16. Storing the plates above 100 °C prevents the silica from binding the water that is in the air. The latter affects the composition of the TLC solvent and lowers the silica's capacity to bind lipids.
17. When the samples are cold, the water in the air can condensate on top of the extracts causing hydrolysis and smearing of the lipid spots.
18. Chloroform evaporates easily and the gas that is building up inside the pipet tends to push out the liquid within the tip. By equilibrating first the pipet and tip with chloroform, this is strongly reduced and increases the handling and loading of the lipid samples.
19. Because of the rapid evaporation of chloroform, the extract tends to be pushed out. Use that drop to touch the pencil-marked spot on the TLC plate. Keep the pipet vertical at all times. It is also no problem to gently touch the silica with the tip. Prevent the spot from becoming too big, by loading small drops of extract at the time.

20. After 1.5 h, the front will still be a few cm from the top. It has, however, no use to run it completely until the end. The speed of the migration front slows down exponentially, while spots will diffuse in all directions, so there is no gain in resolving power by running the TLC longer.

21. If you do not need to use the lipids any further, one can speed up the solvent evaporation by placing the plate in the 115 °C oven for 5 min.

References

1. Welti R, Shah J, Li W, Li M, Chen J, Burke JJ, Fauconnier ML, Chapman K, Chye ML, Wang X (2007) Plant lipidomics: discerning biological function by profiling plant complex lipids using mass spectrometry. Front Biosci 12:2494–2506
2. Arisz SA, Valianpour F, van Gennip AH, Munnik T (2003) Substrate preference of stress-activated phospholipase D in *Chlamydomonas* and its contribution to PA formation. Plant J 34:595–604
3. Munnik T, de Vrije T, Irvine RF, Musgrave A (1996) Identification of diacylglycerol pyrophosphate as a novel metabolic product of phosphatidic acid during G-protein activation in plants. J Biol Chem 271:15708–15715
4. Munnik T, Meijer HJ, Ter Riet B, Hirt H, Frank W, Bartels D, Musgrave A (2000) Hyperosmotic stress stimulates phospholipase D activity and elevates the levels of phosphatidic acid and diacylglycerol pyrophosphate. Plant J 22:147–154
5. Arisz SA, van Himbergen JA, Musgrave A, van den Ende H, Munnik T (2000) Polar glycerolipids of *Chlamydomonas moewusii*. Phytochemistry 53:265–270
6. Munnik T, van Himbergen JAJ, ter Riet B, Braun FJ, Irvine RF, van den Ende H, Musgrave AR (1998) Detailed analysis of the turnover of polyphosphoinositides and phosphatidic acid upon activation of phospholipases C and D in *Chlamydomonas* cells treated with non-permeabilizing concentrations of mastoparan. Planta 207:133–145
7. Meijer HJG, Divecha N, van den Ende H, Musgrave A, Munnik T (1999) Hyperosmotic stress induces rapid synthesis of phosphatidyl-D-inositol 3,5-bisphosphate in plant cells. Planta 208:294–298
8. Meijer HJG, Arisz SA, van Himbergen JAJ, Musgrave A, Munnik T (2001) Hyperosmotic stress rapidly generates lyso-phosphatidic acid in *Chlamydomonas*. Plant J 25:541–548
9. Meijer HJG, Berrie CP, Iurisci C, Divecha N, Musgrave A, Munnik T (2001) Identification of a new polyphosphoinositide in plants, phosphatidylinositol 5-monophosphate (PtdIns5P), and its accumulation upon osmotic stress. Biochem J 360:491–498
10. Bargmann BO, Laxalt AM, ter Riet B, Testerink C, Merquiol E, Mosblech A, Leon-Reyes A, Pieterse CM, Haring MA, Heilmann I, Bartels D, Munnik T (2009) Reassessing the role of phospholipase D in the Arabidopsis wounding response. Plant Cell Environ 32:837–850
11. Welti R, Li W, Li M, Sang Y, Biesiada H, Zhou HE, Rajashekar CB, Williams TD, Wang X (2002) Profiling membrane lipids in plant stress responses: role of phospholipase D alpha in freezing-induced lipid changes in *Arabidopsis*. J Biol Chem 277:31994–32002
12. Zien CA, Wang C, Wang X, Welti R (2001) In vivo substrates and the contribution of the common phospholipase D, PLDα, to wound-induced metabolism of lipids in *Arabidopsis*. Biochim Biophys Acta 1530:236–248
13. Arisz SA, Testerink C, Munnik T (2009) Plant PA signaling via diacylglycerol kinase. Biochim Biophys Acta 1791:869–875
14. Arisz SA, Munnik T (2013) Distinguishing phosphatidic acid pools from de novo synthesis, PLD and DGK. Methods Mol Biol 1009:55–62
15. den Hartog M, Musgrave A, Munnik T (2001) Nod factor-induced phosphatidic acid and diacylglycerol pyrophosphate formation: a role for phospholipase C and D in root hair deformation. Plant J 25:55–65
16. Van der Luit AH, Piatti T, van Doorn A, Musgrave A, Felix G, Boller T, Munnik T (2000) Elicitation of suspension-cultured tomato cells triggers the formation of phosphatidic acid and diacylglycerol pyrophosphate. Plant Physiol 123:1507–1516
17. den Hartog M, Verhoef N, Munnik T (2003) Nod factor and elicitors activate different

phospholipid signaling pathways in suspension-cultured alfalfa cells. Plant Physiol 132:311–317
18. Ramos-Diaz A, Brito-Argaez L, Munnik T, Hernandez-Sotomayor SM (2006) Aluminum inhibits phosphatidic acid formation by blocking the phospholipase C pathway. Planta 225:393–401
19. van Leeuwen W, Vermeer JE, Gadella TW Jr, Munnik T (2007) Visualization of phosphatidylinositol 4,5-bisphosphate in the plasma membrane of suspension-cultured tobacco BY-2 cells and whole Arabidopsis seedlings. Plant J 52:1014–1026
20. Zonia L, Munnik T (2004) Osmotically induced cell swelling versus cell shrinking elicits specific changes in phospholipid signals in tobacco pollen tubes. Plant Physiol 134:813–823
21. Latijnhouwers M, Munnik T, Govers F (2002) Phospholipase D in *Phytophthora infestans* and its role in zoospore encystment. Mol Plant Microbe Interact 15:939–946
22. Frank W, Munnik T, Kerkmann K, Salamini F, Bartels D (2000) Water deficit triggers phospholipase D activity in the resurrection plant *Craterostigma plantagineum*. Plant Cell 12:111–124
23. Monreal JA, Lopez-Baena FJ, Vidal J, Echevarria C, Garcia-Maurino S (2010) Involvement of phospholipase D and phosphatidic acid in the light-dependent up-regulation of sorghum leaf phosphoenolpyruvate carboxylase-kinase. J Exp Bot 61: 2819–2827
24. Darwish E, Testerink C, Khaleil M, El-Shihy O, Munnik T (2009) Phospholipid-signaling responses in salt stressed rice leaves. Plant Cell Physiol 50:986–997
25. Munnik T, Musgrave A, de Vrije T (1994) Rapid turnover of polyphosphoinositides in carnation flower petals. Planta 193:89–98
26. De Vrije T, Munnik T (1997) Activation of phospholipase D by calmodulin antagonists and mastoparan in carnation petal tissue. J Exp Bot 48:1631–1637
27. Munnik T, Laxalt AM (2013) Measuring PLD activity in vivo. Methods Mol Biol 1009: 219–231

Chapter 2

Analysis of D3-,4-,5-Phosphorylated Phosphoinositides Using HPLC

Teun Munnik

Abstract

Detection of polyphosphoinositides (PPIs) is difficult due to their low chemical abundancy. This problem is further complicated by the fact that PPIs are present as various, distinct isomers, which are difficult, if not impossible, to separate by conventional thin layer chromatography (TLC) systems. PPIs in plants include PtdIns3P, PtdIns4P, PtdIns5P, PtdIns(3,5)P_2, and PtdIns(4,5)P_2. Here, a protocol is described analyzing plant PPIs using ^{32}P-orthophosphorus pre-labeled material. After extraction, lipids are deacylated and the resulting glycerophosphoinositol polyphosphates (GroPInsPs) separated by HPLC using a strong anion-exchange column and a shallow salt gradient. Alternatively, PPIs are first separated by TLC, the lipids reisolated, deacylated, and the GroPInsPs then separated by HPLC.

Key words Polyphosphoinositides, Glycerophosphoinositolphosphates, GroPIns, PPI isomers, Deacylation, Mono-methylamine, PtdInsP, PtdInsP_2

1 Introduction

Polyphosphoinositides (PPIs) represent only a minor fraction of the phospholipids of eukaryotic membranes, yet they are extremely important for signal transduction and membrane trafficking. Their concentration is so low and their ionization and chemical properties unfavorable, that they normally do not appear in any of the lipidomic type of analyses that are currently routine (1–4).

The most abundant plant PPI is phosphatidylinositol phosphate (PtdInsP), being present in three different isomers, with the inositol headgroup phosphorylated either at the 3-, 4-, or 5-position (i.e., PtdIns3P, PtdIns4P, PtdIns5P). PtdIns4P is the most abundant, representing ~80 % of the PtdInsP pool, with PtdIns3P and PtdIns5P combined ranging between 5 and 15 % (5–7). Plants exhibit only two phosphatidylinositol bisphosphate (PtdInsP_2) isomers, i.e., PtdIns(4,5)P_2 and PtdIns(3,5)P_2 (8). Compared to mammalian systems, plant PtdInsP_2 levels are extremely low, 30–100-fold

Teun Munnik and Ingo Heilmann (eds.), *Plant Lipid Signaling Protocols*, Methods in Molecular Biology, vol. 1009,
DOI 10.1007/978-1-62703-401-2_2, © Springer Science+Business Media, LLC 2013

lower than PtdInsP, which in animals is present in approx. equal amounts (4, 9). Most (>95 %) of the $PtdInsP_2$ pool consists of $PtdIns(4,5)P_2$ with tiny amounts of $PtdIns(3,5)P_2$ being present and made in response to osmotic stress (8). Earlier, the isomer $PtdIns(3,4)P_2$ had been reported (10) but this lipid was probably mistaken for $PtdIns(3,5)P_2$, which had not yet been described at that time. In animals cells, $PtdIns(3,4)P_2$ is a breakdown product of $PtdIns(3,4,5)P_3$ which is also lacking from plants.

The low amounts of PPIs make them relatively difficult to study. One of the ways to get around this is to use $^{32}P_i$ labeling and exploit the sensitive detection possible with this isotope. $^{32}P_i$-label is rapidly taken up by plant cells and quickly incorporated into ATP and CTP pools to label phospholipids. Because PPIs turn over more rapidly than structural phospholipids, they are preferentially ^{32}P-labeled (5, 6, 11–13). Isomers are, however, difficult (4,5 - 3,5) if not impossible (3-,4-,5P) to separate by thin layer chromatography (6, 8). Hence, deacylation by *mono*-methylamine and HPLC analysis of the resulting glycerolphosphoinostitol phosphates (GroPInsPs) using a strong anion exchanger is required of which a detailed procedure is described here, below.

2 Materials

2.1 Plant Material, ^{32}P-Labeling, Lipid Extraction, and TLC

1. See detailed procedures described in Chapter 1, Subheadings 2.1–2.5.
2. Spray gun with distilled water.
3. Hair dryer.
4. Strong plastic kitchen foil.

2.2 Deacylation

1. Geiger counter.
2. 2 ml Eppendorf Safe-lock tubes.
3. Mono-methylamine reagent: For 100 ml, take 42.8 ml 25 % (w/v) mono-methylamine (Sigma), 45.7 ml MeOH, and 11.5 ml *n*-ButOH (14). Store at 4 °C and keep it cold! (*See* **Note 1**).
4. Deacylation wash solution: H_2O/MeOH/*n*-ButOH (42.8:45.7:11.5).
5. Thermoblock, 53 °C.
6. N_2 (g).
7. Vacuum centrifuge with cold trap.

2.3 Fatty Acid Removal

1. Geiger counter.
2. *n*-ButOH/petroleum ether 40–60°/ethyl formate (20:4:1).
3. 2 ml "Safe-lock" reaction tubes (Eppendorf).
4. 0.5 ml-reaction tubes.

2.4 Desalting

1. Cation exchange resin, Bio-Rad AG 50W ×8 200–400, H^+ form.
2. Empty PD-10 columns.
3. Geiger counter.

2.5 HPLC

1. HPLC.
2. Partisil 10 SAX (Whatman) column with guard; Alternatively, a Zorbax SAX column (DuPont) can be used.
3. Buffer A = water.
4. Buffer B = 1.0 M ammonium phosphate (pH 3.35, phosphoric acid) (*see* **Note 2**).
5. Liquid scintillation vials (5 ml).
6. Liquid scintillation fluid (Packard).
7. Liquid scintillation counter.

3 Methods

3.1 ^{32}P-Labeling of Plant Material and Lipid Extraction

A detailed procedure for lipid labeling and extraction is described in Chapter 1 (Subheadings 3.1–3.5), illustrating methods for suspension-cultured plant cells, seedlings, roots, and leaf disks.

3.2 Deacylation of Phospholipids

To remove the fatty acids of labeled phospholipids, a mono-methylamine deacylation procedure is used, which is based on the method of Clarke and Dawson (15) with modifications as described by Munnik et al. (5, 16) and Meijer et al. (7).

1. Add 400 μL mono-methylamine reagent (careful, DO NOT inhale, keep it cold!) to the dried lipid extract.
2. Incubate for 30 min (NOT longer!) at 53 °C in a fume hood.
3. Evaporate the mono-methylamine vapor using streaming air or N_2 (g) as the vapor would damage the vacuum pump. Be careful to not to inhale the fumes!
4. After 30–60 min, when the mono-methylamine vapor is gone, samples can be dried further using a vacuum centrifuge at room temperature.
5. Dissolve samples in 500 μL distilled H_2O.

To remove the fatty acids:

6. Add 600 μL [*n*-ButOH/petroleum ether 40–60°/ethyl formate] (20:4:1; by vol.).
7. Vortex; remove organic phase (= upper fase).
8. Wash lower water phase with 375 μL [*n*-ButOH/petroleum ether 40–60°/ethyl formate] (20:4:1; by vol.).
9. Transfer the water phase to a clean 0.5 ml-reaction tube.
10. Dry samples in a vacuum centrifuge at 60 °C. (This may take a few hours).

3.3 Deacylation of TLC Purified Phospholipids

In instances when phospholipids were first separated by TLC, HPLC analysis can still be performed. For TLC analysis, see Chapter 1.

1. Dry the TLC plate in a fume hood for 30–60 min, so that all organic-solvent is gone.
2. Wrap the TLC plate in strong kitchen foil.
3. Visualize the ^{32}P-labeled lipids by exposing an autoradiography film (Kodak X-OMAT) to the TLC plate for 30 min to 2 h (*see* **Note 3**).
4. Develop the autoradiography film and dry it with a hairdryer.
5. Fix the film in front of the TLC plate (silica-side) using clips.
6. Hold the TLC plate in the light such that the spots of the film shine through (*see* **Note 4**).
7. Indicate the spots on the back side (glass side) of the TLC plate using a water-resistant marker.
8. When done, remove the film, clips and foil.
9. Use a spray gun with water to wet the silica of the TLC plate. Spray in short intervals (*see* **Note 5**).
10. Upon spraying, the TLC plate becomes transparent so that the marker on the glass-side starts to shine through.
11. Use a small/thin spatula to scrape-off the PPI spots from the plates and collect the silica into "Safe-Lock" reaction tubes (2 ml) (*see* **Note 6**).
12. Dry samples for 15 min in a vacuum centrifuge.
13. Add 400 μL mono-methylamine reagent (careful, DO NOT inhale, keep it cold! *See* **Note 1**).
14. Incubate for 30 min (NOT longer!) at 53 °C in a fume hood.
15. Remove silica by centrifugation (13,000 × *g*) for 2 min and collect the supernatant in a new reaction tube. Wash the silica again with "mono-methylamine *wash* solution" (H_2O/MeOH/n-ButOH (43:46:12; by vol.)).
16. If necessary, repeat step 15, and check whether there is still radioactivity present in the silica remnant.
17. Dry total supernatant under streaming air or N_2 (g) for 60 min, then transfer to a vacuum centrifuge at room temperature. Mono-methylamine fumes are bad for the vacuum pump, too! Therefore, only proceed after all mono-methylamine vapor has been eliminated.
18. Dissolve samples in 500 μL distilled H_2O.
19. To remove the fatty acids, add 600 μL [*n*-ButOH/petroleum ether 40–60 °/ethyl formate] (20:4:1; by vol.).
20. Continue as in Subheading 3.2, step 7.

3.4 Desalting Protocol for HPLC Analysis of Deacylated Phospholipids

1. Dissolve deacylated sample into 100 μL H_2O.
2. Desalt sample with cation exchange column (Bio-Rad AG 50W ×8 200–400, H^+ form):
   (a) Pipet 500 μL of the cation exchange "slurry" into an empty PD-10 column.
   (b) Elute the solvent.
   (c) Add 500 μL H_2O; elute.
   (d) Wash with another 1,000 μL of H_2O (i.e., Total = minimum of 3 column volumes).
   (e) Load 100 μL sample (*see* **Note 4**) onto the column.
   (f) Elute ^{32}P-GroPIns*P*s from column by adding *x*μL H_2O (see below):
   (g) Determine *x* by adding 100 μL aliquots and following the radioactivity with a Geiger counter on the collected fractions and on the column itself. ~500 μL should be sufficient.
   (h) Elute the rest of the samples with the same volumes.
3. Dry by vacuum centrifugation (60 °C).
4. Store at −20 °C in H_2O in a volume appropriate for loading onto the HPLC (100–500 μL).

3.5 HPLC Analysis of Deacylated ^{3}H/^{32}P-Labeled PPIs (Glycerophosphoinositides)

Deacylated lipids are routinely separated by anion-exchange HPLC using a Partisil 10 SAX column and a discontinuous gradient of water (buffer A) and 1.0 M ammonium phosphate, pH 3.35 (phosphoric acid; Buffer B), at a flow rate of 1.0 ml/min.

1. Equilibrate the HPLC column with H_2O prior to sample loading (6 column volumes).
2. Filtrate the deacylated sample over a 0.45 μm filter to remove particles.
3. Load the sample onto the HPLC column using a glass syringe with a blunt-ended needle. Sample loops of 100, 200, or 500 μL can be used.
4. Elute the sample with the following nonlinear gradient of Buffer B: 0–10 min, 0 % B; 10–80 min, 5 % B; 80–140 min, 35 % B; 140–145, 100 % B; 145–160 min 0 % B.
5. Collect 1 ml fractions.
6. Add 4 ml scintillation fluid to each fraction and determine the radioactivity by liquid scintillation counting (*see* **Note 7**).
7. Individual peaks can be identified following periodate treatment by HPLC analysis (17) and/or the use of ^{3}H-standards, including Ins, GroPIns, Ins*1P*, Ins*3P*, Ins*4P*, GroPIns3*P*, GroPIns4*P*, Ins(1,4)P_2, Ins(1,5)P_2, GroPIns(3,4)P_2, GroPIns(4,5)P_2, Ins(1,3,4)P_3, and Ins(1,4,5)P_3.

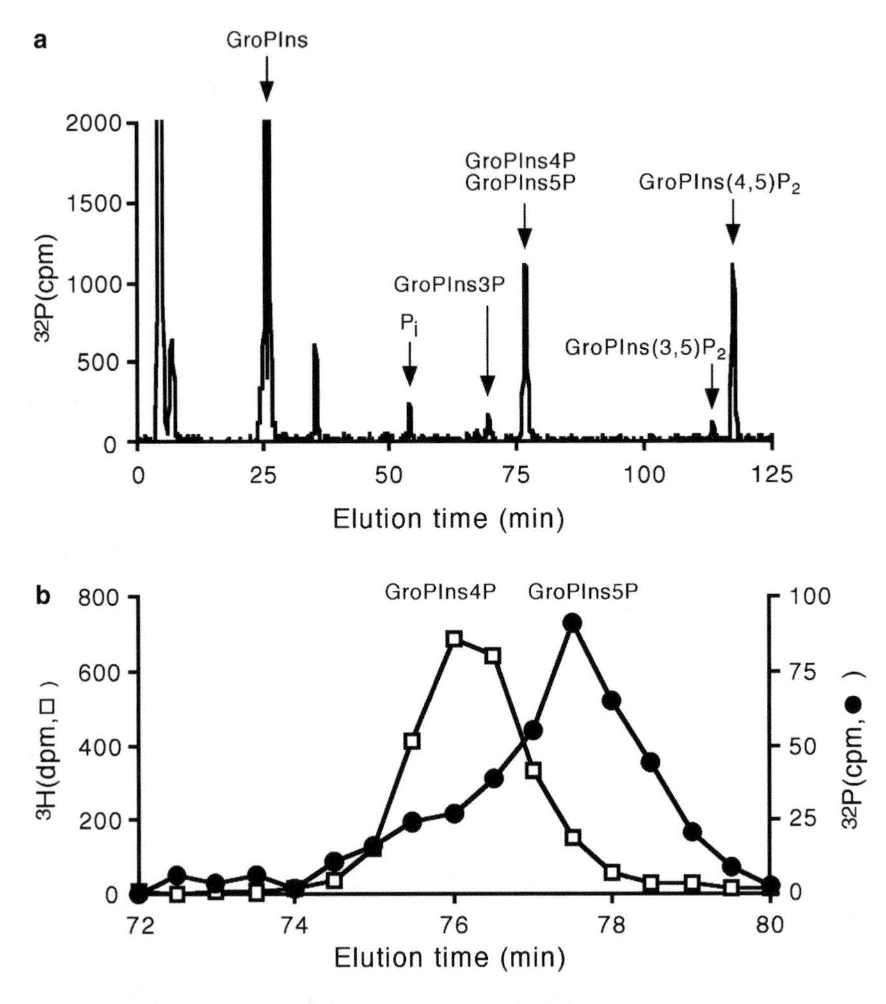

Fig. 1 Anion-exchange HPLC of deacylated phospholipids. (**a**) Chlamydomonas cells were radiolabeled with ^{32}P-P_i for 1 h and then treated for 2.5 min with 150 mM NaCl. Lipids were extracted, deacylated, and the water-soluble products separated by anion-exchange HPLC. (**b**) Separation of GroPIns4*P* and GroPIns5*P*. ^{3}H-GroPIns4P and ^{32}P-GroPIns5P standards were prepared as indicated and separated by HPLC using a more shallow salt gradient and fractions collected at 30 s intervals. Radioactivity was determined by scintillation counting. Adapted with permission from (7)

8. Because the peaks of GroPIns4P and GroPIns5P differ in their retention times by less than 20 s (Fig. 1a), a modified gradient can be utilized, collecting fractions of 0.5 ml: 0–45 min, 0–1.5 % Buffer B; 45–46 min, 1.5–2.4 % Buffer B; 46–80 min, 2.4–4.5 % Buffer B; 80–81 min, 4.5–6.0 % Buffer B, 81–141 min, 6.0–35.0 % Buffer B, 141–142 min, 35–100 % Buffer B, 142–147 min, 100 % Buffer B, 147–150 min 100–0 % Buffer B, 150–180 min 0 % Buffer B wash. The latter gradient produced a 1.5 min peak-to-peak separation of GroPIns4*P* and GroPIns5*P* (see Fig. 1b).

4 Notes

1. Mono-methylamine is dangerous. The fumes are *toxic* and *explosive*! Store at 4–8 °C and keep it cold!
2. Alternatively, 1.25 M NaH_2PO_4, pH 3.7 (with H_3PO_4) can be used as Buffer B. It is the P_i that is important. Obviously, one has to adjust the gradient to separate all GroPIns*P*s.
3. Radioactivity of the PPI spots should be high enough, otherwise you will not be able to get a proper signal after deacylation and HPLC separation (since compounds elute from the column in more than one fraction).
4. Stay behind an 1-cm Perspex screen for radiation safety at all times and wear gloves and safety glasses.
5. Spray in short bursts so that the silica has time to take-up the water (i.e., do not soak it immediately).
6. First try on the side to see how well the silica comes off the glass plate. Too wet will make it too fluffy, if the silica is too dry than it "breaks." When it is right, it scrapes like butter on a knife, sticking together.
7. By adding nonradioactive ADP and ATP and monitoring with a UV detector, specific fractions can be taken rather than the whole run. This saves scintillation fluid and counting.

References

1. Welti R, Li W, Li M, Sang Y, Biesiada H, Zhou HE, Rajashekar CB, Williams TD, Wang X (2002) Profiling membrane lipids in plant stress responses. Role of phospholipase D alpha in freezing-induced lipid changes in *Arabidopsis*. J Biol Chem 277:31994–32002
2. Welti R, Shah J, Li W, Li M, Chen J, Burke JJ, Fauconnier ML, Chapman K, Chye ML, Wang X (2007) Plant lipidomics: discerning biological function by profiling plant complex lipids using mass spectrometry. Front Biosci 12:2494–2506
3. Welti R, Wang X (2004) Lipid species profiling: a high-throughput approach to identify lipid compositional changes and determine the function of genes involved in lipid metabolism and signaling. Curr Opin Plant Biol 7:337–344
4. Munnik T, Nielsen E (2011) Green light for polyphosphoinositide signals in plants. Curr Opin Plant Biol 14:489–497
5. Munnik T, Irvine RF, Musgrave A (1994) Rapid turnover of phosphatidylinositol 3-phosphate in the green alga *Chlamydomonas eugametos*: signs of a phosphatidylinositide 3-kinase signalling pathway in lower plants? Biochem J 298:269–273
6. Munnik T, Musgrave A, de Vrije T (1994) Rapid turnover of polyphosphoinositides in carnation flower petals. Planta 193:89–98
7. Meijer HJG, Berrie CP, Iurisci C, Divecha N, Musgrave A, Munnik T (2001) Identification of a new polyphosphoinositide in plants, phosphatidylinositol 5-monophosphate (PtdIns5P), and its accumulation upon osmotic stress. Biochem J 360:491–498
8. Meijer HJG, Divecha N, van den Ende H, Musgrave A, Munnik T (1999) Hyperosmotic stress induces rapid synthesis of phosphatidyl-D-inositol 3,5-bisphosphate in plant cells. Planta 208:294–298
9. Meijer HJG, Munnik T (2003) Phospholipid-based signaling in plants. Annu Rev Plant Biol 54:265–306
10. Brearley CA, Hanke DE (1993) Pathway of synthesis of 3,4- and 4,5-phosphorylated phosphatidylinositols in the duckweed *Spirodela polyrhiza* L. Biochem J 290:145–150
11. den Hartog M, Musgrave A, Munnik T (2001) Nod factor-induced phosphatidic acid and

diacylglycerol pyrophosphate formation: a role for phospholipase C and D in root hair deformation. Plant J 25:55–65
12. Latijnhouwers M, Munnik T, Govers F (2002) Phospholipase D in *Phytophthora infestans* and its role in zoospore encystment. Mol Plant Microbe Interact 15: 939–946
13. Zonia L, Munnik T (2004) Osmotically induced cell swelling versus cell shrinking elicits specific changes in phospholipid signals in tobacco pollen tubes. Plant Physiol 134:813–823
14. Auger KR, Serunian LA, Cantley LC (1990) Seperation of novel polyphosphoinositides. In: Irvine RF (ed) Methods in inositide research. Raven, New York, pp 159–166
15. Clarke NG, Dawson RM (1981) Alkaline O leads to N-transacylation. A new method for the quantitative deacylation of phospholipids. Biochem J 195:301–306
16. Munnik T, de Vrije T, Irvine RF, Musgrave A (1996) Identification of diacylglycerol pyrophosphate as a novel metabolic product of phosphatidic acid during G-protein activation in plants. J Biol Chem 271:15708–15715
17. Stephens LR, Hughes KT, Irvine RF (1991) Pathway of phosphatidylinositol(3,4,5)-trisphosphate synthesis in activated neutrophils. Nature 351:33–39

Chapter 3

Mass Measurement of Polyphosphoinositides by Thin-Layer and Gas Chromatography

Mareike Heilmann and Ingo Heilmann

Abstract

Phosphoinositides derive from the phospholipid, phosphatidylinositol (PtdIns), by phosphorylation of the inositol ring in the lipid head group. The determination of phosphoinositide species is a particular challenge, because the structurally similar inositolphosphate-head groups must be analyzed as well as the lipid-associated fatty acids. The method presented in this chapter consists of two steps: First phosphoinositides are separated by thin-layer chromatography (TLC) according to their characteristic head groups and the individual lipids are isolated. Second, the fatty acids associated with each isolated lipid are analyzed using a gas-chromatograph (GC). The combination of these two classical methods for lipid analysis, TLC and GC, provides a cost-efficient and reliable alternative to lipidomics approaches requiring more extensive instrumentation.

Key words Phosphoinositides, Thin-layer chromatography, Gas-chromatography, Fatty acids, Quantitative analysis, Molecular lipid species

1 Introduction

Phosphoinositides are a class of structurally related phospholipids that derive from PtdIns by phosphorylation of the inositol ring of the lipid-head group (1). Different phosphorylation signatures of the D3, D4, and D5 positions of the inositol ring of PtdIns are represented as the PtdIns-monophosphates, PtdIns3P, PtdIns4P, and PtdIns5P, the PtdIns-bisphosphates, PtdIns(3,4)P_2, PtdIns(3,5)P_2, and PtdIns(4,5)P_2, and PtdIns-trisphosphate (PtdIns(3,4,5)P_3), which has not so far been detected in plants (2). Because of their structural similarity, the analytic distinction of phosphoinositide classes is difficult, and an additional challenge is presented by the low abundance of these lipids in biological materials. In addition to the separation of phosphoinositides according to the nature of their head groups, recent advances in biochemical research indicate that analytic techniques are required that additionally discriminate the nature of the associated fatty acids. As a cost-effective alternative

Teun Munnik and Ingo Heilmann (eds.), *Plant Lipid Signaling Protocols*, Methods in Molecular Biology, vol. 1009,
DOI 10.1007/978-1-62703-401-2_3, © Springer Science+Business Media, LLC 2013

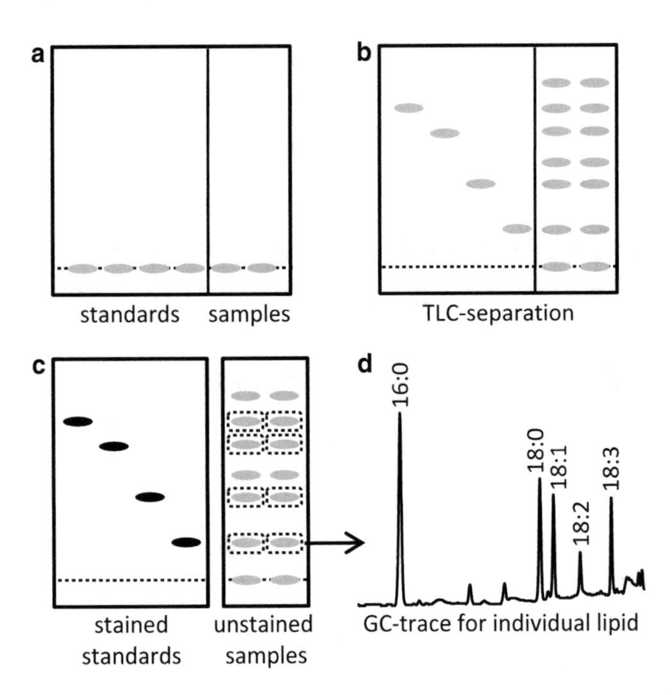

Fig. 1 Schematic overview of phosphoinositide analysis by TLC and GC. (**a**) Lipid extract from biological samples and authentic lipid standards are applied to the same TLC plate. (**b**) The TLC plate is developed. (**c**) The portion holding the standard lipids is removed and stained. The positions of sample lipids are marked according to the migration of the standards. (**d**) Marked sample lipids are isolated from the TLC plate, the associated fatty acids are derivatized and the resulting fatty acid methyl esters (FAMEs) are analyzed using GC

to more advanced high-throughput methods for the analysis of intact lipids, here a comparably simple method is presented which combines the classical techniques of TLC and GC to first separate phosphoinositides according to their head groups and then analyze the fatty acid composition of each collected fraction. A schematic overview of the method described is given in Fig. 1. Using this protocol, biological material from different sources has been successfully analyzed, including from plants (3–5), from yeast (6), and from mammalian tissues (7). Based on previous experiments (6), phosphoinositides were recovered from silica powder to a rate of >90 %, regardless of the TLC-developing solvent used, and the equivalent of as little as 8 pmol lipid has been detected using a GC with flame-ionization detector.

2 Materials

All solvents should be of analytical grade. All solutions should be prepared using filtered and deionized water, here specified as H_2O. Regulations for proper waste disposal of organic solvents should be strictly followed.

2.1 Acidic Extraction of Polyphosphoinositides

1. HCl-solution: 2.4 M HCl in H_2O.
2. EDTA-solution: 0.4 M EDTA in H_2O.
3. Backwashing solution: 0.5 N HCl in 50 % (v/v) methanol.
4. Redissolving solution: chloroform–methanol (1:2, v/v).

2.2 Thin-Layer Chromatography

1. Potassium oxalate-solution: 5 % (w/v) potassium oxalate in H_2O.
2. CDTA-solution: stir 4.55 g of disodium *trans*-1,2-diaminocyclohexane-*N*,*N*,*N*′,*N*′-tetraacetic acid (CDTA) in a mixture of 165 mL H_2O, 330 ml ethanol, and 3 mL of 10 N NaOH until CDTA is fully dissolved (8).
3. Developing solvent I: dissolve 12 g boric acid in 180 ml of chloroform–methanol–pyridine (75:60:45, v/v/v), add 7.5 mL H_2O, 3 ml 88 % (v/v) formic acid, and 75 μL of ethoxyquin (technical grade) (8).
4. Developing solvent II: chloroform–methanol–ammonium hydroxide–H_2O (57:50:4:11, v/v/v/v) (9).
5. Developing solvent III: chloroform–methanol–H_2O–acetic acid (10:10:3:1, v/v/v/v).

2.3 Visualization of Lipid Standards

1. $CuSO_4$-solution: 10 % (w/w) $CuSO_4$, 8 % (w/w) H_3PO_4 in H_2O.

2.4 Fatty Acid Analysis

1. Internal lipid standard: 5 μg of tripentadecanoin or another lipid, e.g., phosphatidylcholine.

3 Methods

Many steps in the described protocol involve organic solvents, which should be handled in a fume hood. TLC and GC should also be set up so that exposure to toxic solvent fumes is minimized. Required safety precautions should always be adhered to, including the wearing of safety glasses for all steps involving acids or liquid N_2.

3.1 Acidic Extraction of Polyphosphoinositides (10)

1. Grind 1–10 g fresh weight of plant tissue in liquid N_2 to a fine powder using a mortar and a pestle. Cultured plant cells or yeast (up to 2 g) can be harvested by centrifugation.
2. To cell material successively add 5–10 ml $CHCl_3$–methanol (1:2), 250 μl EDTA solution, 500 μl HCl-solution, and 500 μL chloroform with intermittent mixing (*see* **Note 1**).
3. Let biological material extract for at least 2 h at 4 °C while slowly shaking (*see* **Note 2**).
4. Spin mixture at 600 × *g* for 2 min to separate phases.

5. Transfer lower (organic) phase to fresh open glass tube.
6. Reextract aqueous phase twice as above and combine organic phases.
7. To organic phase add 1.5 mL of backwashing solvent, mix.
8. Spin mixture at 600 × *g* for 2 min to separate phases.
9. Pull off upper (aqueous) phase and discard.
10. Backwash residual organic phase two more times (steps 7–9).
11. Transfer lower (organic) phase into fresh open glass tube.
12. Evaporate solvent under N_2 flow (*see* **Note 3**).

3.2 Chromatographic Separation of Lipids

Thin-layer chromatography (TLC) is classically performed using standard silica S60 plates in 20 × 20 × 10 cm vertical glass-chambers preequilibrated with 100 mL of developing solvent and containing one sheet (20 × 20 cm) of filter paper to maintain a constant solvent atmosphere. Alternatively, horizontal high-performance (HP)-TLC setups of smaller scale can also be used as specified by the respective manufacturers.

1. Activate TLC-plate. For separation of phosphatidylinositol monophosphates, submerge plate for 10 s in CDTA-solution. Alternatively, for separation of phosphoinositides and other phospholipids, submerge plate for 10 s in potassium oxalate solution. After either treatment, dry plate in oven at 180 °C for at least 1 h.
2. Redissolve dried lipids in 50 μL redissolving solvent.
3. Spot sample lipids onto activated TLC-plate.
4. Also spot authentic lipid standards corresponding to the lipids of interest.
5. Develop TLC-plate in appropriate developing solvent. For separation of phosphatidylinositol monophosphates, use developing solvent I (8), for phosphoinositides use developing solvent II (9), and for generic phospholipids use developing solvent III (*see* **Notes 4–6**).

3.3 Visualization of Lipid Standards

1. After development, dry TLC-plate in fume hood.
2. Sever plate-section containing lipid standards using a glass-cutter (*see* **Note 7**).
3. Submerge dried plate briefly in $CuSO_4$-solution (*see* **Note 8**).
4. Let plate dry upright on paper towels.
5. Heat plate to 180 °C on a heating plate (*see* **Note 9**) until charred lipid bands appear.
6. Let plate cool in fume hood.

3.4 Isolation of Lipids from Silica Plates

1. Place severed part of TLC plate with stained standard lipids next to unstained portion containing biological samples. Mark positions of sample lipids according to migration of standards using a pencil.
2. Scrape areas with sample lipids from the unstained portion of the plate and collect silica powder for each spot into fresh open glass tubes (*see* **Notes 10–12**).
3. Extract lipids from collected silica powder twice with 2 mL of the respective developing solvent. Perform the second extraction in closed glass tubes with screw caps and teflon seals overnight at −20 °C.
4. Combine organic phases from extraction and reextraction and evaporate solvent under streaming N_2.
5. For quantitative analysis, add internal standard to samples before extraction.

3.5 Preparation of $NaOCH_3$-Solution

Prior to GC-analysis the fatty acids associated with isolated lipids must be transesterified to methanol to form fatty acid methyl esters (FAMEs). The transesterification reaction is performed using $NaOCH_3$. The following steps describe how a 1 N solution of $NaOCH_3$ can be prepared. Alternatively, the solution can be obtained from commercial sources.

1. In a fume hood, cut a piece of metallic sodium with a razor. When cutting the sodium, remove all oxidized edges of the sodium block.
2. Immediately add sodium to preweighed 50 mL reaction tube containing 20 mL of methanol. Tightly close screw cap of reaction tube.
3. Weigh reaction tube after sodium has been added and record difference as the weight of the sodium. Reacting 1.15 g of sodium with 50 mL methanol will yield a 1 M solution of $NaOCH_3$.
4. Add methanol to the reaction to bring volume to match desired concentration.
5. Keep reaction tube in the fume hood and let sodium dissolve completely. When the sodium is dissolved and the tube has cooled, the solution is ready to use.

3.6 Transesterification of Lipid-Associated Fatty Acids for GC-Analysis (11)

1. Add 1.35 mL toluene–methanol (1:2, v/v) to dried lipids in a closed glass tube with a screw cap and teflon seal.
2. Add 0.5 mL 1 M $NaOCH_3$ in methanol and mix.
3. Incubate at room temperature for 1 h while slowly shaking.
4. Add 1 mL 5 M NaCl in H_2O.

5. Add 2 mL hexane and mix.
6. Briefly spin mixture at 600 × *g* to separate phases.
7. Pull hexane (upper) phase into fresh closed glass tube with a screw cap and teflon seal.
8. Reextract lower phase twice with hexane.
9. Evaporate solvent under N_2 flow. FAMEs will form a film at the bottom of the tube.

3.7 Fatty Acid Analysis

FAMEs analysis can be carried out using most GC-equipment currently available. The chromatography column should be chosen to enable the separation of hydrophobic compounds. Examples for suitable columns include types DB-23 (Agilent) or Innowax (Hewlett Packard). Other matrices may also be used as exist.

1. Redissolve dried lipids in 5 μL of acetonitrile.
2. Inject 1 μL of the lipid sample in the gas-chromatograph at 220 °C injector temperature.
3. Perform chromatography on a 30 m DB-23 column (Agilent) using helium as a carrier gas as follows: Hold oven temperature at 150 °C for 1 min, then raise temperature to 200 °C at a rate of 8 °C/min, then to 250 °C at 25 °C/min, and then keep temperature at 250 °C for 6 min (*see* **Note 13**).

3.8 Data Analysis

1. Identify FAME-peaks according to coelution with FAME-standards prepared from authentic fatty acid mixes, such as menhaden-oil, or by mass spectrometry (*see* **Note 14**).
2. Integrate FAME-peaks for each lipid sample and calculate molar amounts in relation to the internal standard (*see* **Note 15**).
3. The molar amount of lipid present equals the molar sum of all FAMEs present for lysolipids; the sum of all fatty acids present divided by 2 for phospholipids; or the sum of all fatty acids present divided by 3 for triacylglycerols.

4 Notes

1. Protonation of phosphoinositides is required for successful extraction and no quantitative extraction is achieved with non-acidic protocols.
2. Extraction of cultured cells is aided by first grinding the extraction mixture in a 50 ml glass douncer on ice with 20 slow movements.
3. Dried lipids can be stored overnight at −20 °C in glass tubes under an inert atmosphere of N_2 or argon.

4. Activation of TLC plates is not strictly necessary, but improves resolution.
5. For added resolution use a spotting robot if available.
6. Upon development, dried TLC-plates can be redeveloped in the same or a different developing solvent to further improve resolution as needed.
7. Standards should be run on the same—not an identical!—plate to ensure comigration with lipids from biological material.
8. Dip plate steadily in the $CuSO_4$ solution to avoid tide marks. An automated immersion device can aid uniform dipping.
9. Standard amounts below 1 μg can be visualized using the $CuSO_4$ solution. If no heating plate is available, plates can also be heated more slowly in an oven.
10. To verify linear migration, the scraped remainder of the plate can also be stained by submerging in $CuSO_4$-solution and subsequent heating.
11. The silica powder should not be inhaled, and therefore, make sure to wear protective gear covering your mouth and nostrils. By using sheets of folded cardboard the powder can easily be collected and transferred to glass tubes for extraction.
12. Working with lipids and organic solvents suggests the use of glass tubes wherever possible. While glass is not so critical for all steps prior to TLC separation of lipids, all steps after TLC should be strictly carried out in glass. For lipid extraction ensure that screw-on caps contain absolutely no rubber and that inlays are made from teflon instead. Otherwise, GC analysis will be encumbered by the presence of artifact peaks contaminating the biological samples.
13. The chromatography column and temperature gradient specified are only an example for a method that has worked reliably over the past years. There are numerous published variations for other columns and temperature gradients that can be applied.
14. When using a mass-detector for FAME-analysis, spectra corresponding to many fatty acids of biological relevance can be found at the Web site of the American Oil Chemists' Society (AOCS; http://lipidlibrary.aocs.org/).
15. When calculating lipid concentrations according to the internal standard, be aware that different standards may contain different stoichiometric amounts of fatty acids, e.g., 3 in triacylglycerols versus 2 in glycerophospholipids.

References

1. Palmer S, Hawkins PT, Michell RH, Kirk CJ (1986) The labelling of polyphosphoinositides with [^{32}P]P$_i$ and the accumulation of inositol phosphates in vasopressin-stimulated hepatocytes. Biochem J 238:491–499
2. Heilmann I (2009) Using genetic tools to understand plant phosphoinositide signalling. Trends Plant Sci 14:171–179
3. König S, Ischebeck T, Lerche J, Stenzel I, Heilmann I (2008) Salt stress-induced association of phosphatidylinositol-4,5-bisphosphate with clathrin-coated vesicles in plants. Biochem J 415:387–399
4. König S, Mosblech A, Heilmann I (2007) Stress-inducible and constitutive phosphoinositide pools have distinct fatty acid patterns in *Arabidopsis thaliana*. FASEB J 21:1958–1967
5. Saavedra L, Balbi V, Lerche J, Mikami K, Heilmann I, Sommarin M (2011) PIPKs are essential for rhizoid elongation and caulonemal cell development in the moss *Physcomitrella patens*. Plant J 67:635–647
6. König S, Hoffmann M, Mosblech A, Heilmann I (2008) Determination of content and fatty acid composition of unlabeled phosphoinositide species by thin layer chromatography and gas chromatography. Anal Biochem 278:197–201
7. Goebbels S, Oltrogge JH, Kemper R, Heilmann I, Bormuth I, Wolfer S, Wichert SP, Mobius W, Liu X, Lappe-Siefke C, Rossner MJ, Groszer M, Suter U, Frahm J, Boretius S, Nave KA (2010) Elevated phosphatidylinositol 3,4,5-trisphosphate in glia triggers cell-autonomous membrane wrapping and myelination. J Neurosci 30:8953–8964
8. Walsh JP, Caldwell KK, Majerus PW (1991) Formation of phosphatidylinositol 3-phosphate by isomerization from phosphatidylinositol 4-phosphate. Proc Natl Acad Sci USA 88:9184–9187
9. Perera IY, Davis AJ, Galanopoulou D, Im YJ, Boss WF (2005) Characterization and comparative analysis of Arabidopsis phosphatidylinositol phosphate 5-kinase 10 reveals differences in Arabidopsis and human phosphatidylinositol phosphate kinases. FEBS Lett 579:3427–3432
10. Cho MH, Chen Q, Okpodu CM, Boss WF (1992) Separation and quantification of [^{3}H] inositol phospholipids using thin-layer-chromatography and a computerized ^{3}H imaging scanner. LC-GC 10:464–468
11. Hornung E, Korfei M, Pernstich C, Struss A, Kindl H, Fulda M, Feussner I (2005) Specific formation of arachidonic acid and eicosapentaenoic acid by a front-end Delta5-desaturase from *Phytophthora megasperma*. Biochim Biophys Acta 1686:181–189

Chapter 4

Measurement of Inositol (1,4,5) Trisphosphate in Plant Tissues by a Competitive Receptor Binding Assay

Ingo Heilmann and Imara Y. Perera

Abstract

The phosphoinositide signaling pathway is important for plant responses to many different stresses. As part of the responses to a stimulus, $InsP_3$ levels may increase rapidly and transiently. The receptor binding assay for $InsP_3$ described here is easy to use and an ideal method to monitor and compare $InsP_3$ levels in multiple samples from large scale experiments. The method is based on competitive binding of $InsP_3$ to the mammalian brain $InsP_3$ specific receptor protein. This chapter describes a protocol for extracting and neutralizing plant samples and performing the receptor binding assay (using a commercially available kit). The protocol described has been used effectively to monitor $InsP_3$ levels in plant tissues of different origin and in response to different stresses.

Key words Inositol-1,4,5-trisphosphate, $InsP_3$, Receptor binding assay

1 Introduction

D-*myo*-Inositol 1,4,5-trisphosphate ($InsP_3$) is an important second messenger, produced in response to a stimulus by the hydrolysis of phosphatidylinositol 4,5-bisphosphate ($PtdInsP_2$) in eukaryotic cells. In mammalian cells, $InsP_3$ triggers the release of Ca^{2+} from intracellular stores activating a signaling cascade (1).

In plants, $InsP_3$-mediated Ca^{2+} release may be only one of several Ca^{2+} regulating mechanisms (2). Part of the uncertainty in attributing particular cellular functions to $InsP_3$ is due to the fact that no $InsP_3$-sensitive Ca^{2+}-channel has been discovered in plants (3). Moreover, $InsP_3$ can be further phosphorylated to other inositol polyphosphates, which have been shown to be functional components of plant signaling pathways (4). In any case, $InsP_3$ is an important signaling component either directly or as an intermediate in the biosynthesis of $InsP_5$ or $InsP_6$, and rapid changes in $InsP_3$ have been documented in response to many abiotic and biotic stimuli (5–7).

Teun Munnik and Ingo Heilmann (eds.), *Plant Lipid Signaling Protocols*, Methods in Molecular Biology, vol. 1009,
DOI 10.1007/978-1-62703-401-2_4,

A predominant method used by plant biologists to measure inositol phosphates is HPLC. Numerous HPLC protocols capable of measuring $InsP_3$ have been previously reviewed (8). However, HPLC is time-consuming, requires specialized equipment and is not an ideal method for the analysis of large numbers of samples, such as have to be analyzed for stimulation time-course experiments.

The $InsP_3$ receptor binding assay described here is quick and easy to use and its application enables larger-scale experimental set-ups. The assay is based on the specific binding of Ins(1,4,5)P_3 to the bovine adrenal binding protein (9). While a number of researchers have themselves prepared receptor protein from fresh bovine brains, the commercially available kit reduces exposure to potentially dangerous biological material and provides better consistency from batch to batch. Based on the properties of the bovine brain adrenal binding protein, the assay is highly specific for Ins(1,4,5)P_3 over other inositol phosphates including Ins(1,3,4)P_3, Ins(2,4,5)P_3, Ins(1,3,4,5)P_4, Ins(1,3,4,5,6)P_5, and $InsP_6$.

One limitation of the protocol presented here is that the Ins(1,4,5)P_3 levels in plant cells may be small compared to the total inositol phosphate pool (10), and measurements from unpurified plant extracts may be affected by interference from other compounds present in the sample. Therefore, it may not be valid to compare the absolute values obtained from this assays between different plant tissues and species. Nevertheless, we consider the assay useful to compare relative levels of $InsP_3$, for instance between mutants and wild type or between samples from non-treated vs. treated plants. Several researchers have validated the use of the receptor binding assay for plant $InsP_3$ measurements by comparing their results from the assay with HPLC (11, 12). Additionally, we have previously shown that pretreatment of plant extracts with recombinant mammalian type I inositol 5-phosphatase, an enzyme that specifically hydrolyzes $InsP_3$ and $InsP_4$ (13, 14), removed >90 % of the signal measured by the receptor binding assay (15–17), further supporting the validity of this assay for plant material.

Using the protocol described, we have been able to measure $InsP_3$ levels in a variety of different plant tissues or cells including, maize pulvini (16), suspension cultured tobacco cells (18), Arabidopsis roots, shoots, or inflorescence stems (19), tomato leaves and fruit (20), or the red alga, *Galdieria sulphuraria* (15, 21). When using the assay with previously untested plant material, some optimization particularly with regard to sample preparation might be required. Care should be taken in the neutralization step, since pigments can interfere with accurate neutralization based on color change of a pH-indicator dye.

2 Materials

2.1 Plant Sample and Preparation Components

All solutions should be prepared with deionized, filtered water.

1. Plant material: Samples should be harvested and immediately frozen and ground in liquid N_2. Approximately 80–100 mg of tissue per assay is optimal.
2. 10 % (v/v) perchloric acid.
3. Neutralization buffer: 1.5 M KOH, 60 mM Hepes.
4. Universal pH Indicator dye (Fisher cat # S160-500).

2.2 $InsP_3$ Assay Components

1. Inositol-1,4,5-trisphosphate [^{3}H] radioreceptor assay kit (Cat # NEK064 From PerkinElmer Life Sciences Inc.).

   The kit contains the following components:

   (a) Receptor/[^{3}H] $InsP_3$ tracer (lyophilized).

   (b) $InsP_3$ standard (lyophilized).

   (c) Blanking solution.

   (d) Assay buffer.
2. 0.15 M NaOH.
3. Liquid scintillation cocktail (Atomlight or Pico Fluo).

3 Methods

All regulations concerning safe handling strong acids should be adhered to, including the wearing of safety glasses and protective gloves. The disposal of radioactive material must be carried out according to applying rules and regulations.

3.1 Sample Collection

1. Harvest fresh tissue and immediately freeze in liquid N_2 or if time permits grind the tissue to a fine powder in liquid N_2 and aliquot into ~80–100 mg tissue per sample in microfuge tubes. Keep a record of the exact weight of each sample.
2. Store frozen plant material or frozen ground samples at −80 °C until ready to assay (*see* **Note 1**).

3.2 Sample Extraction and Neutralization

1. Add 0.5 mL universal indicator dye to 10 mL neutralization solution. Mix well and store on ice. Solution will instantly turn dark purple but will eventually become a dark teal color. Also prepare ~5 mL each of 5 % (w/v) and 10 % (w/v) PCA solutions containing the universal dye. These solutions will be pink.
2. Place the frozen ground samples on ice and add 200 μL of 10 % (w/v) PCA to each tube (*see* **Note 2**). Mix the tubes vigorously and incubate them on ice for 20 min.

3. Centrifuge the samples at ~14,000 × *g* in microfuge for 10 min at 4 °C to sediment insoluble material.
4. Carefully transfer the supernatant to a fresh tube without disturbing the pellet.
5. If debris is carried over, do an additional 5 min centrifugation to clarify and transfer supernatant to a new tube.
6. Neutralize the samples to pH 7.5 by adding the neutralization solution containing the dye. Initially add 200 μL of neutralization solution to each tube and vortex (*see* **Note 3**). At pH 7.5 the solution will be an apple green. Vortex well after each addition and allow the precipitate to settle. Additionally check the pH using pH indicator paper.
7. Centrifuge samples at ~14,000 × *g* in microfuge for 10 min at 4 °C to sediment insoluble material.
8. Carefully remove the supernatant to a fresh tube and record the final volume (*see* **Note 4**).
9. Carry out $InsP_3$ assay immediately or store the cleared supernatant at −20 °C for up to 1 week prior to assay.

3.3 $InsP_3$ Assay Using the Inositol-1,4,5-Trisphosphate [^3H] Radioreceptor Assay Kit from PerkinElmer Life Sciences

3.3.1 Preparation of Assay Reagents

1. Working stock of Receptor preparation/Tracer: Reconstitute the receptor preparation/tracer (which is supplied in a lyophilized state) by adding 2.5 mL of cold water. Swirl to mix and store on ice for ~15 min prior to use. Determine the volume of working stock that will be required for your experiment and prepare a working stock by diluting the reconstituted receptor/tracer 1:15 with assay buffer (*see* **Note 5**).
2. $InsP_3$ standard solution: $InsP_3$ standards are provided in a lyophilized state. Add 2 mL of distilled water to 1 vial of $InsP_3$ standard and mix well by inverting. This 120 nM solution will be equivalent to 12 pmol/100 μL. Aliquot into 500 μL aliquots and store reconstituted standard at −20 °C.
3. Serial dilutions of standard: For generating the standard curve you will need to prepare a series of standard solutions ranging from 12 to 0.12 pmol/100 μL by serial dilution.
   (a) Label a set of tubes S2–S7, These will correspond to 4.8, 2.4,1.2,0.6,0.3, and 0.12 pmol, respectively (*see* **Note 6**).
   (b) Prepare the standards as follows:
      - Add 375 μL of water to tubes S2 and S7 and 250 μL of water to tubes S3, S4, S5, and S6.
      - To tube S2 add 250 μL $InsP_3$ standard. Vortex and mix well. Transfer 250 μL from tube S2 to S3. Mix well. Continue serial dilution by adding 250 μL from S3 to S4, S4 to S5, and so on.

3.3.2 $InsP_3$ Assay Protocol

Assays should be performed in duplicate for each sample and incubated on ice. Assays can be carried out in the tubes provided in the kit or alternatively using microfuge tubes. In addition to the samples to be assayed and the $InsP_3$ standards there will be three additional controls, total counts (TC), nonspecific binding (NSB), and B zero (B_0).

Label a duplicate set of assay tubes and set them in a rack in an ice bucket so the tubes are kept cool (*see* **Note 7**). Add 100 μL of the neutralized sample to each of the labeled duplicate tubes.

1. Add 100 μL of blanking solution (provided in the kit) to the tubes labeled NSB.
2. Add 100 μL of distilled water to the tubes labeled B0.
3. Add 400 μL of receptor to each tube and mix (*see* **Note 8**).
4. Incubate the tubes for 1 h at 2–8 °C (*see* **Note 9**).
5. After incubation centrifuge the samples at 2,000 × *g* at 4 °C for 15–20 min (*see* **Note 10**).
6. Decant the supernatant into a radioactive waste container and allow the tubes to drain on absorbent paper for ~3 min. Be careful not to dislodge the pellet.
7. Add 50 μL of 0.15 M NaOH to each tube and mix well.
8. Incubate tubes at room temperature for 10 min and mix again (*see* **Note 11**).
9. Place the entire tube containing the resuspended pellet into a 7 mL glass scintillation vial and add 5 mL of Atomlight or Pico-Fluo. Cap the vials and invert to mix thoroughly. Make sure that the solution is homogenous and that the level of scintillant cocktail is even inside and outside the tube (*see* **Note 12**).
10. Place vial in a scintillation counter and count for 3–5 min using a program appropriate for (^3H).

3.3.3 Calculation of Results

1. First determine the average counts per min (cpm) of all the duplicate samples. If duplicates diverge substantially you may want to recount the samples.
2. Subtract the average of the NSB from the average for each sample and from the average values for each $InsP_3$ standard.
3. Calculate the % bound for each sample and $InsP_3$ standard as follows:

   $$\% \, B/B_0 = (\text{Standard or sample cpm} - \text{NSB})/(B_0 - \text{NSB}) \times 100$$

4. Plot the % B/B_0 for the standard against the standard concentration on a semi-logarithmic paper with the standard concentrations ($InsP_3$ pmol/0.1 mL) on the *x* axis. The resulting graph represents a standard curve.

5. Determine the concentration in each sample by interpolation from the standard curve (*see* **Note 13**).
6. Based on the initially recorded fresh weight and the sample volumes calculate the concentrations of $InsP_3$ in the plant samples (pmol/g fresh weight).

3.4 Comparison of $InsP_3$ Receptor Binding Kits

For many years we had used the Amersham D-*myo*-Inositol 1,4,5-trisphosphate (IP_3) [^3H] Biotrak Assay System sold by GE Healthcare (catalog # TRK1000) for plant tissues with good success. This kit has been discontinued. The Inositol 1,4,5-trisphosphate [^3H] radioreceptor assay kit from PerkinElmer (catalog # NEK064) represents an alternative based on the same principle. We have tested the PerkinElmer-kit and found that some features have been improved. Furthermore, reproducibility and sensitivity of the assay are also improved over those of the discontinued GE-kit.
Improved features:

1. The receptor preparation appears to be more highly purified and comes as a lyophilized preparation. This preparation is easily reconstituted in water and the assay buffer to generate a homogenous suspension, enabling a more uniform addition of receptor to the reaction. Furthermore, the final pellet is easier to resuspend in 0.1 M NaOH and requires minimal mixing effort compared to the TRK1000 kit pellets.
2. The receptor and tracer are lyophilized together and when reconstituted in assay buffer provide a single assay reagent. Therefore, the experimental setup is simpler and only requires addition of two components, sample and receptor, where The TRK1000 kit had four separate components that needed to be added for the final reaction.
3. Rather than transferring the resuspended receptor pellet from each sample into scintillation vials, the entire assay tube containing the resuspended pellet is transferred into the scintillation vials. This reduces sample loss and counting errors.

3.5 Specificity of Assay for $InsP_3$ Versus $InsP_6$

The NEK064 kit uses $InsP_6$ to estimate NSB instead of $InsP_3$. In comparative experiments, we measured comparable NSB values using either $InsP_3$ (as provided in the TRK1000 kit) or $InsP_6$ (as provided in the NEK064 kit). While the bovine brain adrenal binding protein does not have high affinity to $InsP_6$, $InsP_6$ will saturate the receptor if applied at sufficiently high concentrations. In order to determine NSB using $InsP_6$ a concentration of 300 mM $InsP_6$ is used, whereas a concentration of only 4 μM $InsP_3$ is sufficient when using $InsP_3$. In order to test whether saturating the bovine brain adrenal binding protein with $InsP_6$ poses a problem when working with plant samples, we compared $InsP_3$ values detected in the

Table 1
$InsP_3$-levels determined in different plant lines using the PerkinElmer $InsP_3$-assay kit

Arabidopsis line	Ins(1,4,5)P_3 (pmol/g fresh weight)	SD	%
Wild type	217	11	100
ipk1-1	205	23	95
InsP 5-ptase	27	6	13

$InsP_3$ levels in 2-week-old Arabidopsis seedlings were measured using the NEK064 kit (PerkinElmer) and $InsP_6$ to saturate the receptor protein. Data are the average ± SD from four independent biological replicates

Arabidopsis *ipk1-1* mutant (4) Arabidopsis *InsP 5-ptase* plants (2, 19) and wild type Arabidopsis controls (Table 1). Whereas wild type and *ipk1-1* lines exhibited comparable levels of $InsP_3$, $InsP_3$-levels in *InsP 5-ptase* plants were reduced by >90 %. Therefore, the receptor binding assay appears to be specific for $InsP_3$ and not $InsP_6$.

4 Notes

1. If the samples are from a time-course experiment it is best to harvest all the samples and freeze them and grind and process them at a later time.
2. PCA is highly corrosive. Be sure to wear goggles, a lab coat and gloves.
3. If the color is pink or pale yellow the sample is too acidic and will need more neutralization solution. However, it is easy to overshoot, so add extra base in small increments (5–10 μL). If the color is dark blue green or purple it is too basic and will need more acid. Use the 5 % or 10 % PCA solution containing the indicator dye to adjust the pH to 7.5.
4. It is important to note the final volume of your sample. You will need to know this as well as the weight of your sample to finally calculate the total amount of $InsP_{3.}$
5. Determine how much working stock you will need for your experiment. Each assay requires 400 μL of working stock. Prepare enough working stock for your experiment with ~1 mL extra for pipetting errors (i.e., If you plan to run 35 assays you will need 35 × 0.4 mL = 14 mL + 1 mL extra = 15 mL. Therefore, take 1 mL of reconstituted receptor and dilute it to 15 mL with assay buffer).

6. You do not need to prepare a tube for S1 (12 pmol) because this will be the reconstituted $InsP_3$ standard directly without any dilution. Also remember the tracer contains low levels of [^{3}H]-radiolabel, so be sure to wear gloves and a lab coat and work on bench paper. Also set up specially designated waste containers to collect both solid and liquid waste.
7. If using the tubes provided with the kit, label them using a soft pencil.
8. The receptor/tracer working solution needs to be kept on ice at all times and mixed frequently by inversion during addition to the samples to ensure that the receptor preparation is homogenous and to prevent any settling.
9. If using the tubes provided with the kit, make sure to cover the rack with plastic wrap to prevent evaporation. If using microfuge tubes, be sure to cap them.
10. The centrifugation should be carried out in the cold. If using the tubes provided they can be placed in the foam racks provided with the kit and will fit in a Sorvall H100B rotor. Alternatively, you can place the tubes in sleeves that fit in a fixed angle microfuge.
11. Make sure that the membrane pellet is completely dissolved.
12. When adding the 5 mL of scintillant to the vial, start by adding to the inner reaction tube and letting it flow over into the vial. Cap the vial and invert to allow trapped bubbles to rise to the surface. Slowly invert a few times and then shake vials.
13. To get better accuracy, plot the standard curve using Excel and find the best fit regression line. Use the values from the equation to calculate the concentration of $InsP_3$ in each sample.

Acknowledgments

The authors would like to acknowledge funding from the following sources: Agriculture and Food Research Initiative Grant # 2009-65114-06019 from the USDA National Institute of Food and Agriculture to I.Y.P.

References

1. Berridge MJ (1993) Inositol trisphosphate and calcium signalling. Nature 361:315–325
2. Perera IY, Hung CY, Moore CD, Stevenson-Paulik J, Boss WF (2008) Transgenic *Arabidopsis* plants expressing the Type 1 inositol 5-phosphatase exhibit increased drought tolerance and altered abscisic acid signaling. Plant Cell 20:2876–2893
3. Krinke O, Novotna Z, Valentova O, Martinec J (2007) Inositol trisphosphate receptor in higher plants: is it real? J Exp Bot 58:361–376
4. Stevenson-Paulik J, Bastidas RJ, Chiou ST, Frye RA, York JD (2005) Generation of phytate-free seeds in Arabidopsis through disruption of inositol polyphosphate kinases. Proc Natl Acad Sci USA 102:12612–12617

5. Krinke O, Ruelland E, Valentova O, Vergnolle C, Renou JP, Taconnat L, Flemr M, Burketova L, Zachowski A (2007) Phosphatidylinositol 4-kinase activation is an early response to salicylic acid in *Arabidopsis* suspension cells. Plant Physiol 144: 1347–1359
6. Meijer HJ, Munnik T (2003) Phospholipid-based signaling in plants. Annu Rev Plant Biol 54:265–306
7. Stevenson JM, Perera IY, Heilmann I, Persson S, Boss WF (2000) Inositol signaling and plant growth. Trends Plant Sci 5:252–258
8. Im YJ, Phillippy BQ, Perera IY (2010) InsP3 in plant cells. In: Munnik T (ed) Plant lipid signalling. Springer, Berlin, Germany, pp 145–160
9. Challiss RA, Batty IH, Nahorski SR (1988) Mass measurements of inositol(1,4,5)trisphosphate in rat cerebral cortex slices using a radioreceptor assay: effects of neurotransmitters and depolarization. Biochem Biophys Res Commun 157:684–691
10. Brearley CA, Hanke DE (2000) Metabolic relations of inositol 3,4,5,6-tetrakisphosphate revealed by cell permeabilization. Identification of inositol 3,4,5, 6-tetrakisphosphate 1-kinase and inositol 3,4,5,6-tetrakisphosphate phosphatase activities in mesophyll cells. Plant Physiol 122:1209–1216
11. Lee Y, Choi YB, Suh S, Lee J, Assmann SM, Joe CO, Kelleher JF, Crain RC (1996) Abscisic acid-induced phosphoinositide turnover in guard cell protoplasts of *Vicia faba*. Plant Physiol 110:987–996
12. Shigaki T, Bhattacharyya MK (2000) Decreased inositol 1,4,5-trisphosphate content in pathogen-challenged soybean cells. Mol Plant Microbe Interact 13:563–567
13. Laxminarayan KM, Chan BK, Tetaz T, Bird PI, Mitchell CA (1994) Characterization of a cDNA encoding the 43-kDa membrane-associated inositol-polyphosphate 5-phosphatase. J Biol Chem 269:17305–17310
14. Laxminarayan KM, Matzaris M, Speed CJ, Mitchell CA (1993) Purification and characterization of a 43-kDa membrane-associated inositol polyphosphate 5-phosphatase from human placenta. J Biol Chem 268: 4968–4974
15. Heilmann I, Perera IY, Gross W, Boss WF (1999) Changes in phosphoinositide metabolism with days in culture affect signal transduction pathways in *Galdieria sulphuraria*. Plant Physiol 119:1331–1339
16. Perera IY, Heilmann I, Boss WF (1999) Transient and sustained increases in inositol 1,4,5-trisphosphate precede the differential growth response in gravistimulated maize pulvini. Proc Natl Acad Sci USA 96:5838–5843
17. Perera IY, Heilmann I, Chang SC, Boss WF, Kaufman PB (2001) A role for inositol 1,4,5-trisphosphate in gravitropic signaling and the retention of cold-perceived gravistimulation of oat shoot pulvini. Plant Physiol 125:1499–1507
18. Perera IY, Love J, Heilmann I, Thompson WF, Boss WF (2002) Up-regulation of phosphoinositide metabolism in tobacco cells constitutively expressing the human type I inositol polyphosphate 5-phosphatase. Plant Physiol 129:1795–1806
19. Perera IY, Hung CY, Brady S, Muday GK, Boss WF (2006) A universal role for inositol 1,4,5-trisphosphate-mediated signaling in plant gravitropism. Plant Physiol 140: 746–760
20. Khodakovskaya M, Sword C, Wu Q, Perera IY, Boss WF, Brown CS, Winter Sederoff H (2010) Increasing inositol (1,4,5)-trisphosphate metabolism affects drought tolerance, carbohydrate metabolism and phosphate-sensitive biomass increases in tomato. Plant Biotechnol J 8:170–183
21. Heilmann I, Perera IY, Gross W, Boss WF (2001) Plasma membrane phosphatidylinositol 4,5-bisphosphate levels decrease with time in culture. Plant Physiol 126:1507–1518

Chapter 5

Quantification of Diacylglycerol by Mass Spectrometry

Katharina vom Dorp, Isabel Dombrink, and Peter Dörmann

Abstract

Diacylglycerol (DAG) is an important intermediate of lipid metabolism and a component of phospholipase C signal transduction. Quantification of DAG in plant membranes represents a challenging task because of its low abundance. DAG can be measured by direct infusion mass spectrometry (MS) on a quadrupole time-of-flight mass spectrometer after purification from the crude plant lipid extract via solid-phase extraction on silica columns. Different internal standards are employed to compensate for the dependence of the MS and MS/MS signals on the chain length and the presence of double bonds in the acyl moieties. Thus, using a combination of single MS and MS/MS experiments, quantitative results for the different molecular species of DAGs from *Arabidopsis* can be obtained.

Key words Diacylglycerol, Solid-phase extraction, Phospholipase C, *Arabidopsis*, Mass spectrometry, Quadrupole time of flight

1 Introduction

Diacylglycerol (DAG) is an important intermediate in biosynthetic processes and signaling transduction pathways in plants. An accurate method for its quantification is therefore of high importance. Previously, the most common technique for DAG analysis involved the purification via thin-layer chromatography (TLC) and quantification via gas chromatography after conversion into fatty acid methyl esters. However, this method is time consuming and does not provide information on the molecular species composition of DAGs. The development of highly sensitive and accurate mass spectrometers has provided the means to gather information on the molecular species composition and to record changes that occur in mutants or during exposure to environmental or biotic stresses in plants.

DAG contains a glycerol backbone with two fatty acids in ester linkage, and one free hydroxy group. Fatty acyl residues can be linked to the *sn*-1 and *sn*-2 positions (*sn*-1,2-isomer), or the *sn*-1 and *sn*-3 positions (*sn*-1,3-isomer) (Fig. 1). The *sn*-1,2-isomer is the one naturally occurring in organisms. In addition to the position, DAGs can

Teun Munnik and Ingo Heilmann (eds.), *Plant Lipid Signaling Protocols*, Methods in Molecular Biology, vol. 1009,
DOI 10.1007/978-1-62703-401-2_5, © Springer Science+Business Media, LLC 2013

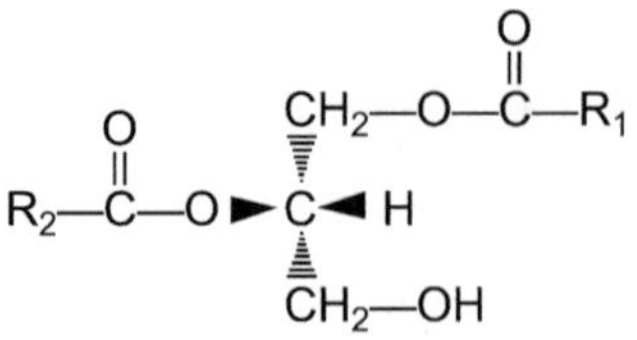

Fig. 1 Structure of *sn*-1,2-isomer of DAG. Because the C-2 of 1,2-DAG is chiral, the molecule exists in two enantiomeric configurations. The *sn*-1,2 isomer is the naturally occurring form

Table 1
Masses of neutral losses of acyl groups from DAGs during MS/MS experiments

	Formula		Molecular mass (*m/z*)
	Fatty acid	Acyl ammonium adduct	Acyl ammonium adduct
14:0	$C_{14}H_{28}O_2$	$C_{14}H_{31}O_2N$	245.2355
16:3	$C_{16}H_{26}O_2$	$C_{16}H_{29}O_2N$	267.2198
16:2	$C_{16}H_{28}O_2$	$C_{16}H_{31}O_2N$	269.2355
16:1	$C_{16}H_{30}O_2$	$C_{16}H_{33}O_2N$	271.2511
16:0	$C_{16}H_{32}O_2$	$C_{16}H_{35}O_2N$	273.2668
18:3	$C_{18}H_{30}O_2$	$C_{18}H_{33}O_2N$	295.2511
18:2	$C_{18}H_{32}O_2$	$C_{18}H_{35}O_2N$	297.2668
18:1	$C_{18}H_{34}O_2$	$C_{18}H_{37}O_2N$	299.2824
18:0	$C_{18}H_{36}O_2$	$C_{18}H_{39}O_2N$	301.2981

Fatty acyl groups are cleaved off the DAG molecular ions as acyl ammonium adducts (neutral loss) during CID in MS/MS experiments. The *m/z* values were calculated using the MassCalculator tool of the Agilent MassHunter Qualitative Software

differ with regard to the chain lengths and degree of desaturation of the acyl residues. The fatty acids occurring in DAGs of plants are mainly 16:0, 18:1, 18:2, and 18:3. The structural variability of DAGs poses difficulties with respect to MS analysis of molecular species. As previously shown for TAGs, MS/MS signal intensities of DAG ions depend on the chain length and degree of unsaturation of the acyl group (1, 2).

Upon collision induced dissociation (CID), fatty acyl chains are cleaved off the glycerol backbone of DAG, while the positive charge remains on the monoacylglycerol (MAG) ion. The *m/z* value of the neutral loss corresponds to the ammonium adducts of the carboxylate (Table 1). The peak intensities of the product ions generated by CID of $(M+NH_4)^+$ ions depend on the degree of unsaturation and on the linkage to the *sn-1* or *sn-2* positions of the

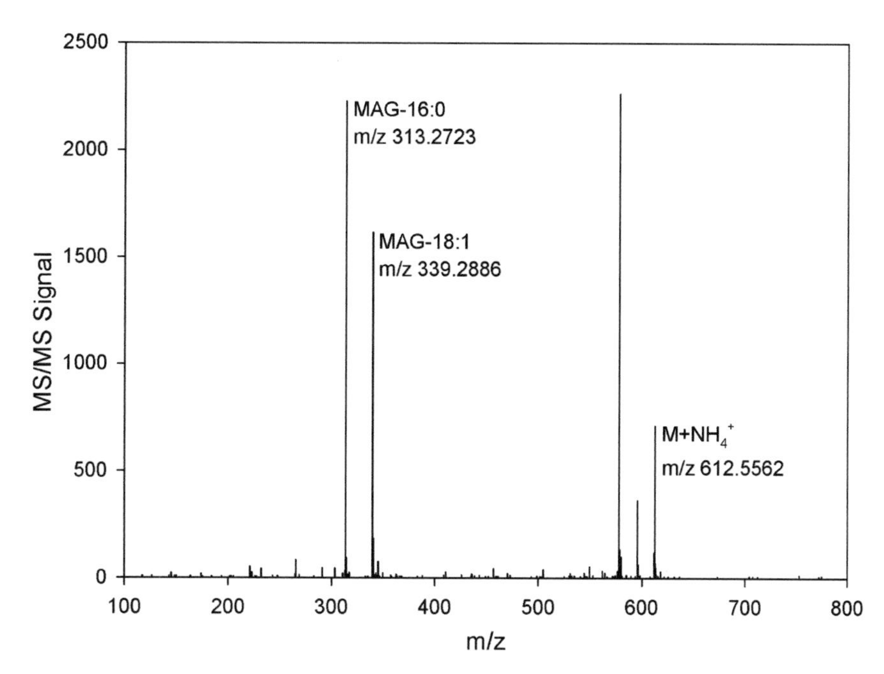

Fig. 2 Q-TOF MS/MS spectrum of *sn*-1-16:0-*sn*-2-18:1-DAG. Collision induced dissociation of the $(M+NH_4)^+$ adduct ion of *sn*-1-16:0-*sn*-2-18:1-DAG produces 16:0-MAG and 18:1-MAG fragment ions. Note that the two MAG fragment peaks show different signal intensities, depending on the degree of unsaturation and position of the fatty acid on the glycerol backbone

acyl groups (Fig. 2) (1, 2). The most abundant product ions resulting from fragmentation of a *sn*-1-16:0-*sn*-2-18:1-DAG are 16:0-MAG and 18:1-MAG after neutral loss of one fatty acid moiety. The signal intensity of the 16:0-MAG is higher than that of 18:1-MAG, even though the acyl groups are present in equal molar ratios (Fig. 2).

In the past, a number of LC-MS based strategies were applied to measure glycerolipids, including TAG and DAG. High-performance liquid chromatography using a reverse phase column coupled to atmospheric pressure chemical ionization MS was applied to quantify the molecular species of TAGs and DAGs in plant oils (3). Separation by normal-phase liquid chromatography coupled to electrospray ionization (ESI) MS was employed for the measurement of nonpolar lipids including DAG and TAG (4).

A major obstacle for DAG quantification in crude lipid extracts is its extremely low abundance. DAG can be enriched and purified from "contaminating" phospholipids by silica column chromatography prior to ESI-triple quadrupole MS (5). Alternatively, DAG was purified by TLC and quantified as Li^+ adducts by ESI-MS (6). Further strategies for DAG quantification include the derivatization of the free hydroxy group prior to fragmentation with the goal to improve ionization and to introduce a good leaving group. For example, DAG was derivatized with N-chlorobetainyl chloride (7) or by acetylation (8).

Quantification of DAG from unfractionated lipid extracts from *Arabidopsis* leaves was achieved by neutral loss scanning of ammonium adducts $(M+NH_4)^+$, using ESI-MS/MS on a triple quadrupole instrument (9). However, only relative amounts for the mass signals of DAG molecular species could be obtained, as the signal intensity in MS/MS experiments of DAG depends on the position on the glycerol backbone and presence of double bonds of the acyl chains (see above).

Here, a method is described for the quantitative determination of DAG via direct infusion nanospray quadrupole time-of-flight mass spectrometry (Q-TOF MS). To this end, lipid fractions purified by solid-phase extraction on silica columns are applied to the Q-TOF MS instrument via direct nanospray infusion. Without further chromatographic separation, DAGs are ionized in the nanospray ion source as ammonium adducts and measured in the positive mode. To compensate for differences in the chain lengths and presence of double bonds in the acyl groups, a set of four internal standards is employed, i.e., one DAG standard each with saturated short chain, saturated long chain, unsaturated short chain, and unsaturated long chain acyl moieties.

The molecular DAG ions are first quantified via TOF MS without fragmentation ("MS only mode") to avoid a bias of signal intensities due to differences in fragmentation efficiencies of the acyl groups. In this aspect, it is advantageous to purify DAG fractions prior to the measurements, and to use accurate mass spectrometers (e.g., TOF) to avoid "contamination" of DAG molecular ion peaks with matrix peaks. Calculation based on the internal standards yields the amounts of $(M+NH_4)^+$ DAG ions in nmol per sample fresh weight (Fig. 3a). Subsequently, MS/MS analysis with neutral loss scanning is employed to identify and quantify the fatty acyl residues in the molecular DAG ions (Table 2, Fig. 3b). The results of these two approaches are in agreement, as the distribution of molecular species obtained via "MS only" mode or via MS/MS are comparable (Fig. 3).

2 Materials

For sample preparation for LC-MS, only solvents of highest purity should be used. Glass vials should be clean and prewashed with $CHCl_3$/methanol (2:1) (*see* **Note 1**).

2.1 Lipid Standards for Q-TOF MS

The following DAGs are used as internal standards for quantification via Q-TOF MS:

1,2/1,3-di-14:0-(28:0)-DAG, 1,2/1,3-di-20:0-(40:0)-DAG, 1,2/1,3-di-14:1-(28:2)-DAG and 1,2/1,3-di-20:1-(40:2)-DAG (Avanti Lipids, Alabaster, Alabama).

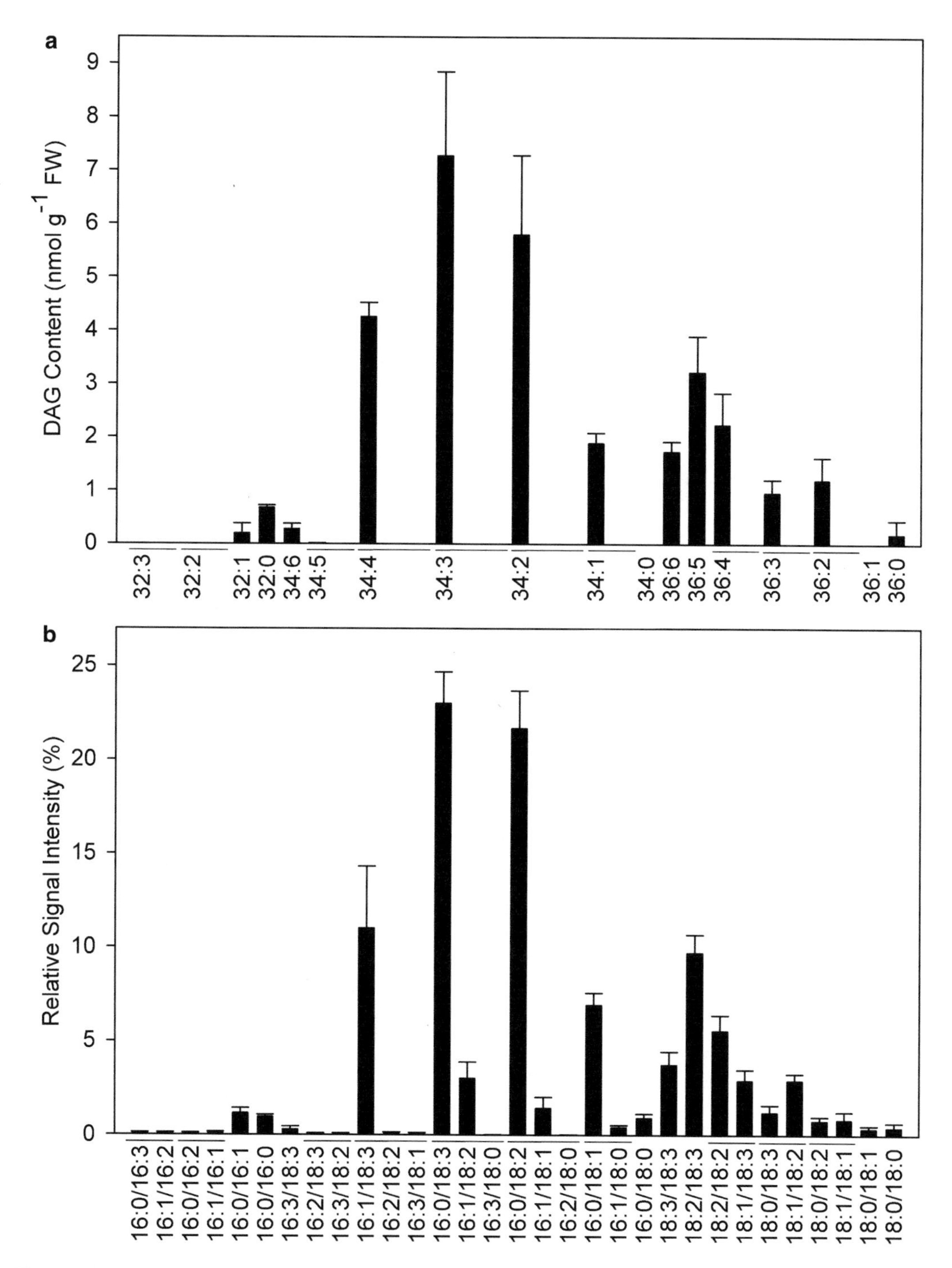

Fig. 3 DAG composition of *Arabidopsis* wild type Col-0 leaves. (**a**) The contents of different DAGs were quantified by MS experiments without fragmentation. The identity of DAG molecular ions was confirmed by CID of the $(M+NH_4)^+$ adducts. (**b**) The relative amounts of DAG molecular species were obtained by MS/MS experiments of the $(M+NH_4)^+$ adducts of molecular ions (see Table 2). The MS/MS signals of 16:0/16:0, 16:0/18:0, and 18:0/18:0 were normalized to saturated standards, and the other molecular species to unsaturated standards

Table 2
Molecular species of DAG in leaves

	Formula	Molecular mass (m/z)	
	$M + NH_4^+$	$M + NH_4^+$	Fatty acid composition[a]
28:2	$C_{31}H_{60}O_5N$	526.4471	14:1/14:1 (internal standard)
28:0	$C_{31}H_{64}O_5N$	530.7790	14:0/14:0 (internal standard)
32:2	$C_{35}H_{68}O_5N$	582.5097	16:1/16:1
32:1	$C_{35}H_{70}O_5N$	584.5249	16:0/16:1
32:0	$C_{35}H_{72}O_5N$	586.5405	16:0/16:0
34:6	$C_{37}H_{64}O_5N$	602.4780	16:3/18:3
34:5	$C_{37}H_{66}O_5N$	604.4936	16:2/18:3, 16:3/18:2
34:4	$C_{37}H_{68}O_5N$	606.5092	16:1/18:3, 16:2/18:2, 16:3/18:1
34:3	$C_{37}H_{70}O_5N$	608.5249	16:0/18:3>>16:1/18:2, 16:3/18:0
34:2	$C_{37}H_{72}O_5N$	610.5405	16:0/18:2>>16:1/18:1, 16:2/18:0
34:1	$C_{37}H_{74}O_5N$	612.5562	16:0/18:1>16:1/18:0
34:0	$C_{37}H_{76}O_5N$	614.5718	16:0/18:0
36:6	$C_{39}H_{68}O_5N$	630.5092	18:3/18:3
36:5	$C_{39}H_{70}O_5N$	632.5249	18:2/18:3
36:4	$C_{39}H_{72}O_5N$	634.5405	18:2/18:2>18:1/18:3
36:3	$C_{39}H_{74}O_5N$	636.5562	18:0/18:3, 18:1/18:2
36:2	$C_{39}H_{76}O_5N$	638.5718	18:0/18:2, 18:1/18:1
36:1	$C_{39}H_{78}O_5N$	640.5875	18:0/18:1
36:0	$C_{39}H_{80}O_5N$	642.6031	18:0/18:0
40:2	$C_{43}H_{84}O_5N$	694.6349	20:1/20:1 (internal standard)
40:0	$C_{43}H_{88}O_5N$	698.6657	20:0/20:0 (internal standard)

">" indicate that one molecular species is more abundant than the other one
[a]The acyl composition was identified by MS/MS experiments
The m/z ratios of NH_4^+ adducts of molecular DAG species were calculated using the MassCalculator tool of the Agilent MassHunter Qualitative Software

2.2 Lipid Extraction and Solid-Phase Extraction

1. Glass tubes with screw caps and Teflon inserts (*see* **Note 1**).
2. 2 mL microfuge tubes.
3. Pasteur pipettes.
4. $CHCl_3$/methanol/formic acid (1:1:0.1, v/v/v).

5. 1 M KCl/0.2 M H_3PO_4.
6. $CHCl_3$/methanol (2:1, v/v).
7. Silica columns for low-pressure normal-phase chromatography (solid-phase extraction; 100 mg silica).
8. Organic solvents: $CHCl_3$, hexane, diethylether.
9. Sample concentrator with nitrogen gas stream.

2.3 Q-TOF MS and Data Analysis

1. Mass spectrometry solvent: methanol/$CHCl_3$/300 mM NH_4^+ acetate (665:300:35, v/v/v) (9, 10).
2. Agilent Series 6530 Accurate Mass Q-TOF LC-MS instrument with Chip Cube source and Agilent Series 1200 LC and infusion chip (Agilent Technologies).
3. Agilent MassHunter Qualitative Analysis software (Agilent Technologies).
4. Microsoft Excel (Microsoft Corp.).

3 Methods

3.1 Preparation of Internal Standards for DAG Quantification

1. Prepare 10× stock: weigh DAG standards (see Subheading 2.1) and dissolve in 100 % $CHCl_3$ to give a final concentration of about 1 µmol/mL. Lipid standards dissolved in $CHCl_3$ can be stored at −20 °C in a glass tube with screw cap and Teflon inlay.
2. Dilute standard stock 1:10 with $CHCl_3$ and mix 100 µL of this dilution with 5 µg of pentadecanoic acid (15:0) internal standard for fatty acid methyl ester synthesis and quantification by GC (see Chapter 8).
3. Add 200 µL of each diluted DAG standard (0.1 nmol/µL) to one vial and add $CHCl_3$ to 1,000 µL final volume. The standard mixture contains 0.02 nmol/µL of each internal DAG standard.

3.2 Lipid Extraction from Arabidopsis thaliana Leaves

Harvest 100 mg of leaves from *Arabidopsis*, quickly determine fresh weight and immediately freeze in liquid nitrogen.

1. Grind sample to fine powder with steel or ceramic balls using the Mixer Mill (Retsch Corp., Haan, Germany) or Precellys homogenizer (Bertin Technol., Saint-Quentin-en-Yvelines, France), or alternatively with mortar and pestle (*see* **Note 2**). Do not allow your samples to thaw during homogenization (*see* **Note 3**)!
2. Quickly add 500 µL of $CHCl_3$/methanol/formic acid (1:1:0.1, v/v/v) to the ground tissue (*see* **Note 4**) (11, 12). Thoroughly vortex the sample.
3. Add 50 µL of internal standard mixture, containing 1 nmol of each DAG standard to the extract and vortex again (*see* **Notes 5** and **6**).

4. Add 250 µL of 1 M KCl/0.2 M H_3PO_4 (13). Then vortex vigorously and centrifuge for 5 min at 5,000 × *g* to obtain phase separation.
5. Transfer the lower (organic) phase into a fresh glass tube with screw cap by carefully pipetting through the aqueous (upper) phase and the tissue layer (interphase) with a Pasteur pipette (*see* **Notes 1** and 7).
6. Extract the residual tissue and aqueous phase two more times with $CHCl_3$/methanol (2:1) and combine the organic extracts in the glass tube.

Evaporate the solvents under N_2 gas flow.

3.3 Lipid Fractionation via Two-Step Solid-Phase Extraction

Nonpolar lipids are first separated from glycolipids and phospholipids on a small silica column (steps 1–3) (14). Subsequently, DAGs are purified from other nonpolar lipids by a second round of step gradient chromatography on a silica column (steps 4–9) (see: http://www.cyberlipid.org) (*see* **Note 8**).

1. Dissolve dried lipids in 400 µL of 100 % $CHCl_3$, close the glass tube with a screw cap with teflon inlay and vortex vigorously.
2. Equilibrate a silica column (100 mg silica) by flushing with 2 mL of $CHCl_3$.
3. Place column onto fresh glass vial (*see* **Note 2**) and apply extracted lipids to the equilibrated column with a Pasteur pipette. Elute nonpolar lipids with 1 mL of $CHCl_3$. Evaporate solvent under N_2 flow.
4. DAGs are separated from other nonpolar lipids with a step gradient elution from a second silica column (100 mg silica) using hexane and diethylether.
5. Add 400 µL of 100 % hexane to dissolve the nonpolar lipids.
6. Equilibrate silica column by flushing with 2 mL of hexane.
7. Place column on clean glass vial and apply nonpolar lipids to equilibrated column. Elute stepwise with the following:
   (a) 3 mL of hexane: hydrocarbons and squalene;
   (b) 6 mL of hexane/diethylether (99:1): sterol esters, waxes, and fatty methyl esters;
   (c) 5 mL of hexane/diethylether (95:5): TAG, alkyl and alkenyl acylglycerols, and tocopherols;
   (d) 5 mL of hexane/diethylether (92:8): free fatty acids and fatty alcohols;
   (e) 8 mL of hexane/diethylether (85:15): sterols and DAG;
   (f) 5 mL of diethylether: some DAG and MAG.
8. Fractions (e) and (f) contain the DAGs and are combined. The other fractions can be discarded.

9. Evaporate solvent from DAG fractions under N_2 gas stream and dissolve in Q-TOF solvent prior to analysis. Lipids can be stored in solvent at –20 °C.

3.4 Q-TOF MS Analysis of DAGs

1. Lipid samples purified by SPE on silica columns are dissolved in Q-TOF solvent (9, 10) and introduced into the Q-TOF instrument via direct nanospray infusion without column chromatography (*see* **Note 8**). The flow rate is 1 μL/min. Instrumental settings are as follows: collision energy (CE), 20 V; drying gas, 8 L/min of nitrogen; fragmentor voltage, 200 V; gas temperature, 300 °C, HPLC chip capillary voltage V_{cap}, 1,700 V; scan rate, 1 spectrum/s.
2. Ionization is done in the positive mode, yielding NH_4^+ adducts of DAG molecules.
3. Total ion chromatograms are recorded by MS experiments without fragmentation ("MS only mode"). The masses of molecular DAG ions $(M+NH_4)^+$ are extracted from the total ion chromatograms using the MassHunter Qualitative Analysis software, and an average intensity from four spectra is calculated. The window for molecular ion detection is set to 100 ppm.
4. In MS/MS experiments, the first quadrupole selects molecular DAG ions $(M+NH_4^+)$ (Table 2) for subsequent fragmentation in the collision cell (collision gas: N_2) (see Fig. 2 for a representative MS/MS spectrum). A "neutral loss scan" for the loss of fatty acids from $M+NH_4^+$ is done using the MassHunter Qualitative Analysis software (Table 1). The window for fragment ion detection is set to 100 ppm.

3.5 Data Analysis of Q-TOF Measurements

1. The peak intensities ("MS only" mode) of specific fragments extracted from the chromatograms using the Agilent MassHunter Qualitative Analysis software are exported to Microsoft Excel (see Subheading 3.4, steps 3 and 4) or equivalent software.
2. The isotope distribution pattern for a given molecule is calculated by the use of the Agilent MassHunter Isotope Distribution Calculator and the contribution of isotopic overlap is subtracted from the peak intensities (*see* **Note 9**).
3. A correction trend line is calculated for the saturated standards (di-14:0-DAG, di-20:0-DAG) and for the unsaturated standards (di-14:1-DAG and di-20:1-DAG) by linear regression taking into account the peak sizes of the molecular DAG ions (15) (*see* **Note 10**).
4. The content of DAG in nmol/g FW is calculated using the internal saturated standards (for 34:0, 36:0, 38:0) and the unsaturated standards (all other DAGs) (see Fig. 3a).

5. The distribution of molecular species is calculated from MS/MS experiments employing a strategy analogous to the one used in "MS only" mode, including correction for isotopic overlap and calculation of a correction trend line for the two saturated and unsaturated standards. The MS/MS signal intensities of the MS/MS peaks are normalized to the internal standards and presented in % of total DAG molecular species (Fig. 3b).

4 Notes

1. To prevent contamination of the purified lipid fraction, glass tubes that are used for storage of lipid extracts or column fractions should be rinsed with $CHCl_3$/methanol 2:1 prior to use.
2. Homogenization with pestle and mortar is more laborious than the use of automatic homogenization mills and therefore should be used for small numbers of samples.
3. When harvesting many plant samples at the same time, make sure that all Eppendorf tubes are completely immersed in liquid nitrogen and no samples are allowed to thaw at the surface of the Dewar container. Transfer frozen samples from the liquid nitrogen container to the −80 °C freezer on a regular basis during harvesting.
4. Never process more than one frozen sample at a time, as it takes some time before all material is dissolved in the extraction solvent. Thawing of sample tissue in the absence of extraction solvent might result in lipid degradation by lipases.
5. Standards should be selected to cover the mass range of naturally occurring DAGs and should to be added to the sample in equimolar ratios.
6. Complete saturation of the fatty acyl chain in DAG decreases the ionization efficiency of a DAG molecule as compared to DAGs with fatty acids containing one or more double bonds. A set of saturated and unsaturated internal standards is used to compensate for these differences: saturated DAG molecular species are quantified using saturated standards and unsaturated standards are used for the quantification of unsaturated DAG molecular species.
7. When transferring the lower organic phase from one vial to another after phase separation, carefully blow small amounts of air through the Pasteur pipette while moving through the aqueous layer to prevent contaminants from being soaked into the pipette tip.
8. As DAGs represent only a minor lipid class in leaves, and DAGs are prone to ion suppression during LC-MS experiments in the

presence of polar lipids, a purification of DAGs using SPE is recommended.

9. Some of the DAG molecular species measured by TOF MS differ by exactly two *m/z* units (A+2 peaks). Therefore, the peak of the molecule A+ 2 exclusively containing ^{12}C atoms might be “contaminated” by the contribution of the $^{13}C_2$ isotope peak (containing two ^{13}C atoms in the molecule) of an A ion carrying one additional double bond in the acyl chain.
10. It has been shown that ionization and fragmentation of different molecular species of a lipid depend on the size of the molecule. In many (but not all) cases, the larger molecular species shows a lower MS/MS peak signal than a smaller molecular species (10, 15). For this reason, it is desirable to include two different molecular species of a given lipid class as internal standards for quantification.

Acknowledgments

We would like to thank Helga Peisker (University of Bonn) for help during sample preparation, method development and data evaluation. This work was supported by an Instrument Grant (Forschungsgrossgeräte-Antrag) and by the Sonderforschungsbereich 645 of Deutsche Forschungsgemeinschaft.

References

1. Li X, Evans J (2005) Examining the collision-induced decomposition spectra of ammoniated triglycerides as a function of fatty acid chain length and degree of unsaturation. I. The OXO/YOY series. Rapid Commun Mass Spectrom 19:2528–2538
2. Han X, Gross R (2001) Quantitative analysis and molecular species fingerprinting of triacylglyceride molecular species directly from lipid extracts of biological samples by electrospray ionization tandem mass spectrometry. Anal Biochem 295:88–100
3. Holčapek M, Jandera P, Zderadička P, Hrubá L (2003) Characterization of triacylglycerol and diacylglycerol composition of plant oils using high-performance liquid chromatography–atmospheric pressure chemical ionization mass spectrometry. J Chromatogr A 1010: 195–215
4. Hutchins P, Barkley R, Murphy R (2008) Separation of cellular nonpolar neutral lipids by normal-phase chromatography and analysis by electrospray ionization mass spectrometry. J Lipid Res 49:804–813
5. Callender H, Forrester J, Ivanova P, Preininger A, Milne S, Brown H (2007) Quantification of diacylglycerol species from cellular extracts by electrospray ionization mass spectrometry using a linear regression algorithm. Anal Biochem 79:263–272
6. Hsu F-F, Ma Z, Wohltmann M, Bohrer A, Nowatzke W, Ramanadham S, Turk J (2000) Electrospray ionization/mass spectrometric analyses of human promonocytic U937 cell glycerolipids and evidence that differentiation is associated with membrane lipid composition changes that facilitate phospholipase A2 activation. J Biol Chem 275:16579–16589
7. Li YL, Su X, Stahl PD, Gross ML (2007) Quantification of diacylglycerol molecular species in biological samples by electrospray ionization mass spectrometry after one-step derivatization. Anal Chem 79:1569–1574
8. Bates P, Durrett T, Ohlrogge J, Pollard M (2009) Analysis of acyl fluxes through multiple pathways of triacylglycerol synthesis in developing soybean embryos. Plant Physiol 150: 55–72

9. Peters C, Li M, Narasimhan R, Roth M, Welti R, Wang X (2010) Nonspecific phospholipase C NPC4 promotes responses to abscisic acid and tolerance to hyperosmotic stress in *Arabidopsis*. Plant Cell 22:2642–2659
10. Welti R, Li W, Li M, Sang Y, Biesiada H, Zhou H-E, Rajashekar C, Williams T, Wang X (2002) Profiling membrane lipids in plant stress responses. Role of phospholipase Dα in freezing-induced lipid changes in Arabidopsis. J Biol Chem 277:31994–32002
11. Bligh EG, Dyer WJ (1959) A rapid method of total lipid extraction and purification. Can J Biochem Physiol 37:911–917
12. Browse J, Warwick N, Somerville CR, Slack CR (1986) Fluxes through the prokaryotic and eukaryotic pathways of lipid synthesis in the '16:3' plant *Arabidopsis thaliana*. Biochem J 235:25–31
13. Hajra A (1974) On extraction of acyl and alkyl dihydroxyacetone phosphate from incubation mixtures. Lipids 9:502–505
14. Wewer V, Dombrink I, vom Dorp K, Dörmann P (2011) Quantification of sterol lipids in plants by quadrupole time-of-flight mass spectrometry. J Lipid Res 52:1–16
15. Brügger B, Erben G, Sandhoff R, Wieland FT, Lehmann WD (1997) Quantitative analysis of biological membrane lipids at the low picomole level by nano-electrospray ionization tandem mass spectrometry. Proc Natl Acad Sci USA 94:2339–2344

Chapter 6

Distinguishing Phosphatidic Acid Pools from De Novo Synthesis, PLD, and DGK

Steven A. Arisz and Teun Munnik

Abstract

In plants, phosphatidic acid (PA) functions as a metabolic precursor in the biosynthesis of glycerolipids, but it also acts as a key signaling lipid in the response to environmental stress conditions (Testerink and Munnik, J Exp Bot 62:2349–2361, 2011). In vivo ^{32}P-radiolabeling assays have shown the level of PA to increase within seconds/minutes of exposure to a stimulus. This response can be due to the activity of diacylglycerol kinase (DGK) and/or phospholipase D (PLD). A method is described to investigate which of the pathways is responsible for PA accumulation under a particular stress condition.

First, a differential ^{32}P-radiolabeling protocol is used to discriminate ^{32}P-PA pools that are rapidly labeled versus those requiring long prelabeling times, reflecting DGK and PLD activities, respectively. Second, to specifically monitor the contribution of PLD, a transphosphatidylation assay is applied, which makes use of the artificial lipid phosphatidylbutanol as an in vivo marker of PLD activity.

Key words Phosphatidic acid, Diacylglycerol kinase, Phospholipase C, Phospholipase D, Phospholipid metabolism, Differential radiolabeling, Transphosphatidylation

1 Introduction

Many environmental and developmental stimuli trigger the accumulation of phosphatidic acid (PA) in plants (1). While PA is known as a metabolic intermediate in glycerolipid biosynthesis and turnover, it is also emerging as a signaling lipid in stress responses. In de novo synthesis, PA is generated through the consecutive acylations of glycerol-3-phosphate (Gro-P) and lysophosphatidic acid (LPA, Fig. 1, I). In stress responses, PA mainly arises through two additional pathways, featuring phospholipase D (PLD) activity and diacylglycerol kinase (DGK). PLD produces PA by hydrolysis of structural phospholipids like PE and PC (Fig. 1, II), while DGK generates it by phosphorylating diacylglycerol (Fig. 1, III). To understand PA's biological significance, it is important to discern these routes, and for this purpose an in vivo differential $^{32}P_i$-labeling protocol has been developed (2, 3).

Teun Munnik and Ingo Heilmann (eds.), *Plant Lipid Signaling Protocols*, Methods in Molecular Biology, vol. 1009, DOI 10.1007/978-1-62703-401-2_6,

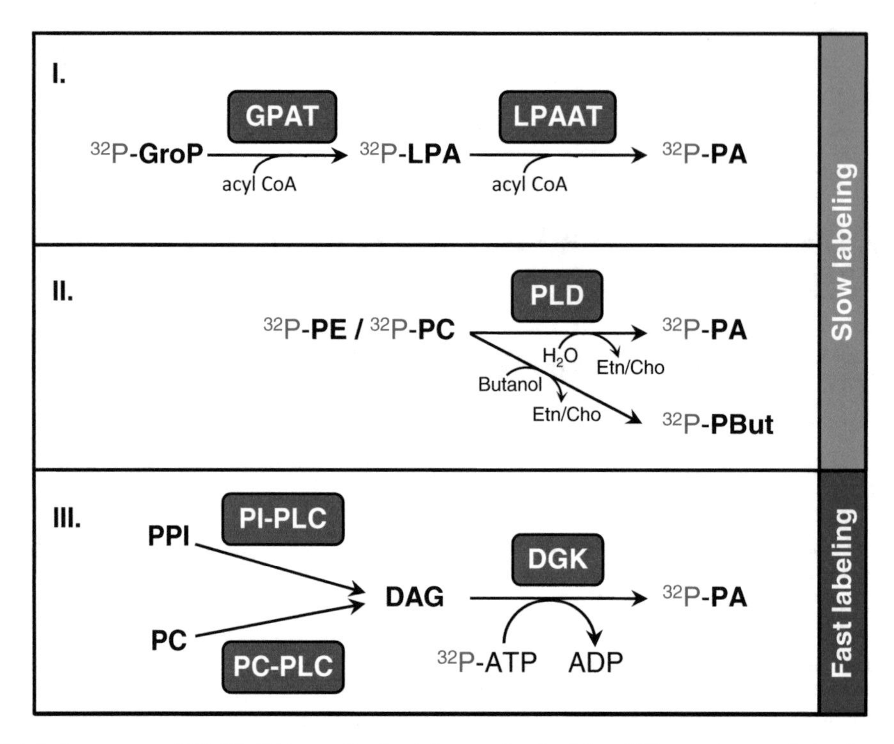

Fig. 1 Metabolic sources of ^{32}P-PA. PA is synthesized de novo in plastid and ER membranes through successive acylations of Gro-P and LPA, by Gro-P acyltransferase (GPAT) and LPA acyltransferase (LPAAT), respectively (*route I*). The activity of PLD hydrolyzes structural phospholipids like PC/PE to generate PA and a free head group (*route II*). In the presence of butanol, this acts as acceptor of the phosphatidyl moiety, generating the artificial lipid PBut, which is used as in vivo reporter of PLD activity. PA can also be formed by phosphorylation of DAG by DGK (*route III*). The three pathways exhibit different kinetics of ^{32}P-labeling. PA species resulting from de novo synthesis and PLD activity display slow ^{32}P-labeling (*blue*). In contrast, PA formed through DGK is very rapidly labeled (*red*) because ATP is one of the first compounds that become labeled when incubating cells with $^{32}P_i$. The *differential labeling protocol* uses this difference to provide evidence for the participation of the DGK pathway

When cells, plants, or plant parts are incubated with ^{32}P-orthophosphate ($^{32}P_i$), it is quickly taken up and incorporated into ATP, and subsequently into macromolecules, such as DNA and phospholipids. Some phospholipids, such as polyphosphoinositides, get heavily labeled within seconds/minutes, whereas others, like phosphatidylethanolamine (PE) or phosphatidylglycerol (PG), require much longer labeling (Fig. 2a, b). This difference is the consequence of the fact that polyphosphoinositides acquire their monoester-linked ^{32}P-phosphates directly from the ATP-pool, which is rapidly loaded with $^{32}P_i$, while PE and PI get their radiolabel indirectly, via precursor pools such as ^{32}P-CDP-ethanolamine and ^{32}P-CDP-DAG. Hence, differences in turnover rates and pool sizes of phospholipids and their precursors result in diverging labeling kinetics.

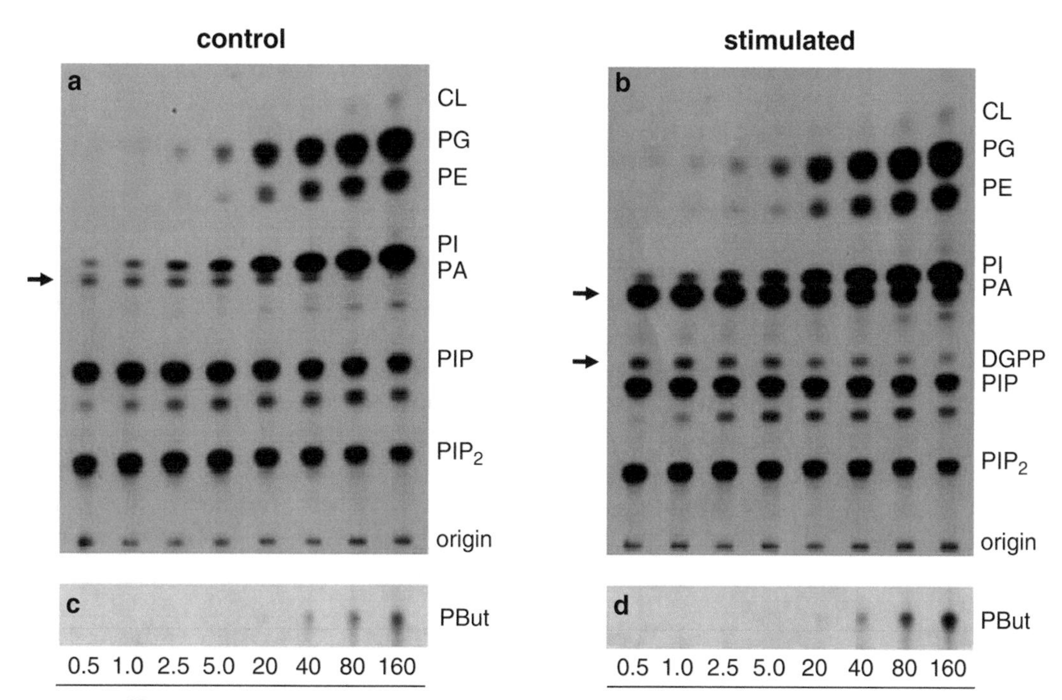

Fig. 2 Differential ^{32}P-labeling and transphosphatidylation. The unicellular green alga *Chlamydomonas moewusii* was metabolically prelabeled with $^{32}P_i$ for the times indicated and subsequently treated for 1 min with either buffer (*control*; **a** and **c**) or 1 μM mastoparan (*stimulated*; **b** and **d**), in the presence of 0.1 % butanol. Lipids were extracted, split into half, and separated either on ethyl acetate TLC for PLD-generated ^{32}P-PBut formation (**c** and **d**), or by alkaline TLC to visualize the rest of the phospholipids (**a** and **b**). After 0.5–5 min of prelabeling, stimulation resulted in large PA (and DGPP) increases (compare **a** and **b**, indicated by *arrows*), while the structural phospholipids were hardly labeled at these time points. Therefore, this ^{32}P-PA could not be derived from PLD activity (no ^{32}P-labelled PBut yet), and most likely is due to DGK. Nevertheless, the increment witnessed in PBut (compare **c** and **d**) is evidence of stimulated PLD activity. Both pathways were concluded to contribute to mastoparan-induced PA accumulation in *Chlamydomonas* (3)

Accordingly, differences are found in PA pools arising from DGK or PLD activity, the latter showing a slow and steady increase in ^{32}P-PA similar to its precursors ^{32}P-PE and ^{32}P-PC, the former a rapid and more transient rise, reflecting the kinetics of ATP-labeling. Hence, short labeling conditions reflect DGK-generated PA, and long labeling times predominantly show PLD-derived PA. Nevertheless, it should be kept in mind that also de novo synthesis of PA displays slow ^{32}P-radiolabeling kinetics (4).

While this so-called differential labeling is useful to discriminate between DGK-derived and PLD-derived PA, we describe another procedure to selectively measure PLD. For this, PLD's unique ability to catalyze a transphosphatidylation reaction is used. In the presence of a primary alcohol, such as *n*-butanol, the phosphatidyl moiety of the phospholipid substrate is transferred to the alcohol in a reaction that produces phosphatidylbutanol (PBut, Fig. 1, II). This artificial phospholipid can be visualized by TLC (Fig. 2c, d) and is used as an in vivo marker of PLD activity (5).

2 Materials

Store all solvents at 4 °C and lipid extracts at −20 °C. Use Milli-Q water for solvents. All ratios (between brackets) are by volume. Organic solvents are stored in dark-glass flasks with screw caps lined with solvent-resistant septa. Procedures involving the transfer of organic solvents (e.g., chloroform or methanol) should be performed in a fume hood. Working with radioactivity requires special precautions which are not discussed here. In general, always take care of proper shielding (e.g., 1 cm perpex) to protect against radiation and check frequently for contaminations.

2.1 In Vivo ^{32}P-Labeling

1. 2 mL Eppendorf polypropylene reaction tubes with safe-lock.
2. 2.56 mM MES buffer, pH 5.7. Dissolve 0.05 g MES (2-(N-morpholino) ethanesulfonic acid) in about 90 mL water. Adjust to pH 5.7 using KOH buffer, and then add water to 100 mL.
3. Diluted $^{32}P_i$-buffer (0.2 μCi/μL). Pipet 8 μL carrier-free $^{32}P_i$ (source 10 μCi/μL) in a 2 mL polypropylene reaction tube and then add 392 μL MES buffer. If 25 μL is added per incubation, this amounts to 5 μCi each. When radioactivity of the stock has declined, the volumes have to be modified accordingly (*see* **Note 1**).
4. 10 % perchloric acid solution. Add 10 mL concentrated perchloric acid (70 % (w/v)) to 60 mL H_2O in a glass flask/beaker.
5. 0.9 % (w/v) NaCl. Dissolve 0.9 g NaCl in 100 mL H_2O.
6. 1 M HCl: as above, except add 8.3 mL concentrated HCl.
7. Extraction mix: chloroform/methanol/HCl (50:100:1, v/v/v).
8. $CHCl_3$.
9. Washing mix: chloroform/methanol/1 M HCl (3:48:47, v/v/v).
10. Flask for liquid radioactive waste.
11. 2-propanol.
12. Table centrifuge for 2 mL reaction tubes, set at 13,000 × *g*.
13. Vortex, preferably suitable for simultaneous shaking of multiple tubes.
14. Temperature-controlled vacuum centrifuge (Speedvac).

15. N_2 gas.
16. TLC and autoradiography equipment.
17. Stopwatch/timer.

2.2 In Vivo Transphosphatidylation

1. Stock solution butanol 2.5 % (v/v). Pipet in a 10 mL glass tube 7.8 mL water; then add 200 μL butanol and vortex (*see* **Note 2**).
2. Stimulus-containing solution. Make this solution in a mixture of MES buffer (Subheading 2.1, item 2) and 2.5 % butanol (Subheading 2.2, item 1) in a ratio of 4:1.
3. Solution for mock treatment. Solutions as in Subheading 2.2, item 2 but without stimulus.

3 Methods

3.1 In Vivo ^{32}P-Labeling Time Course

1. Add to a series (*see* **Note 3**) of 2 mL polypropylene reaction tubes 175 μL MES buffer per tube.
2. Transfer plant material (leaf disks, seedlings, roots, etc.) to tubes.
3. Add 25 μL of diluted $^{32}P_i$-buffer (see Subheading 2.1, item 3).
4. At the first sampling time (20 min), add 200 μL 10 % perchloric acid to each of three tubes (*see* **Note 4**).
5. Mix vigorously for 5 min on a vortex and sediment liquids for 5 s.
6. Carefully remove as much as possible of the liquid with a Pasteur pipet, leaving plant material in the tube.
7. Add 750 μL extraction mix.
8. Vortex for 10 min and sediment liquid in table centrifuge for 5 s.
9. Add 750 μL chloroform and 400 μL 0.9 % NaCl, respectively.
10. Vortex 5 min and sediment material in table centrifuge for 2 min.
11. Transfer lower phase to fresh polypropylene reaction tubes, using a Gilson P1000 pipet (*see* **Note 5**).
12. Add to the new tubes 750 μL washing mix.
13. Vortex 5 min and sediment material in table centrifuge (13,000 × *g*) for 2 min.
14. Discard upper phase, using a normal laboratory pipet or a Pasteur pipet.
15. Add 20 μL 2-propanol to each extract.
16. Vortex 1 min and sediment liquid in a table centrifuge (13,000 × *g*) for 10 s.

17. Dry extracts down in a Speedvac set at a temperature of 50 °C for 1 h.
18. Resuspend the dried lipids in 40 μL $CHCl_3$ using a vortex.
19. Store at −20 °C under N_2 gas until further analysis.
20. Repeat this procedure at each time point.

Separate lipids on Alkaline TLC and make autoradiograph according to the protocol described in Chapter 1 (6).

Based on this timecourse of ^{32}P-labeling, two time points must be selected; the first, where the polyphosphoinositides but not structural phospholipids like PE and PC are well labeled; the second, where PC, PE, PG, and PI are all well labeled. Usually, good sampling times are 15 min and 16 h, respectively. But some systems allow for shorter labeling times (e.g., see Fig. 2).

3.2 Differential Labeling and Stimulation

To investigate the metabolic source of a stimulus-triggered PA response, plant material is prelabeled in parallel for a brief period (e.g., 15 min) and a longer period (16 h), respectively, and subsequently treated with the stimulus or mock-treated. ^{32}P-PA detected after short labeling is likely derived from DGK activity, whereas after long labeling, PLD-generated ^{32}P-PA can be demonstrated.

1. Add to two series (designated 1 and 2) of 2 mL polypropylene reaction tubes 175 μL MES buffer per tube.
2. Transfer plant material (leaf disks, seedlings, roots, etc.) to tubes.
3. Add 25 μL of diluted $^{32}P_i$-buffer (see Subheading 2.1, item 3) to start labeling (*see* **Note 6**).
4. After 15 min add 50 μL stimulus-containing solution to series 1 (*see* **Note 7**).
5. After incubation time, stop the reactions by adding 250 μL 10 % perchloric acid to series 1.
6. Mix vigorously for 5 min on a vortex and sediment material in table centrifuge for 2 min.
7. Extract lipids according to this protocol (see Subheading 3.1, steps 6–20).
8. Repeat this procedure for series 2, after the long labeling time.
9. Separate samples of both series on TLC using alkaline and ethyl acetate solvent systems, as described in Chapter 2 (6).

3.3 In Vivo Transphosphatidylation Assay for PLD

To measure in vivo transphosphatidylation, prelabeled plant material is incubated with 0.5 % *n*-butanol 10–30 min prior to the stimulus/treatment (*see* **Note 8**).

1. Add to a series of 2 mL polypropylene reaction tubes 175 μL MES buffer per tube.

2. Transfer plant material to tubes.
3. Add 25 μL of diluted $^{32}P_i$-buffer (see Subheading 2.1, item 3) to start labeling.
4. Incubate for the long labeling time (following the criteria discussed in Subheading 3.1), e.g., 16 h.
5. Add 50 μL of 2.5 % butanol stock (see Subheading 2.2, item 1) in each tube at 20 s intervals (*see* **Note 6**).
6. 15–30 min later (*see* **Note 9**), add 50 μL stimulus-containing solution (see Subheading 2.2, item 2), or, as control, add the same volume of solution for mock treatment (see Subheading 2.2, item 3).
7. To stop the reactions, add 300 μL 10 % perchloric acid.
8. Mix vigorously for 5 min on a vortex.

Then, perform steps 6–21 of Subheading 3.1 of the general lipid extraction protocol, followed by separation of lipids on TLC using the ethyl acetate solvent system described in Chapter 20 (7).

4 Notes

1. The half-life of ^{32}P is 14.3 days, and radioactivity at the specific activity reference date is 10 μCi/μL. Thus, x days after this date, a diluted ^{32}P-buffer (0.2 μCi/μL) can be made, e.g., by diluting $(8 \times 2^{(x/14.3)})$ μL $^{32}P_i$ in MES buffer (Subheading 2.1, item 2) to a total volume of 400 μL.
2. In the preparation of alcohol stock solutions, the limited miscibility of some alkanols should be kept in mind, e.g., the maximal concentration of *n*-butanol in water is about 9 % and that of 2-butanol 12 %, while tertiary butanol (2-methyl-2-propanol) is fully miscible.
3. Use for example 15 tubes, for five time points, taken in triplicates: e.g., 10 min, 30 min, 2 h, 8 h, 16 h.
4. Always add an equal volume of this solution to the incubation mix, such that the final concentration is 5 % perchloric acid.
5. Leave the interphase containing the insoluble material in the old tube and discard this tube; the volume of the extracted lower organic phase should be approximately 1,060 μL.
6. To obtain equal labeling and treatments, start and stop them in a fixed series order with 20 s intervals, which is generally sufficient time for handling a tube and adding a solution.
7. The controls must be mock-treated by adding the same volume of buffer.
8. There are several caveats using butanol to monitor PLD activity: (a) butanol has pleiotropic effects on membrane fluidity,

transmembrane proteins, heterotrimeric G-proteins, and the cytoskeleton; (b) it potentially inhibits or stimulates phospholipase activities; (c) deep tissues may be less accessible for butanol; (d) non-transphosphatidylating PLDs are missed by this assay. To separately investigate the effects of butanol irrespective of its role as a PLD transphosphatidylation substrate, secondary butanol (2-butanol) is often used instead of *n*-butanol for comparison. This isomer is not used by PLD as a transphosphatidylation substrate.

9. Bigger plants/parts may require longer times for access of butanol to deeper tissues.

References

1. Testerink C, Munnik T (2011) Molecular, cellular, and physiological responses to phosphatidic acid formation in plants. J Exp Bot 62:2349–2361
2. Arisz SA, Testerink C, Munnik T (2009) Plant PA signaling via diacylglycerol kinase. Biochim Biophys Acta 1791:869–875
3. Munnik T, Van Himbergen JAJ, Ter Riet B, Braun F-J, Irvine RF, Van den Ende H, Musgrave A (1998) Detailed analysis of the turnover of polyphosphoinositides and phosphatidic acid upon activation of phospholipases C and D in *Chlamydomonas* cells treated with non-permeabilizing concentrations of mastoparan. Planta 207:133–145
4. Arisz SA, Munnik T. The salt-stress induced lysophosphatidic-acid response in *Chlamydomonas* is produced via phospholipase A_2 hydrolysis of diacylglycerol kinase-generated phosphatidic acid. J Lipid Res 52: 2012–2020
5. Munnik T, Arisz SA, De Vrije T, Musgrave A (1995) G protein activation stimulates phospholipase D signaling in plants. Plant Cell 7:2197–2210
6. Munnik T, Zarza X (2013) Analyzing plant signaling phospholipids through $^{32}P_i$-labeling and TLC. Methods Mol Biol. 1009:3–15
7. Munnik T, Laxalt AM (2013) Phospholipase D activity in vivo. Methods Mol Biol. 1009:219–231

Chapter 7

Use of Phospholipase A_2 for the Production of Lysophospholipids

Steven A. Arisz and Teun Munnik

Abstract

Biological lipid extracts often contain small amounts of lysophospholipids (LPLs). Since different functions are emerging for LPLs in lipid metabolism and signalling, there is need for a reliable and cost-effective method for their identification. For this purpose, authentic LPL standards have to be synthesized from phosphoglycerides by PLA_2 digestion in vitro. PLA_2 specifically hydrolyzes the fatty acid ester linkage in the *sn*-2-position of phospholipids to liberate *sn*-2-linked fatty acids and the corresponding LPL. Due to this specificity, the reaction is also useful to analyze the positional distribution of fatty acids within membrane phospholipids. This chapter describes the in vitro generation of LPLs from diacyl-phosphoglycerides and their TLC analysis.

Key words Phospholipase A_2, Lysophospholipids, Positional analysis

1 Introduction

Diacyl-phosphoglycerides are characterized by the presence of a phosphate-ester bond at the *sn*-3-position of the glycerol, and two fatty acids, at the *sn*-1- and *sn*-2-hydroxyl groups, respectively. Lysophospholipids (LPLs) contain only one fatty acid linked to the glycerol backbone. LPLs have come into focus because of their functions in signal transduction (1, 2), vesicle transport (3), and lipid biosynthesis (4). In lipid extracts, LPLs may also occur as products of specific degradation during extraction and analysis.

To identify and quantify LPLs by TLC (or LC), appropriate standards are required. This chapter describes a method to make LPLs through PLA_2-catalyzed hydrolysis of purified commercial or authentic phospholipids (*see* **Note 1**). The enzyme selectively hydrolyzes *sn*-2-linked fatty acids from phospholipids. The method may also be used to prepare LPLs with odd carbon chain length as internal standards for quantitation of LPLs in natural extracts. Moreover, due to the positional specificity of PLA_2, the assay may be used, in combination with GC/LC-MS to analyze the positional

Teun Munnik and Ingo Heilmann (eds.), *Plant Lipid Signaling Protocols*, Methods in Molecular Biology, vol. 1009,
DOI 10.1007/978-1-62703-401-2_7,

distribution of fatty acids on the glycerol backbone, as analysis of the resulting LPLs yields the *sn*-1-linked fatty acyl complement.

Bee venom PLA_2 is used to generate LPLs in a two-phase water–ethylether system. The resulting lipid digest is chromatographed on TLC to monitor the completeness of the digestion and to purify the products. After their localization on the TLC plate, LPLs are recovered from the silica. Since the turnover of phospholipids is regularly studied in vivo using metabolic ^{32}P-phosphate-labelling, a ^{32}P-labelled lipid is taken as a substrate. The radiolabelling allows for visualization of the substrate lipid and LPL product on TLC by autoradiography (*see* **Note 2**).

2 Materials

Store all solvents at 4 °C. Use Milli-Q water for solutions. All ratios (between brackets) are by volume. Organic solvents are stored in dark-glass flasks with screw caps lined with solvent-resistant septa. Procedures involving the transfer of organic solvents (e.g., chloroform or methanol) should be performed in a fume hood. Working with radioactivity requires special precautions which are not discussed here. In general, always take care of proper shielding to protect against radiation, and check frequently for contaminations.

General required lab equipment:

1. Table centrifuge for 2 mL polypropylene reaction tubes, set at 13,000 × *g*.
2. Vortex, suitable for simultaneous shaking of multiple tubes.
3. Temperature-controlled vacuum concentrator (Speedvac).
4. N_2 gas.
5. TLC and autoradiography equipment (see Chapter 1, this book).
6. Stopwatch/timer.
7. Sonication bath.
8. 2 mL polypropylene reaction safe-lock tubes.

2.1 PLA_2 Digestion

1. 500 mM Tris–HCl, pH 8.9: Add about 20 mL of water to a graduated glass beaker. Weigh 6.06 g Tris and transfer to the cylinder. Add water to a volume of 90 mL. Mix using a magnetic stirrer and adjust the pH to 8.9 with HCl. Remove stirring bar and add water to a final volume of 100 mL.
2. 100 mM $CaCl_2$: Dissolve 1.47 g $CaCl_2 \cdot 2H_2O$ in 100 mL water.
3. PLA_2 solution 20 units/mL. Use PLA_2 from bee venom, lyophilized powder, salt-free, 1,220 units/mg (cat # P-9279, Sigma, St. Louis, MO, *see* **Note 3**). Dissolve 100 μg PLA_2 in 6 mL Tris–HCl buffer; this corresponds with 0.4 units per 100 μL (=one digestion).

4. Digestion mix, freshly prepared: Pipet in a 2 mL polypropylene reaction tube 300 μL each of 500 mM Tris–HCl buffer, 100 mM $CaCl_2$, and the PLA_2 solution. Add 600 μL water, and mix gently. This volume is calculated for 15 digestions. Adjust the volumes according to the desired number of digestions (100 μL per digestion).
5. Ethylether/methanol (98:2).
6. EDTA solution 0.5 M. Dissolve 14.6 g EDTA in 100 mL water.
7. Isopropanol.

2.2 Solvents and Solutions for Lipid Extractions

1. Extraction mix: Chloroform/methanol/HCl (50:100:1, v/v/v).
2. 2 M HCl: Carefully add, using a glass pipet, 16.7 mL concentrated HCl to about 70 mL water in a graduated cylinder; then add water to a final volume of 100 mL.
3. 1 M HCl: As above, except add 8.3 mL concentrated HCl.
4. $CHCl_3$.
5. 0.9 % NaCl: Dissolve 0.9 g NaCl in 100 mL H_2O.
6. Washing mix: Chloroform/methanol/1 M HCl (3:48:47).
7. Temperature-regulated vacuum concentrator (Speedvac).
8. 2 mL polypropylene reaction tubes with safe-lock.

2.3 TLC Analysis

1. Alkaline solvent system: Chloroform/methanol/25 % NH_4OH/H_2O (90:70:4:16, v/v/v/v); this mixture is used for the separation of all major phospholipid classes.
2. Glass silica gel 60 plates; with a pencil draw a baseline 2 cm from the bottom; store plates for heat-activation in an oven at 120 °C for at least 16 h before use.
3. TLC tank, the back side of which is lined with filter paper, the lid secured using clamps or weights.
4. Saran wrap.
5. Controlled stream of hot air.
6. Sonication bath.
7. Cassette and film (Kodak X-Omat S) suitable for autoradiography of TLC plate.
8. N_2 or Ar gas.

3 Methods

3.1 PLA_2 Digestion

1. Dissolve 1–10 mg lipid in 1 mL ethylether/methanol (98:2, v/v) using a sonication bath at room temperature for 5 min. Centrifuge for a few seconds to bring all liquid down.

2. Add 100 μL PLA_2 digestion mix (see Subheading 2.1, step 4) and mix gently using a vortex at low speed; briefly sediment liquid in tabletop centrifuge.
3. Incubate for 2 h at room temperature (20–25 °C); regularly shake the tube to remix the two phases (*see* **Note 4**).
4. Add 20 μL 0.5 M EDTA; vortex briefly; sediment liquid in tabletop centrifuge.
5. For quantitative analyses, add an aliquot of an appropriate LPL standard, dissolved in $CHCl_3$; vortex briefly; sediment liquid in tabletop centrifuge.
6. Evaporate upper phase in a Speedvac set at a temperature of 50 °C for 1 h.
7. Add 100 μL 0.9 % NaCl to the remaining lower phase.
8. Extract lipids by adding 750 μL extraction mix (see Subheading 2.2, step 1); vortex for 5 min and centrifuge for 5 s to remove the liquid from the lids.
9. Add 750 μL $CHCl_3$ and 200 μL 2.0 M HCl; vortex for 5 min and centrifuge for 2 min (*see* **Note 5**).
10. Pipet as much as possible of the lower, organic phase to a fresh tube.
11. Add 750 μL washing mix (see Subheading 2.2, step 6) to the latter; vortex for 2 min and spin down.
12. Discard the upper phase with a Pasteur pipet.
13. Add 20 μL isopropanol (*see* **Note 6**).
14. Evaporate solvents to dryness in a Speedvac set at 50 °C for 1 h.
15. Dissolve the residue in 20 μL $CHCl_3$ by sonication.

3.2 TLC of LPLs

1. Prepare a TLC tank by rinsing it with the alkaline solvent system (see Subheading 2.3, step 1); cover the bottom of the tank with approximately 1 cm of alkaline solvent mixture (see Subheading 2.3, step 1) which usually takes about 80 mL.
2. Carefully apply, using a 20 μL pipet, the lipid extracts on the baseline of a TLC plate (*see* **Note 7**).
3. After all samples have been applied, including radiolabelled and/or commercial unlabelled lipid markers to help identify the lipids, put the plate in the TLC tank, in a near-upright position.
4. Let the TLC run until the solvent front is just below the top of the plate (it takes about 2 h).
5. Take the plate out of the tank and let it dry in the fume hood for at least 2 h.
6. Wrap the plate in Saran wrap and, in the darkroom, put it in an autoradiography cassette with a film covering the front surface of the plate.

7. Depending on the amount of ^{32}P-labelling of the lipids of interest, the required exposure time ranges from 1 to 16 h.
8. Develop the film to obtain an autoradiograph. Identify the LPLs using the markers as reference and available data on chromatographic behavior. If required, a phosphoimaging screen can be exposed simultaneously for quantification.

3.3 Recovering LPLs from the Silica

1. Localize LPLs on the TLC plate by putting the autoradiographic film on the wrapped plate in its original position; use the origins on the baseline as reference points.
2. Mark the silica areas corresponding with the LPL spots with a soft pencil.
3. Wet the marked silica areas (with pipet and Milli-Q water) and scrape the silica off into 2 mL polypropylene reaction tubes with a spatula (*see* **Notes 8** and **9**).
4. Add 400 μL extraction mix (see Subheading 2.2, step 1) to each tube.
5. Vortex for 10 min, and then sediment the silica powder by centrifuging in a tabletop centrifuge for 5 min at 13,000 × *g*.
6. Transfer the supernatant to a fresh 2 mL polypropylene reaction tube.
7. Again, add 400 μL extraction mix to the silica-containing tubes, vortex and sediment the silica powder as before, and pool the supernatants.
8. Add 750 μL chloroform and 200 μL 2 M HCl to the extracts.
9. Vortex for 2 min, and centrifuge for 5 min.
10. Discard upper phase with a Pasteur pipet.
11. Dry the extracted lipids in a Speedvac at 50 °C for 1 h.
12. Dissolve the residue, containing purified LPLs by sonication, in 20 μL $CHCl_3$; store under N_2 or Ar gas at −20 °C.

4 Notes

1. The method is based on a protocol described by Kates (5).
2. In the absence of radiolabelled lipids, visualization is achieved by the use of iodine vapor or nondestructive lipid stains. The application of large quantities of lipid may require the use of TLC plates with loading regions.
3. This enzyme has a preference for the zwitterionic phospholipids PC and PE, but also hydrolyzes other phospholipids.
4. PLA_2 works at the water/ether interface; remixing brings new substrate to the interface.

5. The decreased hydrophobicity of LPLs compared to their diacyl-containing parents often results in losses during lipid two-phase extraction. Nevertheless, most LPLs are recovered when a strong acid, HCl, is used to acidify the water phase.
6. Isopropanol promotes the evaporation of remaining water in the extract.
7. To aid rapid evaporation of the solvent on the plate, the application zone is kept under a continuous, modest stream of warm air (e.g., provided by a hairdryer). To further minimize the diffusion of lipids from the baseline, it is important to avoid pipetting the whole sample volume at once. Instead, pipet a small volume, wait for the solvent to evaporate, then pipet the next, etc. To obtain optimal chromatographic resolution, apply each extract on a short (usually 4–8 mm) line segment rather than on one spot.
8. Use a metal spatula with a short (3–5 mm) squared end.
9. For fatty acid methyl ester analysis, the silica may be directly transferred to a tube containing the transmethylation reagent.

References

1. Ryu SB (2004) Phospholipid-derived signaling mediated by phospholipase A in plants. Trends Plant Sci 9:229–235
2. Scherer GF (2010) Phospholipase A in plant signal transduction. In: Munnik T (ed) Plant cell monographs—lipid signaling in plants, vol 16. Springer, Berlin, pp 3–22
3. Weigert R, Silletta MG, Spano S, Turacchio G, Cericola C, Colanzi A, Senatore S, Mancini R, Polishchuk EV, Salmona M, Facchiano F, Burger KN, Mironov A, Luini A, Corda D (1999) CtBP/BARS induces fission of Golgi membranes by acylating lysophosphatidic acid. Nature 402:429–433
4. Bates PD, Ohlrogge JB, Pollard M (2007) Incorporation of newly synthesized fatty acids into cytosolic glycerolipids in pea leaves occurs via acyl editing. J Biol Chem 282: 31206–31216
5. Kates M (1986) Techniques of lipidology: isolation, analysis and identification of lipids. Elsevier, Amsterdam, pp 405–408

Chapter 8

Analysis and Quantification of Plant Membrane Lipids by Thin-Layer Chromatography and Gas Chromatography

Vera Wewer, Peter Dörmann, and Georg Hölzl

Abstract

Galactolipids represent the predominant membrane lipid class in plants. In general, galactolipids are restricted to plastids, but during phosphate deficiency, they also accumulate in extraplastidial membranes. Two groups of plants can be distinguished based on the presence of a specific fatty acid, hexadecatrienoic acid (16:3), in chloroplast lipids. Plants that contain galactolipids with 16:3 acids are designated "16:3-plants"; the other group of plants which lack 16:3 contain mostly 18:3 in their galactolipids ("18:3-plants"). The methods in this chapter describe the extraction of membrane lipids from whole leaves, or from subcellular fractions, and their analysis via thin-layer chromatography (TLC) with different staining methods. Furthermore, a protocol for membrane lipid quantification is presented starting with the separation via TLC, transmethylation of the isolated lipids to fatty acid methyl esters, and their quantitative analysis via gas chromatography (GC).

Key words MGDG, DGDG, Chromatography, FAME, Chloroplast, Arabidopsis, Lotus, Galactolipid, Phospholipid

1 Introduction

The two galactolipids, monogalactosyldiacylglycerol (MGDG) and digalactosyldiacylglycerol (DGDG), represent the predominant lipids in green tissues of plants where they establish the bulk of the bilayer membrane in thylakoids and envelopes of the chloroplasts. Thus, MGDG and DGDG comprise about 50 and 30 %, respectively, of total chloroplast lipids, the remainder being sulfoquinovosyldiacylglycerol (SQDG), phosphatidylglycerol (PG), and minor amounts of phosphatidylcholine (PC) (1, 2). Galactolipids are synthesized in the chloroplast envelopes. Two pathways provide the precursors for galactolipid biosynthesis (3–5). The procaryotic pathway is exclusively localized to the chloroplasts leading to the synthesis of galactolipids with a 16:3 fatty acid at the *sn*-2 position of the glycerol. The eucaryotic pathway includes enzymes from the endoplasmic reticulum and the chloroplasts

Teun Munnik and Ingo Heilmann (eds.), *Plant Lipid Signaling Protocols*, Methods in Molecular Biology, vol. 1009,
DOI 10.1007/978-1-62703-401-2_8,

resulting in the production of galactolipids with only 18:3 fatty acids at their *sn*-2 position (5). In the so-called 16:3-plants like *Arabidopsis* or spinach, the two pathways are active. In "18:3-plants" like pea, soybean, or *Lotus japonicus*, the galactolipid precursors are synthesized via the eucaryotic pathway. Thus, mostly 18:3, but not 16:3, fatty acids occur in these plants. When grown under phosphate deficiency, DGDG and SQDG accumulate in the plant cell (6). These two lipids serve as surrogates for phospholipids, which are decreased when phosphate is limiting. During phosphate deprivation, DGDG is exported to extraplastidial membranes. Therefore, DGDG can be found in different membranes of plant cells including chloroplast and plasma membrane. Because of the predominance of the galactolipids (>65 % of total lipids) (7), they can easily be isolated and quantified from whole leaf lipid extracts. The methods listed here describe the extraction of membrane lipids from leaves and other plant organs such as roots, and from subcellular fractions obtained by phase separation or centrifugation. Furthermore, lipid analysis via thin-layer chromatography (TLC) and quantification in the form of fatty acid methyl esters (FAMEs) via gas chromatography (GC) using a flame ionization detector (FID) is described.

2 Materials

2.1 General Equipment for Working with Lipids

1. Glass tubes with screw thread (12 × 100 mm, Schott).
2. Screw-caps with Teflon septa for glass tubes (Schott).
3. Glass tubes (test tubes) without screw thread (12 × 100 mm, VWR).
4. Pasteur pipettes.
5. Sample concentrator with N_2 gas or air.

2.2 Lipid Extraction

1. Extraction solvent: $CHCl_3$/methanol (1:2, v/v).
2. Extraction solvent: $CHCl_3$/methanol (2:1, v/v).
3. Extraction solvent: $CHCl_3$/methanol/formic acid (1:1:0.1, v/v/v).
4. NaCl solution: 0.9 % NaCl (w/v) in H_2O.
5. KCl, H_3PO_4 solution: 1 M KCl, 0.2 M H_3PO_4 in H_2O.

2.3 Thin-Layer Chromatography and Visualization of Lipids

1. TLC-developing solvent: Acetone/toluene/H_2O (91:30:8, v/v/v).
2. Ammonium sulfate solution for impregnating TLC plates: 0.15 M $(NH_4)_2SO_4$.
3. TLC plates with concentration zone (*see* **Note 1**).
4. Glass tank with lid for TLC.

5. Micropipettes.
6. Glass tank with lid containing about 5 g of crystalline iodine, which sublimates into the gas phase, for lipid staining.
7. Aniline naphthalene sulfonic acid (ANS) reagent: 0.2 % (w/v) ANS in methanol, for staining of lipids. Protect ANS reagent from light by wrapping the reagent flask in aluminum foil.
8. α-Naphthol–H_2SO_4 reagent for staining of glycolipids: Dissolve 8 g of α-naphthol in 250 mL methanol. Add 30 mL concentrated H_2SO_4 dropwise while stirring on ice (*see* **Note 2**).
9. Glass sprayers.
10. UV light table or lamp (366 nm).
11. Heating plate.

2.4 Preparation of Fatty Acid Methyl Esters and Analysis by Gas Chromatography

1. 1 N HCl in methanol: 100 mL 3 N methanolic HCl + 200 mL methanol. Store at 4 °C.
2. Hexane for FAME extraction.
3. NaCl solution: 0.9 % NaCl (w/v) in H_2O.
4. Pentadecanoic acid (15:0): Working stock, 50 μg/mL in methanol (*see* **Note 3**).
5. Glass vials for GC auto sampler.
6. Water bath set at 80 °C.
7. Gas chromatograph with auto injector (injector temperature, 220 °C) and FID (detector temperature, 250 °C; detector gases: hydrogen—flow rate 30 mL/min, synthetic air—flow rate 400 mL/min, helium makeup flow—19 mL/min).
8. GC column: SP 2380, fused silica capillary column, 30 m × 0.53 mm, 0.20 μm film (Supelco), flow rate: 7 mL/min of helium, splitless injection of 2 μL.
9. Low erucic rapeseed oil FAME standard.

3 Methods

Galactolipids can be directly extracted from whole leaves. Analysis of lipids from different cell compartments requires fractionation of the membranes via two-phase partitioning or by centrifugation, which leads to membrane pellets or aqueous membrane lipid mixtures. This chapter describes three different methods allowing the extraction of lipids from whole organs, or subcellular fractions. Lipids can be handled at ambient room temperature, but they should be stored at −20 °C. The use of glassware and wearing of gloves and goggles for safety reasons are recommended when working with organic solvents.

3.1 Extraction of Lipids with Chloroform/Methanol/Formic Acid

This method allows the rapid extraction of lipids from ground plant material or from membrane pellets. Lipid-degrading enzymes (in particular phospholipase D) are inactivated by acidic extraction conditions (5, 8, 9).

1. Grind tissue (e.g., one Arabidopsis leaf) under liquid nitrogen. Membrane pellets obtained after centrifugation can be extracted after removal of supernatant. Add 1 mL of $CHCl_3$/methanol/formic acid (1:1:0.1) to the plant material.
2. Add 0.5 mL of 1 M KCl/0.2 M H_3PO_4 to the extract and shake vigorously.
3. Centrifugation at 4,000 × *g* for 3 min facilitates phase separation with the lower organic phase containing the lipids.
4. Go through the upper aqueous phase with a Pasteur pipette and transfer the lower phase to a new glass vial. Repeat extraction by adding one volume of $CHCl_3$/methanol (2:1, v/v).
5. Evaporate the solvent of the combined chloroform phases in a sample concentrator by a stream of nitrogen gas or air, add 100 μL of $CHCl_3$/methanol (2:1, v/v), and store the lipids at −20 °C. If larger quantities of plant material and solvent were used for lipid extraction, the use of a rotation evaporator for the final concentration of the lipid extracts is recommended.

3.2 Extraction of Lipids with Chloroform/Methanol After Boiling in Water

This method allows the use of whole leaves without grinding. Lipid-degrading enzymes are destroyed by boiling the plant material in water (9–11). Tough plant material (e.g., roots) should be ground prior to extraction.

1. Prior to lipid extraction, boil the leaf or other plant material in water for 10 min.
2. Remove the water and add an excess of $CHCl_3$/methanol (1:2, v/v) to the plant material. 2 mL of solvent is sufficient for one *Arabidopsis* leaf. Extract for 30 min at RT or at least 4 h at 4 °C.
3. Collect the first extract in a new vial with screw thread, add 2 mL $CHCl_3$/methanol (2:1, v/v) to the leaf or other plant material, and continue extraction as described in step 2.
4. Combine the two extracts, adjust the ratio of chloroform/methanol/water to 2:1:0.75 by adding more chloroform (2 mL) and NaCl solution (1.5 mL), and shake vigorously.
5. Centrifugation at 4,000 × *g* for 3 min facilitates phase separation.
6. Transfer the lower chloroform phase with a Pasteur pipette to a new glass vial.
7. Evaporate the solvent with a sample concentrator by a stream of N_2 or air, add 100 μL of $CHCl_3$/methanol (2:1, v/v), and

store the lipid extract at −20 °C. For larger amounts of solvent, the use of a rotation evaporator is recommended.

3.3 Extraction of Lipids from Aqueous Membrane Fractions

1. Add two volumes of $CHCl_3$/methanol/formic acid (1:1:0.1, v/v/v) to the aqueous membrane fraction and shake vigorously. The acidic conditions lead to inactivation of lipid-modifying or -degrading enzymes (5).
2. Centrifuge at 4,000 × *g* for 3 min.
3. Transfer the lower $CHCl_3$ phase with a Pasteur pipette to a new glass vial.
4. Evaporate the solvent with a sample concentrator by a stream of N_2 or air, add 100 μL of $CHCl_3$/methanol (2:1), and store the extract at −20 °C.

3.4 Quantitative and Qualitative Analysis of Lipids via TLC

The TLC system described here is a common method to separate phospho- and glycolipids from plants (5, 12, 13).

1. Activate an ammonium sulfate-impregnated plate by heating at 120 °C for 2.5 h (*see* **Note 4**).
2. Load the lipids as single spots onto the concentration zone of the cooled plate using a micropipette or a yellow-tip pipette. Reference lipids may be loaded in extra lanes.
3. After drying the spots, the plate is developed in a closed glass tank with acetone/toluene/H_2O (91:30:8, v/v/v) as solvent.
4. Subsequently, the plate is dried in the hood and ready for further analysis.
5. For qualitative analysis, all lipids can be stained with iodine (*see* **Note 5**) or glycolipids can be specifically stained with α-naphthol–H_2SO_4 (*see* **Note 6**). Figure 1 shows TLC plates with lipids from *Arabidopsis* leaves and roots stained with iodine or α-naphthol–H_2SO_4.
6. For quantitative analysis, the TLC plate is sprayed with ANS reagent, and lipids are visualized by exposure to UV light (*see* **Note 7**), and marked with a pencil. The lipids can be identified by co-migration with reference lipids (*see* **Note 8**).
7. The spots of the marked lipids are individually scraped off the plate using a razor blade, and the lipid-containing silica material is completely transferred into screw thread glass vials using a glass funnel.
8. The lipids can be extracted from the silica gel by adding 1 mL of $CHCl_3$/methanol (2:1, v/v) and 0.25 mL NaCl-solution (final ratio of $CHCl_3$/methanol/water: 2:1:0.75, v/v/v) in a similar way as described (Subheading 3.2). Alternatively, lipids on silica gel can directly be used for preparation of FAMEs (Subheading 3.5).

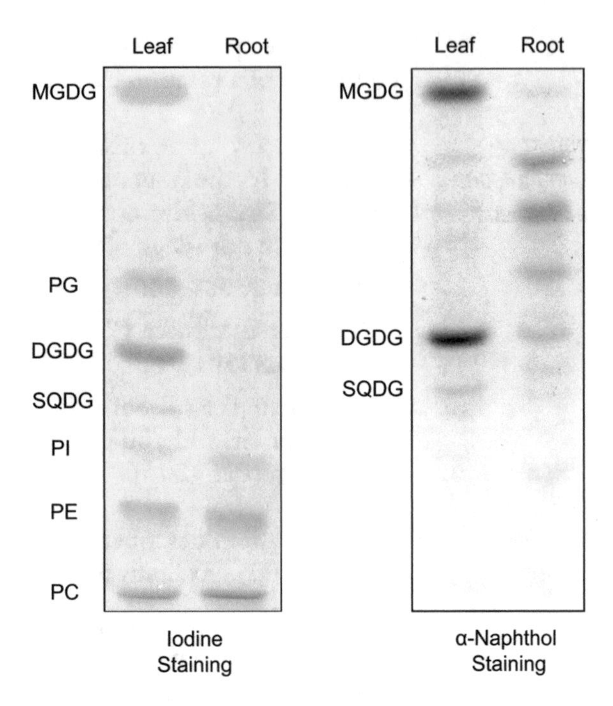

Fig. 1 Separation of membrane lipids from *Arabidopsis* leaves and roots. Lipids were isolated from whole leaves or roots with $CHCl_3$/methanol/formic acid solution (Subheading 3.1), separated by TLC and stained with iodine (*left panel*) or α-naphthol (*right panel*). Note the predominance of the galactolipids in the leaf samples which is due to the high abundance of chloroplasts in leaves

3.5 Preparation of FAMEs and Analysis via GC-FID

The high polarity of glycolipids, phospholipids, and free fatty acids impairs their direct analysis via GC. Therefore, the lipids and fatty acids need to be derivatized to the more volatile FAMEs, which can be easily detected by GC. GC is often employed for lipid quantification or for the determination of the fatty acid composition. The addition of a defined amount of internal standard (e.g., pentadecanoic acid, 15:0) to the samples is required for quantitative analysis of fatty acids. FAMEs are synthesized by transesterification of lipid-bound or free fatty acids to methanol under acidic conditions (14). It is not necessary to isolate lipids prior to FAME synthesis. Instead, plant material or lipid-containing silica gel can directly be used for the FAME reaction. Tough plant material (e.g., roots, seeds) should be ground first.

1. Fill the plant material or lipid-containing silica gel to a glass vial with screw thread (*see* **Note 9**).
2. Add 1 mL of 1 N methanolic HCl to the sample. For quantification, add 100 μL of pentadecanoic acid (working stock, 50 μg/mL) to each sample (yellow-tip pipette can be used).
3. Incubate the closed glass vials in a water bath at 80 °C for 20 min.

4. Remove vials from water bath and wait until they have cooled down to room temperature. Add 1 mL of hexane and 1 mL of NaCl-solution to the sample and shake vigorously, by vortexing or inverting the tubes.
5. Centrifuge for 3 min at 4,000 × *g* to facilitate phase separation.
6. Transfer the upper hexane phase containing the FAMEs with a Pasteur pipette to a new test tube without screw thread.
7. Evaporate the solvent in a sample concentrator under a stream of nitrogen gas, add 50–100 μL of hexane to the sample, and transfer it to a GC glass vial. The samples can be analyzed immediately or stored at −20 °C.
8. The FAMEs are separated by GC with the following temperature gradient: from initial 100 °C to 160 °C with 25 °C/min, to 220 °C with 10 °C/min, and to 100 °C with 25 °C/min.
9. The fatty acids are identified according to their retention time in comparison to the retention times of rapeseed FAME standard mixture. Quantification is based on the integration of the peak areas of the fatty acids in relation to the amount of the internal standard. For exact quantification, the detector response factors for the individual FAME peaks should be calculated using standard fatty acids (e.g., from the rapeseed standard; see Subheading 2.4). Figure 2 shows the GC chromatograms of FAMEs derived from total leaf and root lipid extracts from *Arabidopsis* and *Lotus*.

4 Notes

1. The use of TLC plates with a concentration zone facilitates loading of lipid extracts. The extract can be loaded as a single spot, which is concentrated down to a line upon entering the separation zone of the plate.
2. The addition of H_2SO_4 to methanolic α-naphthol solution leads to heat development. Therefore, keep the vial containing the solution in an ice bath, and use safety goggles.
3. The working stock (50 μg/mL) of pentadecanoic acid is prepared from a super stock (10 mg/mL in methanol) as a 1:200 dilution in methanol. Store methanolic solution of pentadecanoic acid at −20 °C. Make sure that all 15:0 fatty acid has dissolved after taking out the stock from −20 °C freezer.
4. For impregnation, submerge the TLC plates briefly in an ammonium sulfate solution and dry it for at least 2 days at a dust-protected place. Plate activation at 120 °C leads to conversion of $(NH_4)_2SO_4$ to NH_3 and H_2SO_4. This treatment leads

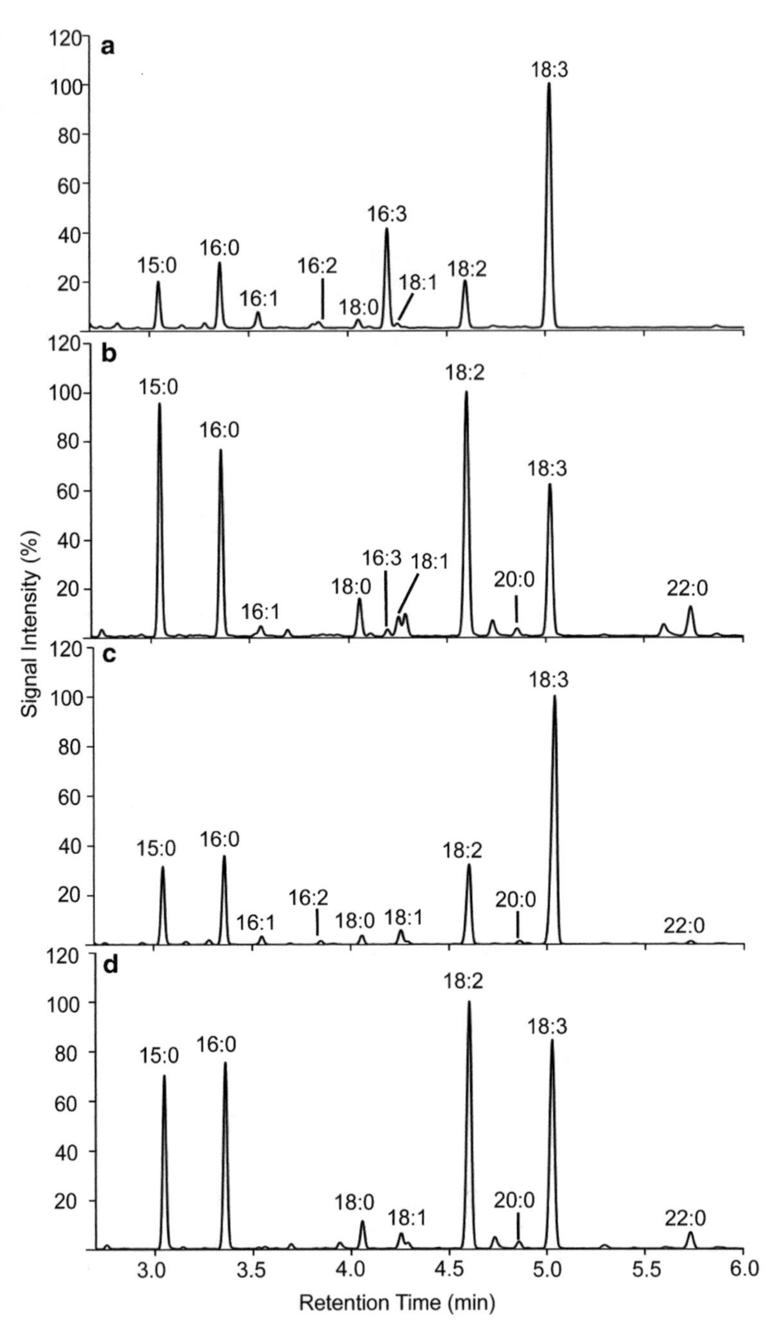

Fig. 2 Gas chromatography of fatty acid methyl esters from leaves and roots of *Arabidopsis* and *Lotus japonicus*. FAMEs were synthesized from whole organs of *Arabidopsis thaliana* leaves (**a**), roots (**b**), or *Lotus japonicus* leaves (**c**), or roots (**d**) and separated and quantified by GC and flame ionization detection. The leaf samples are generally rich in polyunsaturated fatty acids which are abundant in the galactolipids of the chloroplasts. Note that hexadecatrienoic acid (16:3) accumulates in *Arabidopsis* leaves ("16:3 plant"), but is absent from *Lotus japonicus* leaves ("18:3 plant"). The signal intensities were normalized to the peak with the highest intensity (100 %). Pentadecanoic acid (15:0) was used as internal standard

to evaporation of ammonia and to acidification of the silica material by sulfuric acid. Thus, anionic lipids alter their migration resulting in an improved separation of the phospho- and glycolipids.

5. For staining with iodine, expose the dried TLC plate to iodine vapor in a sealed TLC tank containing iodine crystals. All lipids will be stained after 5–10 min in a yellowish-brown color. Prolonged exposure to iodine results in a covalent modification of lipids containing polyunsaturated fatty acids.
6. The α-naphthol–H_2SO_4 reagent specifically stains sugar-containing lipids (MGDG, DGDG, SQDG) in a reddish color. The TLC plate is sprayed with α-naphthol–H_2SO_4 and heated on a heating plate to 137 °C until the color develops. After longer heat exposure, the color of the lipid bands changes from red to dark gray. This staining method leads to destruction of the lipids.
7. ANS reagent stains all lipids. Lipid bands are visualized by fluorescence light (exposure with UV light at 366 nm). ANS reversibly associates with lipids.
8. Reference lipids can be selectively stained with iodine or α-naphthol–H_2SO_4. Protect the area of the lipids to be analyzed by covering with a glass plate or a sheet of paper. For staining of reference lipids with iodine, fill some iodine crystals into a Pasteur pipette where they are fixed between two layers of glass wool. The Pasteur pipette is attached to tubing connected to compressed air. Then a gentle stream of iodine vapor is directed onto the lanes containing the reference lipids for selective staining. For staining with α-naphthol–H_2SO_4, spray the lanes containing the reference lipids with this reagent and heat up only this area on a heating plate.
9. The presence of residual amounts of water in the reaction leads to hydrolysis of membrane lipids instead of trans-esterification. The amount of water should therefore be below 5 % of the final volume of the trans-esterification reaction.

Acknowledgments

We would like to acknowledge the help of Anna Moseler (University of Bonn) with TLC and GC of leaf and root samples. This work was supported by the grants Ho3870/1-2 and Do520/9 (Forschungsschwerpunkt 1212) of Deutsche Forschungsgemeinschaft.

References

1. Siegenthaler P (1998) Molecular organization of acyl lipids in photosynthetic membranes of higher plants. In: Siegenthaler P, Murata N (eds) Lipids in photosynthesis: structure, function and genetics. Kluwer Academic, Dordrecht, pp 119–144
2. Benson AA (1971) Lipids of chloroplasts. In: Gibbs M (ed) Structure and function of chloroplasts. Springer, Berlin, pp 129–148
3. Heinz E, Roughan PG (1983) Similarities and differences in lipid metabolism of chloroplasts isolated from 18:3 and 16:3 plants. Plant Physiol 72:273–279
4. Browse J, Somerville C (1991) Glycerolipid synthesis: biochemistry and regulation. Annu Rev Plant Physiol Plant Mol Biol 42: 467–506
5. Browse J, Warwick N, Somerville CR, Slack CR (1986) Fluxes through the prokaryotic and eukaryotic pathways of lipid synthesis in the '16:3' plant *Arabidopsis thaliana*. Biochem J 235:25–31
6. Benning C, Ohta H (2005) Three enzyme systems for galactoglycerolipid biosynthesis are coordinately regulated in plants. J Biol Chem 280:2397–2400
7. Joyard J, Maréchal E, Miège C, Block MA, Dorne A-J, Douce R (1998) Structure, distribution and biosynthesis of glycerolipids from higher plant chloroplasts. In: Siegenthaler PA, Murata N (eds) Lipids in photosynthesis: structure, function and genetics. Kluwer Academic Publishers, Dordrecht, pp 21–52
8. Hajra A (1974) On extraction of acyl and alkyl dihydroxyacetone phosphate from incubation mixtures. Lipids 9:502–505
9. Bligh EG, Dyer WJ (1959) A rapid method of total lipid extraction and purification. Can J Biochem Physiol 37:911–917
10. Roughan P, Slack C, Holland R (1978) Generation of phospholipid artefacts during extraction of developing soybean seeds with methanolic solvents. Lipids 13:497–503
11. Hölzl G, Leipelt M, Ott C, Zähringer U, Lindner B, Warnecke D, Heinz E (2005) Processive lipid galactosyl/glucosyltransferases from *Agrobacterium tumefaciens* and *Mesorhizobium loti* display multiple specificities. Glycobiology 15:874–886
12. Khan M-U, Williams J (1977) Improved thin-layer chromatographic method for the separation of major phospholipids and glycolipids from plant lipid extracts and phosphatidyl glycerol and bis(monoacylglyceryl) phosphate from animal lipid extracts. J Chromatogr 140:179–185
13. Dörmann P, Hoffmann-Benning S, Balbo I, Benning C (1995) Isolation and characterization of an *Arabidopsis* mutant deficient in the thylakoid lipid digalactosyl diacylglycerol. Plant Cell 7:1801–1810
14. Browse J, McCourt PJ, Somerville CR (1986) Fatty acid composition of leaf lipids determined after combined digestion and fatty acid methyl ester formation from fresh tissue. Anal Biochem 152:141–145

Chapter 9

Lipidomic Analysis of Plant Membrane Lipids by Direct Infusion Tandem Mass Spectrometry

Sunitha Shiva*, Hieu Sy Vu*, Mary R. Roth, Zhenguo Zhou, Shantan Reddy Marepally, Daya Sagar Nune, Gerald H. Lushington, Mahesh Visvanathan, and Ruth Welti

Abstract

Plant phospholipids and glycolipids can be analyzed by direct infusion electrospray ionization triple-quadrupole mass spectrometry. A biological extract is introduced in solvent by continuous infusion into the mass spectrometer's electrospray ionization source, where ions are produced from the lipids. For analysis of membrane lipids, a series of precursor and neutral loss scans, each specific for lipids containing a common head group, are obtained sequentially. The mass spectral data are processed and combined, using the Web application LipidomeDB Data Calculation Environment, to create a lipid profile.

Key words Phospholipid, Galactolipid, Lipidomics, Precursor scan, Neutral loss scan, Electrospray ionization, Triple quadrupole, Mass spectrometry

1 Introduction

"Lipidomics" refers to profiling of lipids, often in relation to genotype, treatments, and/or the developmental state of a biological system; the analyses are typically performed via mass spectrometry. Lipids from whole plants, plant tissues, or fractionated material may be profiled. Here we focus on basic analysis of membrane (polar) lipids extracted from plant tissues. The analyzed compounds include those with structural and signaling roles in membranes. Lipids derived from relatively small amounts of tissue (0.2 mg of tissue dry weight) are sufficient for lipid profiling, as described herein.

Introducing the sample to a mass spectrometer (MS) by direct infusion is simple and effective, particularly for analysis of polar lipids. Direct infusion analysis is best performed on a triple-quadrupole

*Sunitha Shiva and Hieu Sy Vu have contributed equally.

Teun Munnik and Ingo Heilmann (eds.), *Plant Lipid Signaling Protocols*, Methods in Molecular Biology, vol. 1009,
DOI 10.1007/978-1-62703-401-2_9, © Springer Science+Business Media, LLC 2013

mass spectrometer; its three quadrupoles include two (tandem) mass analyzers and a collision cell located in between. An ion detector is located at the end of the ion path, after the second mass analyzer. For direct infusion, the sample is loaded into a syringe pump or into a sample loop from which it is continuously infused into the electrospray ionization source. The ions available for analysis depend on the solution in which the lipids are dissolved and on the source voltage. In the method described herein, an autosampler is used, and lipids are analyzed as $(M+H)^+$, $(M+NH_4)^+$, or $(M-H)^-$. After ionization, solvent is stripped from the ions, which move into and through the MS analyzers and collision cell in the gas phase.

Inside a quadrupole mass analyzer, the ions are in an electrical field. At a particular electrical field strength, ions of a particular mass/charge (*m/z*) ratio move straight toward the detector at the end of the tandem mass spectrometer ion path. Scanning, or systematically varying the electrical field strength, produces a plot (spectrum) of signal (ion hits at the detector) versus *m/z* (correlated with electrical field strength). In the case of lipids, the charge is typically 1, so *m/z* is equal to the ion mass. Scanning of the ions in a tissue or cell extract with a single-quadrupole mass analyzer would show many peaks, representing all the compounds in the sample. A single mass analyzer's spectrum is not easily interpretable because many peaks represent multiple compounds, which have similar *m/z* values. (A quadrupole mass analyzer typically resolves ions only to approximately unit *m/z*).

Precursor (Prec) and neutral loss (NL) scans, performed with the tandem analyzers, allow more specific compound detection. Each Prec or NL scan is typically used to profile membrane lipids specific to a particular lipid class; some classes are analyzed with Prec scans and others with NL scans. The Prec fragment *m/z* or NL fragment mass is that of a head group fragment formed in the collision cell. In a Prec spectrum (scan), the second mass analyzer is set at an (constant) electrical field that allows only ions with the *m/z* of a particular polar lipid head group fragment to move to the detector. Thus, while the first mass analyzer is scanning, the second analyzer acts as a "filter" so that a "hit" at the detector occurs only when an intact ion in the first mass analyzer produces the characteristic head group fragment. Thus, a Prec spectrum displays the lipids that have one head group fragment, typically lipids in one head group class. An NL spectrum is obtained when the charge does not localize to the lipid head group after fragmentation. In this case, the two mass analyzers are scanned in sync with an electrical field offset corresponding to the mass of the neutral head group fragment. Similar to precursor scanning, a hit at the detector occurs only when the first mass analyzer is at a field strength corresponding to an intact ion *m/z* that generates the characteristic neutral loss of the head group. Each NL scan also reveals the ions in one class. Another way to describe Prec or

NL scans would be to say that each peak in each scan is due to a particular intact ion-fragment pair. The Prec and NL scans for particular lipid classes are listed in Table 1.

The *m/z* of each detected intact ion, known to contain a fragment characteristic of a specific head group, is interpreted to identify the rest of the ion, for example, to identify the number of acyl carbons and the total number of carbon–carbon double bonds. Compounds identified using the directions herein are shown in Table 2. An example of the level of compound identification intrinsic to Prec or NL scanning is "PC(34:2)." The head group, here PC, is determined by the presence of a positively charged fragment of *m/z* 184.07 (the phosphocholine head group fragment) used for a Prec scan. The acyl composition, here "34:2," is determined from the *m/z*, 758.5, which, for PC, is consistent with 34 acyl carbons and 2 carbon–carbon double bonds.

Compounds identified by intact ion-fragment pairs, at the level of head group, total acyl carbons, and total acyl carbon–carbon bonds can correspond to one or a few true molecular species. To discern the possible individual acyl combinations of the detected compounds, it is helpful to have knowledge of the organism's fatty acid composition. Going back to the example of PC(34:2), and considering that most commonly occurring plant fatty acids are even-numbered and contain ≥16 acyl carbons per chain, PC(34:2) might contain 16:0/18:2, 16:1/18:1, or 16:2/18:0 combinations. Although the method is not described herein, individual fatty acid chains can be identified by mass spectral product ion analysis, but this may not clarify the acyl chain positions on the glycerol backbone (4, 5). In wild-type *Arabidopsis thaliana*, the acyl combinations for many common membrane lipids have been determined, and PC(34:2) has shown to comprise a combination of PC(16:0/18:2) and PC(16:1/18:1) (5). Another type of ambiguity can occur if unusual fatty acids are present. "MGDG(38:6)" and "MGDG(36:8-2O)" have the same intact ion *m/z* and head group fragment mass. "MGDG(36:8-2O)" is a compound with 2 oxophytodienoyl chains (i.e., oxidized acyl chains, each with the formula 18:4-O, where 4 is the number of double-bond equivalents and -O indicates an additional oxygen atom in the acyl chain).

For each detected lipid molecular species, the size of mass spectral signal allows quantification. To achieve accurate quantification by direct infusion electrospray ionization triple-quadrupole mass spectrometry, internal standard compounds, related to the biological compounds in each class or group, are combined with the plant samples or extracts before their analysis (1, 3–5). Optimally at least two non-naturally occurring internal standards are used for each class. A handy approach to the calculation of lipid profiles from mass spectral data is to use LipidomeDB Data Calculation Environment, a Web application that identifies

Table 1
Mass spectral scanning and acquisition parameters for analysis of polar plant membrane lipids

	Phospholipids						Glycolipids	
Class	**PA**	**PC and LPC**	**PE and LPE**	**PG**	**PI**	**PS**	**DGDG**	**MGDG**
Intact ion analyzed (adduct)	$(M+NH_4)^+$	$(M+H)^+$	$(M+H)^+$	$(M+NH_4)^+$	$(M+NH_4)^+$	$(M+H)^+$	$(M+NH_4)^+$	$(M+NH_4)^+$
Positive ion scan mode	NL of 115.00	Pre of *m/z* 184.07	NL of 141.02	NL of 189.04	NL of 277.06	NL of 185.01	NL of 341.13	NL of 179.08
m/z range	500–850	450–960	420–920	650–1,000	790–950	600–920	890–1,050	700–900
References		(1)	(1)	(2)	(2)	(1)	(3)	(3)
Acquisition parameters for API 4000 (Applied Biosystems) triple-quadrupole mass spectrometer								
Typical scan time (min)	3.51	1.28	3.34	3.21	4.00	4.01	1.67	1.67
Depolarization potential (V)	100	100	100	100	100	100	90	90
Exit potential (V)	14	14	14	14	14	14	10	10
Collision energy (V)	25	40	28	20	25	26	24	21
Collision exit potential (V)	14	14	11	14	14	14	23	23

Table 2
Plant membrane lipids determined by the procedure described herein and included in data processing through preformulated lists in LipidomeDB Data Calculation Environment

Class	Compound designations	Target list in LipidomeDB Data Calculation Environment
Lysophosphatidylcholine (LPC)	16:0, 16:1, 18:0, 18:1, 18:2, 18:3	Plant LPC
Lysophosphatidylethanolamine (LPE)	16:0, 16:1, 18:1, 18:2, 18:3	Plant LPE
Phosphatidic acid (PA)	32:0, 34:6, 34:5, 34:4, 34:3, 34:2, 34:1, 36:6, 36:5, 36:4, 36:3, 36:2	Plant PA
Phosphatidylcholine (PC)	32:0, 34:4, 34:3, 34:2, 34:1, 36:6, 36:5, 36:4, 36:3, 36:2, 36:1, 38:6, 38:5, 38:4, 38:3, 38:2, 40:5, 40:4, 40:3, 40:2	Plant PC
Phosphatidylethanolamine (PE)	32:3, 32:2, 32:1, 32:0, 34:4, 34:3, 34:2, 34:1, 36:6, 36:5, 36:4, 36:3, 36:2, 36:1, 38:6, 38:5, 38:4, 38:3, 40:3, 40:2, 42:4, 42:3, 42:2	Plant PE
Phosphatidylglycerol (PG)	32:1, 32:0, 34:4, 34:3, 34:2, 34:1, 34:0, 36:6, 36:5, 36:4, 36:3, 36:2, 36:1	Plant PG
Phosphatidylinositol (PI)	32:3, 32:2, 32:1, 32:0, 34:4, 34:3, 34:2, 34:1, 36:6, 36:5, 36:4, 36:3, 36:2, 36:1	Plant PI
Phosphatidylserine (PS)	34:4, 34:3, 34:2, 34:1, 36:6, 36:5, 36:4, 36:3, 36:2, 36:1, 38:6, 38:5, 38:4, 38:3, 38:2, 38:1, 40:4, 40:3, 40:2, 40:1, 42:4, 42:3, 42:2, 42:1, 44:3, 44:2	Plant PS
Digalactosyldiacylglycerol (DGDG)	34:6, 34:5, 34:4, 34:3, 34:2, 34:1, 36:6, 36:5, 36:4, 36:3, 36:2, 36:1, 38:6, 38:5, 38:4, 38:3	Plant DGDG
Monogalactosyldiacylglycerol (MGDG)	34:6, 34:5, 34:4, 34:3, 34:2, 34:1, 36:6, 36:5, 36:4, 36:3, 36:2, 36:1, 38:6, 38:5, 38:4, 38:3	Plant MGDG

The *m/z* of each detected intact ion, known to contain a fragment characteristic of a specific head group (because of the specific Pre or NL scan employed), is used to designate the number of acyl carbons and the total number of carbon–carbon double bonds. Note that these designations are based on nominal *m/z* and could, in some cases, be otherwise interpreted as described in the text. The indicated compound coverage was developed originally for *Arabidopsis thaliana*, but additions have been made to make the list applicable to other plant species

the signals of interest, isotopically deconvolutes the mass spectral data, and compares the plant signals to those of the internal standards for quantification. LipidomeDB Data Calculation Environment and complete instructions are available on the Web at http://lipidome.bcf.ku.edu:9000/Lipidomics.

2 Materials

***2.1 For Internal Standards* (See Note 1)**

1. LPC(13:0) (Avanti), 5 mM in $CHCl_3$.
2. LPC(19:0) (Avanti), 5 mM in $CHCl_3$.
3. LPE(14:0) (Avanti), 5 mM in $CHCl_3$/methanol/water (65:35:8).
4. LPE(18:0) (Avanti), 5 mM in $CHCl_3$/methanol/water (65:35:8).
5. PA(28:0) (di14:0) (Avanti), 5 mM in $CHCl_3$/methanol/water (65:35:8).
6. PA(40:0) (diphytanoyl) (Avanti), 5 mM in $CHCl_3$.
7. PC(24:0) (di12:0) (Avanti), 5 mM in $CHCl_3$.
8. PC(48:2) (di 24:1) (Avanti), 5 mM in $CHCl_3$.
9. PE(24:0) (di12:0) (Avanti), 5 mM in $CHCl_3$/methanol/water (65:35:8).
10. PE(46:0) (di23:0) semi-synthesized by transphosphatidylation of di23:0 PC from Avanti (6), 5 mM in $CHCl_3$.
11. PG(28:0) (di14:0) (Avanti), 5 mM in $CHCl_3$/methanol/water (65:35:8).
12. PG(40:0) (diphytanoyl), 5 mM in $CHCl_3$ (Avanti).
13. Hydrogenated PI (Avanti), containing PI(34:0) (16:0,18:0) and PI(36:0) (di18:0), 5 mM in $CHCl_3$.
14. PS(28:0) (di14:0) (Avanti), 5 mM in $CHCl_3$/methanol/water (65:35:8).
15. PS(40:0) (diphytanoyl) (Avanti), 5 mM in $CHCl_3$.
16. Hydrogenated DGDG (Matreya), containing DGDG(34:0) (18:0,16:0) and DGDG(36:0) (di18:0), 5 mM in $CHCl_3$/methanol/water (40:10:1).
17. Hydrogenated MGDG (Matreya), containing MGDG(34:0) (18:0,16:0) and MGDG(36:0) (di18:0), 5 mM in $CHCl_3$/methanol/water (40:10:1).

2.2 For Extraction

1. Isopropanol with 0.01 % BHT (w/v).
2. HPLC-grade $CHCl_3$.
3. HPLC-grade water.
4. $CHCl_3$/methanol (2:1, v/v) with 0.01 % BHT (w/v).
5. KCl, 1 M in water.
6. Glass tubes, 50 mL (25 × 150 mm) with Teflon-lined screw caps (Fisher, 14-930-10J).
7. Pasteur pipettes, 9-in.
8. Dry block heater that can accommodate 50 mL tubes.

9. Vortex mixer.
10. Orbital shaker.
11. Vacuum concentrator, vented to hood, or nitrogen gas stream evaporator, in hood.
12. Centrifuge (*see* **Note 2**).
13. Oven, vented to hood.
14. Balance that determines mass at least to tenths of milligram, but preferably to micrograms.

2.3 For Mass Spectrometry

1. Methanol/300 mM ammonium acetate in water (95:5, v/v).
2. $CHCl_3$.
3. Internal standard mixture (*see* Subheading 3.1).
4. Pre-slit, target Snap-it 11 mm Snap Caps (MicroLiter, 11-0054DB).
5. Amber vials, 12 × 32 mm (MicroLiter, 11-6200).
6. Autosampler, such as CTC Mini-PAL (LEAP), with 1 mL sample loop.
7. Sample trays to hold vials, such as VT54 (LEAP).
8. Large reservoir (e.g., 500 mL) syringe pump with pump controller to provide continuous infusion.
9. Methanol/acetic acid (9:1, v/v) for washing between samples.
10. Methanol/$CHCl_3$/water (665:300:35, v/v/v) to fill the wash reservoirs on the autosampler for washing the syringe and sample loop.
11. Triple-quadrupole mass spectrometer, such as API 4000 (Applied Biosystems, Foster City, CA), with electrospray ionization source.

3 Methods (See Note 3)

3.1 Internal Standard Mixture (see Note 4)

1. Using the 5 mM stock solutions, mix 120 μL (600 nmol) of each LPC and PC; 60 μL (300 nmol) of each LPE, PA, PE, and PG; 80 μL (400 nmol) of PI; 40 μL (200 nmol) of each PS; 240 μL (1,200 nmol) of DGDG; and 480 μL (2,400 nmol) of MGDG.
2. Bring this mixture to 10 mL by adding 8.16 mL $CHCl_3$.

3.2 Lipid Extraction

This is a modification of Bligh and Dyer's procedure (7). The instructions below are for Arabidopsis leaves (*see* **Note 5**). The procedure is described here for one sample; up to 24 samples can be conveniently processed together.

1. Drop up to eight leaves (*see* **Note 6**) into 3 mL of isopropanol with 0.01 % BHT, preheated to 75 °C (*see* **Note 7**), in a 50 mL glass tube with a Teflon-lined screw cap, fully immersing the tissue. Continue heating at 75 °C for 15 min.
2. Add 1.5 mL of $CHCl_3$ and 0.6 mL of water, and vortex (*see* **Note 8**); then agitate on an orbital shaker at 100 rpm at room temperature for 1 h. Use a Pasteur pipette to transfer the lipid extract to a new 50 mL tube. The pipette may be left in the recipient tube for use in transferring additional extractions of the same tissue (step 3).
3. Add 4 mL of $CHCl_3$/methanol (2:1) with 0.01 % BHT; agitate on an orbital shaker for 1 h. Remove the solvent from the tissue and combine with the extract previously transferred to the recipient tube. Repeat this extraction procedure three more times (*see* **Note 9**).
4. Add 1 mL of 1 M KCl to the combined extract, vortex or shake, and centrifuge at 500 × *g* for 10 min; discard upper phase. Add 2 mL water, vortex or shake, and centrifuge at 500 × *g* for 10 min; discard upper phase (*see* **Note 10**).
5. Use a vacuum concentrator or nitrogen gas stream to evaporate the combined extract and dissolve the lipid extract in 1 mL of $CHCl_3$ for storage.
6. Dry the leaf materials in oven (105 °C, 12 h). Weigh the remaining dry leaf materials.

3.3 Mass Spectrometry

1. Add a volume of the total extracted sample (in $CHCl_3$) equivalent to 0.2 mg dry tissue weight, and 10 μL of internal standard mix (*see* **Note 11**). Use $CHCl_3$ to bring the volume to a total of 360 μL. Add 840 μL of the methanol/300 mM ammonium acetate (95:5, v/v) so that the final ratio of solvent is $CHCl_3$/methanol/300 mM ammonium acetate (300:665:35, v/v/v) (*see* **Note 12**).
2. Prepare additional "Standards-only" samples with 10 μL of internal standard mix, 350 μL of $CHCl_3$, and 840 μL of the methanol/300 mM ammonium acetate (95:5, v/v).
3. Place the samples in the vial tray with a "Standards-only" sample in position 1 and as every 11th sample (*see* **Note 13**).
4. An additional set of vials should be filled with methanol/acetic acid (9:1) for washing the tubing and source between samples (*see* **Note 14**).
5. Infuse each sample at 30 μL/min and acquire a series of spectra for each sample using the *m/z* ranges, scan modes, acquisition times, and parameters shown in Table 1. Multiple channel analyzer (MCA) mode should be used. Typical scan speed is 100 U (i.e., amu)/s (*see* **Note 15**).

6. Perform baseline subtraction, smoothing, and centroiding (peak integration) of the spectrum obtained using mass spectrometer software (*see* **Note 16**).
7. Export spectral data from the proprietary mass spectrometer software into Excel files (*see* **Note 17**).
8. For identification and quantification of lipid species, use LipidomeDB Data Calculation Environment at http://lipidome.bcf.ku.edu:9000/Lipidomics/. Follow the instructions in the tutorial at the site, choosing the preformulated plant lipid target lists indicated in Table 2 (*see* **Note 18**).

4 Notes

1. Hydrogenated mixtures should be analyzed by mass spectrometry to assure that the compounds are fully hydrogenated. Concentrations of the phospholipids should be determined by phosphate assay (8). PI, DGDG, and MGDG total concentration and concentration of individual components should be determined by gas chromatography of the fatty acid methyl esters. The concentration of 16:0 methyl ester is equal to the concentration of the 16:0, 18:0 diacyl species; the total fatty acid concentration divided by 2 acyls per diacyl species, minus that of the 16:0, 18:0 species, is the concentration of the di18:0 species. Stock solutions of lipids should be stored at −20 °C in a freezer that does not self-defrost.
2. A low-speed, clinical-type centrifuge is adequate. Swinging buckets are optimal but not required.
3. It is critical that the solvents come in contact only with glass or Teflon. Solvent mixtures should be stored in glass bottles (preferably amber) with ground glass stoppers. If plastic is used at any step, components are dissolved, resulting in sample contamination and possible ion suppression in the mass spectrometer. Addition and transfer of all solvents and lipid extracts should be done in a fume hood.
4. The internal standard mixture is enough for 1,000 samples using 10 μL/sample. 10 μL of the mixtures contains 0.6 nmol of each LPC and PC; 0.3 nmol of each LPE, PA, PE, and PG; and 0.2 nmol of each PS. The total PI amount in 10 μL of the mixture is 0.4 nmol, the total amount of DGDG is 1.2 nmol, and the total amount of MGDG is 2.4 nmol. Amounts of individual PI, DGDG, and MGDG molecular species may vary somewhat in different batches of hydrogenated compounds (*see* **Note 1**).

5. Notes on extraction of plant tissues other than leaves:

   Siliques, *flowers*, or *whole rosettes* (*from small plants*) can be extracted as described for leaves.

   Roots are easily damaged, resulting in very high levels of PA. Harvest quickly, moving directly from growing medium into hot solvent. Growing plants on culture plates or in liquid culture will make it easier to harvest roots without damaging them.

   Stems, *stalks*, or *grass leaves*, which have thicker, more fibrous tissues than Arabidopsis leaves, can be cut from the plant, and snipped into 0.5 cm lengths directly into the hot isopropanol. Lightly crushing stems and stalks with a glass rod after heating will be helpful in extracting lipid thoroughly.

   For *seeds*, weigh and/or count. Twenty-five Arabidopsis seeds are sufficient; more are convenient. Lyophilization of seeds may be optimal for obtaining dry weights. Drop seeds into 1 mL 75 °C isopropanol with 0.01 % BHT. Heat for 15 min. Let cool. A Dounce (ground glass to ground glass) homogenizer should be used to grind the seed in isopropanol after heating. Grind thoroughly. Glass Pasteur pipettes should be used for transferring seeds and solvent to the homogenizer and back to 17 mL glass tube for extraction. Rinse homogenizer with 1 mL of $CHCl_3$ and 1 mL of methanol to recover all the seed parts/lipids; add to isopropanol/seed mix in tube. Add 0.8 mL water to solvent/seed mix. Shake well; the solution should be one phase. Add 1 mL $CHCl_3$ and 1 mL water. Shake well. Centrifuge at $500 \times g$ for 10 min to split phases. Remove lower layer ($CHCl_3$ and lipids). Save to clean glass tube. Add 1 mL $CHCl_3$ to original tube, shake, and centrifuge at $500 \times g$. Combine the lower layer with saved extract. Repeat addition of 1 mL $CHCl_3$, shaking, centrifuging, and removing lower layer two more times, combining with previous extracts. Add 0.5 mL of 1 M KCl to combined lower layers. Shake well; centrifuge at $500 \times g$ for 10 min. Remove upper phase, and discard. Add 1 mL water to $CHCl_3$/lipid layer. Shake well; centrifuge. Remove upper phase, and discard. Evaporate solvent and store as for leaves.

6. Only a fraction of an Arabidopsis leaf (0.2 mg) is required. Use of more tissue, and particularly use of tissue from multiple leaves and/or multiple plants, reduces biological variability among samples. Dry weights of 5–30 mg are convenient.

7. It is extremely important that the plants be extracted immediately after sampling and that the isopropanol be fully preheated. Many plant tissues have very active phospholipase D, which is activated upon wounding; failure to place the sampled tissue *immediately* in hot isopropanol will result in generation of PA.

8. After addition of $CHCl_3$ and water, there should be one phase. If there is phase separation or cloudiness, it is likely to be due to too much water in the mixture, caused by too much plant material or wet plant material. In this case, add 0.1 mL methanol to each tube to clear the cloudiness. If this is not enough to create a clear solution, 0.1 mL methanol can be added a second time.
9. Some tissues from some plants may require more rounds of extraction. Leaves should be extracted until they are white, but all samples in an experiment should be extracted the same number of times. An acceptable variation is to extend one round of extraction for a longer time, such as overnight.
10. The protocol is optimized for polar glycerolipids. If extraction and analysis of sphingolipids are also desired, an additional extraction step, based on the procedure outlined by Markham et al. (9), is performed, alternative to Subheading 3.2, step 4.

    Alternative step 4 requires "Solvent H" (9). Prepare Solvent H by mixing isopropanol/hexane/water (55:20:25, v/v/v) for at least 30 min with a magnetic stirrer. Let the mixture settle for 30 min, remove and discard the upper phase, measure the volume of the lower phase, and add BHT to the concentration of 0.01 %. The clear lower phase with BHT 0.01 % is solvent H.

    Alternative step 4: Add 3 mL of Solvent H to extracted tissues. Incubate Solvent H–tissue mixture at 60 °C for 15 min. Use Pasteur pipette to transfer the solvent, combined with extract from Subheading 3.2, step 3. Repeat this step three more times and continue with Subheading 3.2, step 5.

    Leaf tissues extracted with Solvent H tend to strongly adhere to glass tubes after being dried completely, which will be troublesome for dry tissue weighing (step 6). In order to prevent this inconvenience, before putting samples in oven to dry, use the tip of the Pasteur pipette to roll extracted leaf tissues into a "ball," and gently press this "ball" against the tube side wall (not the bottom of the tube).
11. The internal standards are used to determine the concentration of lipids in the extract by comparison of their signals with the signals of the analyzed compounds (Table 2). In many protocols, internal standards are added prior to extraction. Here the internal standard mixture is added during sample preparation. The rationale for the late addition of the internal standard mixture is to conserve the internal standard mixture, because its assembly is complex, and it is not currently commercially available.
12. A total volume of 1.2 mL per vial is used to assure complete filling of a 1 mL syringe.

13. "Standards-only" samples are used to evaluate instrument background and sample-to-sample carryover. "Standards-only" samples should have very low signals for analyte peaks. "Standards-only" sample data can be processed with other samples and the resulting quantified data for each analyte can be subtracted from data for the same analyte in other samples.
14. The sample syringe is used to fill the 1 mL loop. With the CTC Mini-PAL autosampler, the syringe and injection port are washed with a mixture of methanol/$CHCl_3$/water (666:300:35) from the autosampler's wash reservoirs right after the loop is filled and sample infusion begins. Between each sample (including the "Standards-only" samples), 1 mL of a wash solution (from the wash vials), containing methanol/acetic acid (9:1, v/v), is used to fill the loop and is infused for 8 min. The infusion of this acidic solvent washes the loop, the tubing from the loop to the mass spectrometer, and the electrospray needle. The electrospray source is on during this inter-sample wash cycle. It is important to use an acidic wash solvent between samples to reduce carryover of acidic lipids such as PA and PS.
15. Parameters, such as scanning time, collision energy, source voltages, source temperature, lens voltages, and collision gas pressure for each scan, may need to be optimized for your instrument. You may request a "data acquisition method" file for the API 4000 (Applied Biosystems)/CTC LEAP Mini-PAL from welti@ksu.edu.
16. In Analyst version 1.4.2, the baseline subtraction parameter should be set at "20 amu," and the smooth parameter should be set at 0.4 for previous and next point weight and 1.0 for current point weight.
17. Spectra can be checked against the *m/z* of the designed compounds to assure that all observed compounds (i.e., spectral peaks) are quantified.
18. For diacyl or monoacyl phospholipids, the response of each compound is very close (within 5 or 10 %) to the response of an internal standard of the same class. Thus, the signal for diacyl or monoacyl phospholipids, quantified in comparison to internal standards of known molar amounts, can be determined directly in molar amounts, without correction (response) factors. However, there may be variation in molar response among compounds within the DGDG and MGDG classes; comparison of the signal to that of the internal standards may not provide an accurate indicator of the molar content. It is suggested that you present glycolipid data as "normalized signal"/(tissue metric); this approach allows for sample-to-sample comparison in levels of components.

Acknowledgments

The development of plant lipidomics methodology was supported by the National Science Foundation (MCB 0455318 and MCB 0920663) to R.W., Xuemin Wang, Jyoti Shah, Todd Williams, and Gary Gadbury. We are very grateful to these, other colleagues, and lab members who have contributed to plant lipidomics development. Contribution no. 11-300-B from the Kansas Agricultural Experiment Station.

References

1. Brügger B et al (1997) Quantitative analysis of biological membrane lipids at the low picomole level by nano-electrospray ionization tandem mass spectrometry. Proc Natl Acad Sci USA 94:2339–2344
2. Taguchi R et al (2005) Focused lipidomics by tandem mass spectrometry. J Chromatogr B Analyt Technol Biomed Life Sci 823:26–36
3. Isaac G et al (2007) New mass spectrometry-based strategies for lipids. Genet Eng (N Y) 28:129–157
4. Welti R et al (2002) Profiling membrane lipids in plant stress responses: role of phospholipase D-α in freezing-induced lipid changes in Arabidopsis. J Biol Chem 277:31994–32002
5. Devaiah SP et al (2006) Quantitative profiling of polar glycerolipid species and the role of phospholipase Dα1 in defining the lipid species in Arabidopsis tissues. Phytochemistry 67: 1907–1924
6. Comfurius P, Zwaal RF (1977) The enzymatic synthesis of phosphatidylserine and purification by CM-cellulose column chromatography. Biochim Biophys Acta 488:36–42
7. Bligh EG, Dyer WJ (1959) A rapid method of total lipid extraction and purification. Can J Physiol Pharmacol 37:911–917
8. Ames BN (1966) Assay of inorganic phosphate, total phosphate and phosphatases. In: Neufeld E, Ginsburg V (eds) Methods in enzymology: complex carbohydrates, vol VIII. Academic, New York, pp 115–118
9. Markham JE et al (2006) Separation and identification of major plant sphingolipid classes from leaves. J Biol Chem 281: 22684–22694

Chapter 10

Detection and Quantification of Plant Sphingolipids by LC-MS

Jennifer E. Markham

Abstract

Sphingolipids generate signals in plants in response to a variety of biotic and abiotic stresses. Measuring these signaling compounds is complicated by the heterogeneity of structures within the sphingolipid family and the comparatively low concentration of their metabolites in plant tissues. To date, the only method with the sensitivity, dynamic range, and specificity to measure all sphingolipids in a plant extract is liquid chromatography coupled to mass spectrometry. The drawback of this method is the cost of the hardware, the expertise in mass spectrometry required to critically assess the outcome and the lack of suitable standards for accurate quantitative analysis. The goal of this chapter is to assist researchers in setting up experiments to measure sphingolipids and explain some of the pitfalls and solutions along the way.

Key words HPLC, Mass spectrometry, Long-chain base, Long-chain base phosphate, Ceramide, Sphingosine, Sphingolipid

1 Introduction

There are numerous ways to detect and measure sphingolipids (1–4) but only mass spectrometry of intact molecules provides complete information about their structure and abundance (5, 6). Mass spectrometry may either be performed directly on lipid-containing samples, the so-called shotgun approach, or performed in conjunction with HPLC. The advantages of combining HPLC with mass spectrometry are improved sensitivity and compound identification, reduced interfering compounds, and reduced contamination of the mass-spectrometer by nonvolatile components of the extract. Sphingolipids are detected upon elution from the HPLC column by using the mass-spectrometer in multiple-reaction monitoring (MRM) mode. In this mode the mass spectrometer monitors parent and product ion combinations that are specific for each sphingolipid compound. Quantification of the results can be achieved by including internal standards at the time of lipid extraction and weighing the amount of tissue used. Typically, 10–30 mg

Teun Munnik and Ingo Heilmann (eds.), *Plant Lipid Signaling Protocols*, Methods in Molecular Biology, vol. 1009,
DOI 10.1007/978-1-62703-401-2_10, © Springer Science+Business Media, LLC 2013

dry weight of Arabidopsis tissue is used, although it is possible to work with less material. In total, 168 sphingolipid compounds have been measured in Arabidopsis, including ceramide (Cer), 2-hydroxyceramide (hCer), glucosylceramide (GlcCer), glycosylinositolphosphoceramide (GIPC), and free long-chain bases (LCB) and long-chain base phosphates (LCBP) (6).

2 Materials

All solvents must be HPLC grade or above which typically means ≥99.9 % purity. Measure solvents in a flow hood and wear suitable protective clothing. Keep glassware, including solvent bottles, measuring cylinders, and pipettes, free from detergent, dust, and skin oils. Always measure the required amount of pure solvent before mixing with other solvents in a bottle. Pipette smaller amounts of solvent with glass pipettes or Hamilton syringes.

2.1 Lipid Extraction

1. Extraction solvent: Prepare a mixture of 2-propanol (55 mL), water (20 mL), and hexane (25 mL), mix well and allow the phases to separate, use only the lower phase (*see* **Note 1**).
2. Duall, All-Glass Tissue grinder, size 21 (3 mL) Kimble Chase. Use once and clean with water and methanol after each use. If many samples are to be extracted have up to five available (*see* **Note 2**).
3. Conical glass centrifuge tubes, 16×110 mm, screw thread cap.
4. Round bottom glass culture tubes, 16×100 mm, screw thread cap.
5. Nitrogen drying apparatus or centrivap concentrator for 16×100 mm tubes.
6. Methylamine reagent: Combine 7 mL of methylamine solution (33 % (w/w) in ethanol, SigmaAldrich, St Louis, MO) with 3 mL of water.

2.2 Lipid Standards

1. Lipid standards are available from Avanti Polar Lipids, Alabaster, AL (see Table 1). Dissolve all standards at a concentration of 1 mg/mL in $CHCl_3$/methanol/water (16:16:5, v/v/v) and store at −20 °C.
2. Combine lipid standards in a 2 mL autosampler vial as outlined in Table 1. Dry standards carefully under streaming N_2. Redissolve in 1 mL of extraction solvent and seal vial. Store at −20 °C until use.

2.3 HPLC and Mass Spectrometry

1. AB Sciex 4000 (QTRAP) mass spectrometer fitted with a Turbo V ion source and Turbo Ion Spray (TIS) probe attached to an HPLC system (e.g., Shimadzu) with two binary pumps, low volume mixer, auto-injector, and column oven (*see* **Note 3**).

Table 1
Use of internal standards for quantification of different sphingolipid classes

Sphingolipid class	Internal standard	μL stock per mL	Amount per 10 μL (nmol)
GIPCs	GM1	312.5	2
Glucosylceramide	Glucosyl-C12-ceramide	64.5	1
Ceramide	C12-ceramide	4.83	0.1
Long chain base	C17-sphingosine	2.85	0.1
Long chain base phosphate	C17-sphingosine-1-phosphate	3.65	0.1

Each standard is dissolved individually as a stock solution of 1 mg/mL and then combined in the quantities indicated to produce a standard solution

2. XDB-C18 HPLC column, 3.0 × 100 mm, 3.5 μm particle size fitted with 12 × 5 mm guard column (Agilent Technologies, Santa Clara, CA) (*see* **Note 4**).
3. Ammonium formate solution: 5 mM solution in water (dilute from 50 mM stock).
4. Solvent A: Combine 300 mL of tetrahydrofuran with 200 mL of methanol and 500 mL of ammonium formate solution. Add 1 mL of formic acid.
5. Solvent B: Combine 700 mL of tetrahydrofuran with 200 mL of methanol and 100 mL of ammonium formate solution. Add 1 mL of formic acid.
6. Sample solvent: Combine 400 mL of tetrahydrofuran with 200 mL of methanol and 400 mL of water. Add 1 mL of formic acid.

3 Methods

3.1 Extraction of Lipid Samples

1. Freeze-dry samples overnight to ensure complete removal of water. Once freeze dried, sphingolipids are stable at −80 °C for years or at room temperature for several days (*see* **Note 5**).
2. Weigh 10–30 mg of tissue at room temperature and place in Duall all-glass tissue grinder. Record the weight.
3. Add 10 μL of the internal standard solution and 3 mL of extraction solvent and grind tissue until all tissue is completely disrupted.
4. Vortex and pour sample into a glass centrifuge tube. Rinse tissue grinder with 3 mL of extraction solvent and add to sample.

5. Cap the tube and incubate the sample at 60 °C for at least 15 min.
6. Centrifuge the sample at 500 × *g* for 10 min. Decant the supernatant into a fresh, round bottom tube.
7. Resuspend the pellet in 3 mL of extraction solvent (sonication is usually required—*see* **Note 6**) and incubate at 60 °C for an additional 15 min. Centrifuge as before and combine with the supernatant from step 6.
8. Dry the sample under streaming N_2-gas or in vacuum.
9. Add 2 mL of Methylamine reagent and incubate at 50 °C for 1 h.
10. Dry the sample under N_2-gas and dissolve in 1 mL of Sample solvent (gentle sonication and sample heating is usually required). Transfer to a 2 mL autosampler vial and cap tightly. Store at −20 °C until required.

3.2 HPLC-MS Detection of Sphingolipids

1. Set up five methods for HPLC separation of the different sphingolipid classes. Ceramides and hydroxyceramides use the same HPLC methods but different MRM parameters. All use a flow rate of 1 mL/min, oven temperature of 40 °C and a post-run equilibration of 20 s. The flow for the first minute after sample injection is directed to waste, the gradient program starts immediately following injection. HPLC gradient parameters are described in Table 2.
2. Set up the source parameters for each method as follows: Curtain Gas (CUR) 20; Collision Gas (CAD), Medium; IonSpray Voltage (IS) 5,000; Temperature (TEM) 650; Ion Source Gas 1 (GS1), 60; Ion Source Gas 2 (GS2), 50.
3. Input MRM parameters for the mass-spectrometer using Tables 3 and 4. Dwell time is the same for each MRM, 19.4 ms, which provides a data point for each MRM once per second (allowing a 5 ms settle time between each MRM).
4. Inject 10 μL of lipid extract for each LC-MS method. After injection rinse the needle in sample solvent.
5. Once the data has been gathered, it will be necessary to create quantitation methods to integrate the peak areas and determine the amount of each compound (*see* **Note 7**). The formula is as follows:

$$\frac{A_{\text{analyte}} / A_{\text{standard}} \times \text{nmol of standard} \times R}{\text{grams tissue}}$$

where A_{analyte} and A_{standard} are the peak areas for each analyte and its respective standard, nmol of standard is the amount of each standard added for the respective sphingolipid class (shown in Table 1), R is a response factorial (see Table 5) to account for

Table 2
HPLC parameters for separation of sphingolipid species by HPLC prior to detection by mass spectrometry

	Cer and hCer	GlcCer	GIPC	LCB(P)
[B] start	55 %	55 %	26 %	10 %
Gradient [B], time	85 %, 7 min	80 %, 5 min	58 %, 5 min	20 %, 3.5 min
	100 %, 7.2 min	100 %, 5.2 min	100 %, 5.2 min	100 %, 4 min
Stop	8 min	6 min	6 min	5 min

Table 3
MRM parameters for detection of LCB and LCBPs by mass spectrometry

LCB(P)	Q1	Q3	DP	CE
d17:1	286.3	268.3	55	19
d18:0	302.3	284.3	75	21
d18:1	300.3	282.3	65	18
t18:0	318.3	300.4	70	21
t18:1	316.3	298.4	60	18
d17:1P	366.2	250.3	60	23
d18:0P	382.3	266.3	65	19
d18:1P	380.3	264.3	60	25
d18:2-P	378.3	262.3	60	25
t18:0P	398.3	300.3	65	22
t18:1P	396.3	298.3	60	25

Q1 and Q3 are the parent and product ion *m/z* settings, and DP and CE indicate the declustering potential and collision energies for each compound

different responses between sphingolipids with different chemical structures compared to the internal standard (*see* **Note 8**), and *grams tissue* is the weight of tissue measured in Subheading 3.1, step 2. The formula gives a value for the amount of sphingolipid in a certain sample in nmol per gram dry weight.

4 Notes

1. The bottom phase is 2-propanol/hexane/water at approximately (50:20:5, v/v/v) while the upper phase is 2-propanol/hexane (30:70, v/v). If this extraction is performed with fresh

Table 4
MRM parameters for ceramide (Cer), hydroxyceramide (hCer), glycosylceramide (GlcCer), and glycosylinositolphosphoceramide (GIPC) from Arabidopsis

		Cer				hCer				GlcCer				GIPC			
LCB	FA	Q1	Q3	DP	CE	Q1	Q3	DP	CE	Q1	Q3	DP	CE	Q1	Q3	DP	CE
Standard		482.5	264.3	60	35	482.5	264.3	60	35	644.5	264.3	90	50	1,546.9	366.3	145	50
t18:0	16:0	556.5	300.3	100	35	572.5	300.3	100	36	734.6	300.3	80	68	1,152.6	554.5	145	60
t18:0	18:0	584.6	300.3	100	35	600.6	300.3	100	38	762.6	300.3	80	68	1,180.6	582.5	145	60
t18:0	20:0	612.6	300.3	100	37	628.6	300.3	100	38	790.6	300.3	80	72	1,208.7	610.6	145	60
t18:0	20:1	610.6	300.3	100	37	626.6	300.3	100	44	788.6	300.3	80	75	1,206.7	608.6	145	61
t18:0	22:0	640.6	300.3	100	43	656.6	300.3	100	45	818.7	300.3	80	60	1,236.7	638.6	145	62.5
t18:0	22:1	638.6	300.3	100	43	654.6	300.3	100	45	816.7	300.3	80	63	1,234.7	636.6	145	61
t18:0	24:0	668.7	300.3	100	43	684.7	300.3	100	45	846.7	300.3	80	60	1,264.7	666.6	145	62.5
t18:0	24:1	666.7	300.3	100	43	682.7	300.3	100	45	844.7	300.3	80	65	1,262.7	664.6	145	62
t18:0	26:0	696.7	300.3	100	43	712.7	300.3	100	46	874.7	300.3	80	63	1,292.8	694.7	145	63
t18:0	26:1	694.7	300.3	100	43	710.7	300.3	100	45	872.7	300.3	80	65	1,290.8	692.7	145	63
t18:1	16:0	554.5	298.3	100	38	570.5	298.3	100	36	732.6	298.3	88	49	1,150.6	552.5	145	56
t18:1	18:0	582.6	298.3	100	38	598.6	298.3	100	36	760.6	298.3	70	54	1,178.6	580.5	145	58
t18:1	20:0	610.6	298.3	100	40	626.6	298.3	100	38	788.6	298.3	70	55	1,206.7	608.6	145	61
t18:1	20:1	608.6	298.3	100	40	624.6	298.3	100	38	786.6	298.3	75	60	1,204.7	606.6	145	60
t18:1	22:0	638.6	298.3	100	42	654.6	298.3	100	43	816.7	298.3	88	57	1,234.7	636.6	145	61
t18:1	22:1	636.6	298.3	100	42	652.6	298.3	100	43	814.7	298.3	75	60	1,232.7	634.6	145	60
t18:1	24:0	666.7	298.3	100	42	682.7	298.3	100	45	844.7	298.3	100	57	1,262.7	664.6	145	62

t18:1	24:1	664.7	298.3	100	44	680.7	298.3	100	45	842.7	298.3	100	59	1,260.7	662.6	145	63
t18:1	26:0	694.7	298.3	100	44	710.7	298.3	100	45	872.7	298.3	100	57	1,290.8	692.7	145	63
t18:1	26:1	692.7	298.3	100	44	708.7	298.3	100	45	870.7	298.3	100	62	1,288.8	690.7	145	65
d18:0	16:0	540.5	266.3	40	42	556.5	266.3	80	43	718.6	266.3	85	56	1,136.6	538.5	145	57
d18:0	18:0	568.6	266.3	40	43	584.6	266.3	80	46	746.6	266.3	85	80	1,164.7	566.6	145	57
d18:0	20:0	596.6	266.3	42	43	612.6	266.3	90	48	774.6	266.3	93	80	1,192.7	594.6	145	57
d18:0	20:1	594.6	266.3	40	48	610.6	266.3	88	49	772.6	266.3	93	75	1,190.7	592.6	145	57
d18:0	22:0	624.6	266.3	39	48	640.6	266.3	95	47	802.7	266.3	93	80	1,220.7	622.6	145	58
d18:0	22:1	622.6	266.3	40	48	638.6	266.3	85	44	800.7	266.3	93	75	1,218.7	620.6	145	58
d18:0	24:0	652.7	266.3	39	44	668.7	266.3	92	50	830.7	266.3	93	100	1,248.7	650.6	145	61
d18:0	24:1	650.7	266.3	37	43	666.7	266.3	81	50	828.7	266.3	100	95	1,246.7	648.6	145	61
d18:0	26:0	680.7	266.3	43	48	696.7	266.3	98	50	858.7	266.3	100	100	1,276.8	678.7	145	63
d18:0	26:1	678.7	266.3	46	48	694.7	266.3	88	52	856.7	266.3	100	95	1,274.8	676.7	145	63
d18:1	16:0	538.5	264.3	40	39	554.5	264.3	62	37	716.6	264.3	78	53	1,134.6	536.5	145	57
d18:1	18:0	566.6	264.3	38	39	582.6	264.3	62	41	744.6	264.3	80	56	1,162.7	564.6	145	57
d18:1	20:0	594.6	264.3	44	39	610.6	264.3	68	42	772.6	264.3	80	60	1,190.7	592.6	145	57
d18:1	20:1	592.6	264.3	42	42	608.6	264.3	56	43	770.6	264.3	80	58	1,188.7	590.6	145	57
d18:1	22:0	622.6	264.3	44	46	638.6	264.3	68	47	800.7	264.3	80	62	1,218.7	620.6	145	58
d18:1	22:1	620.6	264.3	39	44	636.6	264.3	65	45	798.6	264.3	80	66	1,216.7	618.6	145	58
d18:1	24:0	650.7	264.3	38	49	666.7	264.3	75	45	828.7	264.3	90	60	1,246.7	648.6	145	61
d18:1	24:1	648.7	264.3	42	43	664.7	264.3	69	45	826.7	264.3	95	63	1,244.7	646.6	145	61
d18:1	26:0	678.7	264.3	38	46	694.7	264.3	83	48	856.7	264.3	90	67	1,274.8	676.7	145	63
d18:1	26:1	676.7	264.3	46	48	692.7	264.3	78	49	854.7	264.3	85	63	1,272.8	674.7	145	63

The LCB and fatty acid (FA) pairings are shown on the left followed by the parent (Q1) and product ion (Q3) *m/z*, the declustering potential (DP) and collision energy (CE) for each compound

Table 5
Response factor for each chemical structure compared to internal standard

Class/LCB	d18:0	d18:1	t18:0	t18:1
Ceramides	3	4	6	5
Hydroxyceramides	3	4	6	5
Glucosylceramides	4	6	4	3
GIPCs	0.16	0.45	0.12	0.08
LCBs	2	1	5	4
LCB(P)s	1.8	1	1	1

Response factors were calculated by comparing standard curve for the internal standard with that for purified Arabidopsis sphingolipids (6)

tissue the water in the tissue will cause phase separation already during the extraction. This can be circumvented by adding the propanol water and hexane separately and reducing the amount of water by the weight of tissue used (assuming 1 g tissue contains 0.9 mL of water).

2. These grinders can be attached to a power tool for processing many samples. Push a 5 cm length of vinyl hose firmly over the end of the grinder handle leaving 2 cm of open tube. Insert the free end of the hose into a screwdriver bit installed on a variable speed power tool. Control the power tool with one hand while holding the grinding vessel with the other.
3. The API 4000 is the instrument of choice for lipidomics due to the sensitivity, mass-range, and triple quadropole design. The data and settings provided here are intended for use with that instrument, other instruments may also be used with adaptations where necessary.
4. There are many manufacturers and types of reversed phase C18 columns. The choice of an Agilent XDB-C18 is made based on the reproducibility of the columns, the scalability from smaller to larger columns and the large number of injections the column can handle. Shifts in retention time or peak shape can usually be resolved by replacing the guard column.
5. Freeze drying is a very effective way of preserving samples, provided no mistakes are made in the freeze drying process. Tissue samples can easily be collected and frozen in liquid N_2 inside coin envelopes. Transfer these to a pre-cooled freeze drying container and store at −80 °C for an hour to allow the liquid N_2 to evaporate then attach to the freeze dryer and leave overnight.

6. The judicious use of an ultrasonic water bath can greatly aid in resuspending pellets or dissolving lipid samples. In general a mix of heating in the water bath, vortexing, and short periods of sonication are sufficient.
7. Elution times for specific compounds have not been provided as they will vary from system to system. Elution times for internal standards are easily identified by including a sample containing only standards. Elution times for specific sphingolipid compounds are easily identified for abundant compounds, less abundant compounds such as d18:0 containing GlcCer will be harder to identify. One approach is to use mutants that contain higher levels of these compounds (7).
8. The response factor is necessary to compensate for the lack of chemically identical internal standards. Changing the position of desaturation or the addition of extra hydroxyl groups has a significant effect on the signal from the mass-spectrometer however there are no synthetic or purified standards for plant sphingolipids, hence the need for this signal compensation or response factor.

Acknowledgments

This work was supported by NSF grant MCB-0843312 and the University of Nebraska-Lincoln.

References

1. Lester RL, Dickson RC (2001) High-performance liquid chromatography analysis of molecular species of sphingolipid-related long chain bases and long chain base phosphates in Saccharomyces cerevisiae after derivatization with 6-aminoquinolyl-*N*-hydroxysuccinimidyl carbamate. Anal Biochem 298:283–292
2. Merrill AH Jr, Caligan TB, Wang E et al (2000) Analysis of sphingoid bases and sphingoid base 1-phosphates by high-performance liquid chromatography. Methods Enzymol 312:3–9
3. Cahoon EB, Lynch DV (1991) Analysis of glucocerebrosides of rye (Secale cereale L. cv Puma) leaf and plasma membrane. Plant Physiol 95:58–68
4. Imai H, Morimoto Y, Tamura K (2000) Sphingoid base composition of monoglucosylceramide in Brassicaceae. J Plant Physiol 157:453–456
5. Sullards MC (2000) Analysis of sphingomyelin, glucosylceramide, ceramide, sphingosine, and sphingosine 1-phosphate by tandem mass spectrometry. Methods Enzymol 312:32–45
6. Markham JE, Jaworski JG (2007) Rapid measurement of sphingolipids from Arabidopsis thaliana by reversed-phase high-performance liquid chromatography coupled to electrospray ionization tandem mass spectrometry. Rapid Commun Mass Spectrom 21:1304–1314
7. Chen M, Markham JE, Dietrich CR et al (2008) Sphingolipid long-chain base hydroxylation is important for growth and regulation of sphingolipid content and composition in Arabidopsis. Plant Cell 20:1862–1878

Chapter 11

Analysis of Defense Signals in *Arabidopsis thaliana* Leaves by Ultra-performance Liquid Chromatography/Tandem Mass Spectrometry: Jasmonates, Salicylic Acid, Abscisic Acid

Nadja Stingl, Markus Krischke, Agnes Fekete, and Martin J. Mueller

Abstract

Defense signaling compounds and phytohormones play an essential role in the regulation of plant responses to various environmental abiotic and biotic stresses. Among the most severe stresses are herbivory, pathogen infection, and drought stress. The major hormones involved in the regulation of these responses are 12-oxo-phytodienoic acid (OPDA), the pro-hormone jasmonic acid (JA) and its biologically active isoleucine conjugate (JA-Ile), salicylic acid (SA), and abscisic acid (ABA). These signaling compounds are present and biologically active at very low concentrations from ng/g to μg/g dry weight. Accurate and sensitive quantification of these signals has made a significant contribution to the understanding of plant stress responses. Ultra-performance liquid chromatography (UPLC) coupled with a tandem quadrupole mass spectrometer (MS/MS) has become an essential technique for the analysis and quantification of these compounds.

Key words Ultra-performance liquid chromatography/tandem mass spectrometry, Jasmonates, Salicylic acid, Abscisic acid

1 Introduction

Plant defense signals and hormones are low molecular weight molecules that are rapidly synthesized de novo in response to internal and external cues. Notably, synthesis of these compounds may take place within a few seconds after onset of a particular stress and increase the endogenous signal compound concentration by two orders of magnitude (1). Due to their low endogenous concentrations, analysis techniques have to be extremely sensitive and specific to observe and quantify these signaling compounds both under basal and stress conditions. Until recently, gas chromatography-mass spectrometry (GC-MS) was most commonly used for the analysis of many plant hormones (2). However, GC-MS requires a time consuming pre-purification process as well as derivatization steps to enhance volatility. Liquid chromatography-tandem

Teun Munnik and Ingo Heilmann (eds.), *Plant Lipid Signaling Protocols*, Methods in Molecular Biology, vol. 1009,
DOI 10.1007/978-1-62703-401-2_11,

quadrupole mass spectrometry has emerged as an effective, equally sensitive, and specific method for phytohormone analysis (3, 4). The presented method needs 50 mg of Arabidopsis leaf tissue to determine phytohormone profiles and enables high-throughput multiparallel microscale ball-mill-extraction followed by centrifugation and direct analysis of the supernatant without further purification. The use of adequate internal standards that are similar to the analytes and partially co-elute with the signal compounds and thus display similar ionization and fragmentation properties is critical in establishing a reliable method. This is crucial because molecules originating from the sample matrix or the solvent that co-elute with the compounds of interest can interfere in the ionization process in the mass spectrometer, causing ionization suppression or enhancement, which is the so-called matrix effect. For quantification, isotope-labeled internal standards are the first choice. Unfortunately, not all internal standards are commercially available. Oxygen-18-labeled internal standards can be easily prepared in the case of acidic phytohormones by oxygen-18 exchange of the two oxygen atoms of carboxyl groups (2) and these internal standards are very suitable for LC-MS/MS analysis.

2 Materials

All solvents and chemicals utilized in this protocol are of LC grade or better.

2.1 General Equipment

1. Centrifuge for 1.5 ml polypropylene reaction tubes.
2. Vacuum freeze-drier.
3. Small bead mill (Type MM201, Retsch, Haan); ceramic beads with a diameter of 64 mm.
4. Centrifugal vacuum evaporator (Type 2-18 CD, Christ, Osterode).

2.2 Plant Material

Arabidopsis plants were grown in soil under short day conditions (9 h light 100–120 μmol photons/m^2 s 22 °C, 15 h dark). Fully expanded leaves of 6-week-old plants have been used for analysis.

2.3 Signal Compound Extraction

1. Extraction solution: Ethylacetate:formic acid, 99:1 (v/v).
2. Internal standard solution: This solution contains each of the following internal standards at 1 ng/μl in acetonitrile: dhJA, [$^{18}O_2$]OPDA, JA-Nval, [D_4]SA, and [D_6]ABA (see Subheading 3.2 for preparation of internal standards).
3. Reconstitution solution: Acetonitrile: water, 1:1 (v/v).
4. Mass spectrometer tuning solution: This solution contains each of the following internal standards: dhJA, [$^{18}O_2$]OPDA,

JA-Nval, [D_4]SA, and [D_6]ABA, and analytes: JA, OPDA, JA-Ile, SA, and ABA, at 1 ng/µl in acetonitrile.

2.4 LC-MS/MS Analysis and Quantification

1. Separation and quantification of signal compounds was performed by ultra-performance liquid chromatography/tandem mass spectrometry (UPLC-MS/MS) analyses using a Waters Quattro Premier triple-quadrupole mass spectrometer with an electrospray interface (ESI) coupled to a Waters Acquity ultra-performance liquid chromatography (UPLC®) setup (Milford, MA, USA).
2. Acquity UPLC BEH RP C18 column (2.1 × 50 mm, 1.7 µm particle size, Waters, Milford, MA, USA) with an in-line filter (0.2 µm pore size).
3. Solvents used for the LC: *Eluent A*: water:formic acid, 99.9:0.1 (v/v) and *Eluent B*: acetonitrile.
4. Calibration solutions: These solutions contain each analyte in equal concentration and varying concentrations of the analytes between solutions, over the range expected in samples (100 pg/µl, 1 ng/µl, 10 ng/µl, 30 ng/µl, and 50 ng/µl), along with 1 ng/µl of each internal standards in acetonitrile. These solutions are to be prepared as samples: solvent is removed by using a centrifugal vacuum evaporator, and then the residues are taken up in Reconstitution solution to reach at the final desired concentrations.

3 Methods

This protocol is dedicated to investigate phytohormone levels in leaf tissues and requires at least 50 mg fresh weight or 5 mg dry weight of plant tissue. The extraction and preparation of each sample takes at least 2 h and the UPLC-MS/MS analysis takes 12 min per sample. Nevertheless, the preparation of internal standards and the complexity of the procedure, initial setup, and optimization of the UPLC-MS/MS system may take a few days to be completed, and should be performed by experienced workers.

3.1 Preparation of Internal Standards

All internal standards were checked for purity and isotopic enrichment by UPLC-MS/MS and quantified against the unlabeled analytes. For quantification, the analyte (1 ng/µl) and an about equal amount of the internal standard were injected (5 µl injection volume) into the LC-MS/MS system and the MRM transitions (Table 1) were recorded. From the ratio of the peak areas of the analyte and the internal standard compound, the concentration of the internal standard was determined assuming a response factor of one (*see* **Note 1**).

Table 1
Optimized MS/MS conditions utilized for Quattro Premier triple-quadrupole mass spectrometer and multiple reaction monitoring (MRM) with a dwell time of 0.025 s per transition

Compound	Transition	Cone voltage (V)	Collision energy (eV)	Retention time (min)
SA	137→93	17	17	2.97
$[D_4]$SA	141→97	17	17	2.95
ABA	263→153	26	14	3.10
$[D_6]$ABA	269→159	26	14	3.09
JA	209→59	19	17	3.58
dhJA	211→59	19	17	3.91
JA-Ile	322→130	24	18	4.13
JA-Nval	308→116	24	18	3.83
OPDA	291→165	22	26	5.11
$[^{18}O_2]$OPDA	295→165	22	26	5.11

1. *Dihydro-Jasmonic Acid (dhJA)*: Synthesis of dhJA was performed by catalytic hydrogenation of racemic JA (obtained from Sigma-Aldrich, Steinheim, Germany) (5). The entire procedure must be performed in a fume hood. Dissolve JA (25 mg) in 1 ml methanol, add 5 mg of Pd/Al_2O_3 catalyst (Degussa Type E207 R/D; 5 % Pd, 5–10 % H_2O; Sigma-Aldrich, Steinheim, Germany) and stir slowly with a magnetic bar. Hydrogen gas is bubbled through the mixture for 30 min with a flow of 5 ml/min. The catalyst is filtered off by passing the suspension through a pipette tip packed with a homemade glass fiber filter. When the solution has passed the filter, the hydrogen-loaded catalyst should not run dry since the catalyst may start to glow in the presence of air. Therefore, highly flammable solvents should be removed from the hood beforehand and the filter should be stored in an empty flask in the hood for at least 5 h before it is discarded. The sample is taken to dryness under a stream of nitrogen. The residue is dissolved in acetonitrile and stored at −20 °C.
2. *Jasmonic Acid-Norvaline Conjugate (JA-Nval)*: Synthesis of JA-norvaline is performed by conjugation of norvaline to racemic JA (both compounds obtained from Sigma-Aldrich, Steinheim, Germany) (6). JA (1 mmol) is dissolved in 5 ml anhydrous tetrahydrofurane containing 1.08 mmol of triethyl amine. Add isobutyl chloroformate (1.07 mmol) and incubate for 60 min at 0 °C. Filter the solution and mix the clear filtrate with water containing

3.5 mmol norvaline lithium salt and incubate for 5 h. Evaporate the mixture under a stream of nitrogen, add 15 ml water, adjust the pH with 1 M HCl to pH 3, and extract ten times with 10 ml of $CHCl_3$. Dry the collected $CHCl_3$-phases under a stream of nitrogen and dissolve the residue in 500 μl hexane:ethyl acetate:acetic acid, 75:25:1 (v/v/v). The solution is loaded on a 500 mg silica solid phase extraction column (Sepra silica with a particle size of 50 μm and a pore size of 65 Å, Phenomenex, Aschaffenburg, Germany) which was washed before with two column volumes of methanol and afterwards equilibrated with three column volumes of hexane: ethyl acetate: acetic acid, 75:25:1 (v/v/v). The column is eluted with three column volumes of hexane:ethyl acetate:acetic acid, 75:25:1 (v/v/v), three column volumes of hexane:ethyl acetate:acetic acid, 50:50:1 (v/v/v) and two column volumes of ethyl acetate:acetic acid, 100:1 (v/v). To determine the fractions in which JA-Nval is present, each fraction is analyzed by LC-MS/MS. JA-Nval is mostly recovered with hexane:ethyl acetate:acetic acid, 50:50:1 (v/v/v) and ethyl acetate:acetic acid, 100:1 (v/v). JA-Nval containing fractions are combined and taken to dryness under a stream of nitrogen. Dissolve the residue in acetonitrile and store at −20 °C.

3. [1,1-$^{18}O_2$]12-oxo-phytodienoic acid ([$^{18}O_2$]OPDA): Unlabeled racemic *cis*-12-Oxo-10,15(Z)-phytodienoic acid was obtained from Larodan (Malmö, Sweden). [$^{18}O_2$]OPDA was prepared by enzyme-catalyzed [^{18}O]-exchange (2). Dissolve unlabeled OPDA (0.5 mg) in 50 μl of anhydrous ethylene glycol and suspend 1 mg of Enzyme (Lipase II, crude from porcine pancreas, 100–400 units/mg, Sigma, Buchs, Switzerland) in this solution. Add 25 μl [^{18}O] water containing buffer salts (dried solution of 2.5 μl 1 M Tris–HCl buffer, pH 7.4) and incubate for 15 h at 40 °C. Extract with 1 ml of hexane:diethyl ether, 2:1 (v/v), containing 1 % of acetic acid and dry the organic phase under a stream of nitrogen. Dissolve the residue in acetonitrile and store at −20 °C.

4. [3,5,5,7,7,7-D_6]abscisic acid ([D_6]ABA): Synthesis of [D_6]ABA was performed by exchange of the ring protons of unlabeled 2-*cis*, 4-*trans*-ABA (Sigma-Aldrich, Steinheim, Germany) (7). Dissolve 5 mg of unlabeled abscisic acid in 10 ml D_2O (isotopic purity 99.8 %, Merck, Darmstadt, Germany) containing 0.1 M sodium methanolate and incubate for 48 h at room temperature. Stop the reaction by acidifying the solution with 1 M HCl to pH 3 and extract the mixture three times with 10 ml of diethyl ether. Dry the combined organic phases with anhydrous sodium sulfate and remove the salt by filtration. Take the sample to dryness under a stream of nitrogen and dissolve the residue in acetonitrile and store at −20 °C.

5. [3,4,5,6-D_4]salicylic acid ([D_4]SA): For the synthesis of [D_4]salicylic acid, [D_6]salicylic acid was purchased from Sigma-Aldrich

(Steinheim). Since the carboxylic acid and hydroxyl deuterium atoms readily exchange with hydrogen in protic solvents, stable [D_4]SA is produced by dissolving dry [D_6]SA in methanol. Thereafter, the sample is dried under a stream of nitrogen. Dissolve the residue in acetonitrile and store at –20 °C.

3.2 Signal Compound Extraction

1. Collect plant tissue and immediately shock freeze the tissue in liquid nitrogen and store until analysis below –20 °C. Grind plant material with a mortar and pestle (*see* **Note 2**).
2. Weigh out 50 mg of homogenized frozen tissue (or 5 mg freeze dried material) into 2 ml reaction tubes with screw cap (suitable for subsequent ball milling), add 0.950 ml of cold Extraction solution and mix rapidly until the tissue is completely suspended in the Extraction solution.
3. Add 50 μl of the internal standard solution and mix (*see* **Note 3**).
4. Add a ceramic bead and perform the extraction with a small bead mill for 3 min at 20 Hz.
5. Centrifuge at 16,000 × *g* for 10 min at room temperature.
6. Transfer supernatant to a new tube and remove solvent by using a centrifugal vacuum evaporator at 30 °C.
7. Dissolve the residue in 40 μl of Reconstitution solution (*see* **Note 4**).

3.3 LC-MS/MS Analysis and Quantification

1. ESI-parameters: To establish the appropriate MS conditions for the individual compounds and internal standards, the Mass spectrometer tuning solution is directly infused into the MS and the cone voltage (CV) is adjusted to maximize the intensity of the deprotonated molecular species (*see* **Note 5**). The ESI source was operated in negative ionization mode with a capillary voltage of 3.0 kV and a source temperature of 120 °C. Use nitrogen as the desolvation and cone gas with a flow rate of 800 and 50 l/h, respectively.
2. MRM-parameters: The optimized fragmentation settings for each of the analytes and internal standards are shown in Table 1, although final tuning is suggested (*see* **Note 5**). Set the time windows for the MRM signals according to the retention time (±0.5 min).
3. UPLC-parameters: The following conditions using a BEH RP C18 column were applied: 5 μl injection volume through partial loop, linear solvent strength gradient using Eluent A and Eluent B (3 % Eluent B at 0 min, 100 % at 7 min, 100 % at 9 min; re-equilibrate the column with 3 % Solvent B for 3 min) at a flow rate of 0.25 ml/min, column temperature set to 40 °C and autosampler temperature at 10 °C. The determined retention times of each compound and corresponding internal standard are listed in Table 1 and Fig. 1 (*see* **Note 5**).

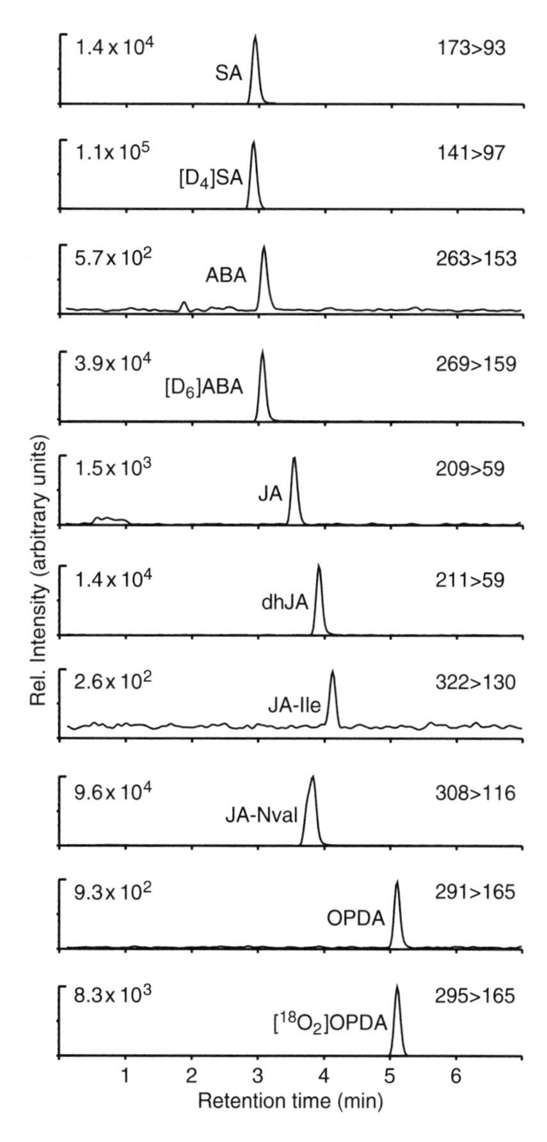

Fig. 1 Representative LC chromatograms of five plant defense signals and their internal standards from an extract of 50 mg fresh weight of *Arabidopsis thaliana* plants ecotype Columbia 0

4. Quantification: After optimization of all LC-MS/MS parameters, analyze all Calibration solutions and the plot peak area ratio of analyte and internal standard (Peak area ratio (analyte/IS)) in function of the concentration ratio of analyte and internal standard [ratio (analyte/IS) injected] for each analyte/internal standard pair. An example calibration curve is shown in Fig. 2 (*see* **Note 6**). The calibration curves are used to confirm a linear detector response over the chosen analyte concentration range and for quantify the phytohormone concentration in real samples (*see* **Note 7**).

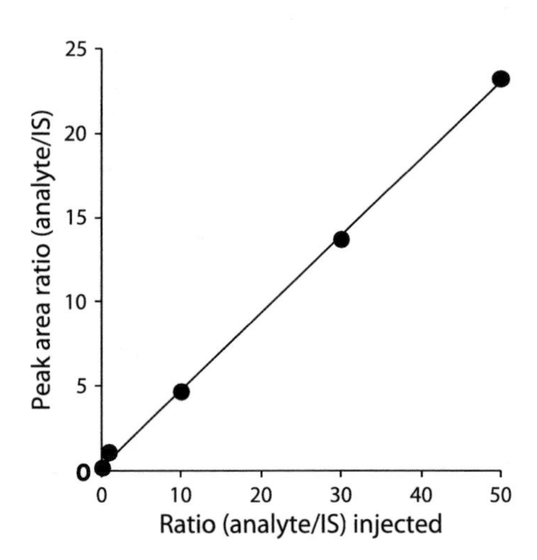

Fig. 2 Example calibration curve for ABA. The peak area ratio of analyte and internal standard was plotted in function of the concentration ratio of analyte and internal standard

5. Sample analysis: Analyze plant extracts containing the internal standards and quantify the analytes using the calibration curves constructed above (see Fig. 2) by calculating the ratio (analyte/IS) injected via the following formula:

$$\text{Ratio (analyte/IS) injected} = \frac{\text{Peak area ratio(analyte / IS)} - a}{b}$$

where a, intercept; b, slope

In addition blank samples containing only internal standards should be analyzed as the first and last sample in all batch of samples to ensure that no contamination, carry over effect, or retention time shift occurred.

4 Notes

1. A detector response factor of one means that the ratio of the peak areas of the internal standard and the analyte is one after injecting equal moles of internal standard and analyte into the LC-MS/MS system. The true response factor for the IS/analyte pairs is close to but not exactly one and therefore the internal standard concentration cannot exactly be determined. However, it is not important to know the exact internal standard concentration since calibration curves of the target components are used for quantification.
2. After harvest, plant material has to be immediately shock frozen in liquid nitrogen and must remain frozen during grinding with a precooled mortar and pestle. During grinding liquid

nitrogen has to be added to prevent thawing of plant material. These precautions are necessary since some plant signals such as JA can be formed within seconds after wounding (1). The plant material has to be stored at least at −20 °C until the extraction solvent is added.

3. [^{18}O]-isotope-labeled standards are stable for several days in aqueous solutions in the pH range of 3–7. However, prolonged storage of samples may lead to some loss of [^{18}O]-label. Therefore storage in anhydrous and aprotic solvents such as acetonitrile is recommended. In plant extracts, plant lipases and esterases (8) can convert the internal standard into the target compound by back-exchange of the [^{18}O]-label. Some plant lipases may be very resistant to denaturation and are even active in organic solvents. The stability of the internal standard [$^{18}O_2$]OPDA during extraction, storage of the sample for 5 days at room temperature and analysis was tested in triplicate. No back-exchange of the [^{18}O]-label could be observed. Nevertheless plant extracts should be generally stored at least at −20 °C.
4. Sonication may help to dissolve or to suspend the residue. While internal standards and analytes are readily soluble, however, not all extract components may dissolve.
5. The optimal ionization parameters and fragmentation settings are instrument dependant thus final tuning might be needed. The retention times might differ using different types of UPLC instruments, and thus, the MRM window has to be set individually. The LC and the MS were controlled by MassLynx v. 4.1 software (Waters Cooperation, Mississauga, Ontario, Canada).
6. Within this protocol, 500 pg to 250 ng on column corresponds to a concentration range of 40 ng to 20 μg/g of plant fresh weight. The five point calibration curve was linear over the analyte range of 500 pg to 250 ng per 5 μl injection volume. It is suggested to calculate the regression coefficient R^2 that should be above 0.9. As a further quality control, it is suggested to determine the precision of the method. Analyze three replicates of different Calibration solutions and determine the relative standard deviation (RSD) which should be below 10 %.
7. The performance characteristics (precision, linearity, and sensitivity) of the extraction and analysis were determined. To determine the precision of the analysis three replicates of the Calibration solutions (100 pg/μl, 1 ng/μl, 10 ng/μl, 30 ng/μl, and 50 ng/μl) were injected and the relative standard deviations of the peak area were calculated. For determination of the precision of the extraction method three extracts of equal plant tissues were analyzed and the relative standard deviation

Table 2
LOD and LOQ for the analytes SA, ABA, JA, JA-Ile, and OPDA derived from chromatograms of plant extracts (50 mg fresh material)

Analyte	SA	ABA	JA	JA-Ile	OPDA
LOD/fresh weight (ng/g)	1.7	0.7	0.5	0.1	0.7
LOQ/fresh weight (ng/g)	5.6	2.3	1.7	0.7	2.2

of the peak area ratio (analyte/IS) was calculated. With the setup used in this method RSD of 2–8 % for precision of the analysis and RSD of 8–12 % for precision of the extraction were achieved. To study the linearity of this method all Calibration solutions were analyzed and peak area ratio of analyte and internal standard was plotted in function of the concentration ratio of analyte and internal standard for each analyte/internal standard pair, and R^2 calculated. The following R^2 values were determined with the setup used in this procedure: 0.93 (SA), 0.97 (OPDA and JA) and 0.99 (JA-Ile and ABA) in the range of 100 and 50,000 pg/μl. Sensitivity was assessed by calculating LOD and LOQ of three extracts of equal plant tissue with the following formulas:

$$\mathrm{LOD} \rightarrow \frac{\mathrm{Signal}}{\mathrm{Noise}} \approx 3, \qquad \mathrm{LOQ} \rightarrow \frac{\mathrm{Signal}}{\mathrm{Noise}} \approx 10.$$

Within this procedure achieved LOD and LOQ are shown in Table 2.

Acknowledgments

This work was supported by grants of the Graduiertenkolleg 1342 "Lipid signaling" and the SFB 567 of the Deutsche Forschungsgemeinschaft (DFG).

References

1. Glauser G, Dubugnon L, Mousavi SA, Rudaz S, Wolfender JL, Farmer EE (2009) Velocity estimates for signal propagation leading to systemic jasmonic acid accumulation in wounded Arabidopsis. J Biol Chem 284:34506–34513
2. Mueller MJ, Mene-Saffrane L, Grun C, Karg K, Farmer EE (2006) Oxylipin analysis methods. Plant J 45:472–489
3. Giannarelli S, Muscatello B, Bogani P, Spiriti MM, Buiatti M, Fuoco R (2010) Comparative determination of some phytohormones in wild-type and genetically modified plants by gas chromatography-mass spectrometry and high-performance liquid chromatography-tandem mass spectrometry. Anal Biochem 398: 60–68

4. Pan X, Welti R, Wang X (2008) Simultaneous quantification of major phytohormones and related compounds in crude plant extracts by liquid chromatography–electrospray tandem mass spectrometry. Phytochemistry 69: 1773–1781
5. Gundlach H, Muller MJ, Kutchan TM, Zenk MH (1992) Jasmonic acid is a signal transducer in elicitor-induced plant cell cultures. Proc Natl Acad Sci USA 89:2389–2393
6. Kramell R, Schmidt J, Schneider G, Sembdner G, Schreiber K (1988) Synthesis of *N*-(jasmonoyl) amino acid conjugates. Tetrahedron 44: 5791–5807
7. Gomez-Cadenas A, Pozo OJ, Garcia-Augustin P, Sancho JV (2002) Direct analysis of abscisic acid in crude plant extracts by liquid chromatography–electrospray/tandem mass spectrometry. Phytochem Anal 13:228–234
8. Pollard M, Ohlrogge J (1999) Testing models of fatty acid transfer and lipid synthesis in spinach leaf using in vivo oxygen-18 labeling. Plant Physiol 121:1217–1226

Chapter 12

Analysis of Fatty Acid Amide Hydrolase Activity in Plants

Sang-Chul Kim, Lionel Faure, and Kent D. Chapman

Abstract

N-Acylethanolamines (NAEs) are fatty acid derivatives amide-linked to ethanolamine. NAEs vary in chain lengths and numbers of double bonds and generally reflect the fatty acids found in membrane lipids in the tissues in which they reside. NAEs are present naturally in trace amounts and occur in a wide range of organisms including plants, animals, and microbes. Some NAE types are known to be involved in the endocannabinoid signaling system of vertebrates, and in plants they may play important regulatory roles in several physiological processes, such as root growth, seedling development, stress responses, and pathogen interactions. The biological effects of NAEs are terminated through their hydrolysis into the ethanolamine and free fatty acid by a membrane enzyme known as the fatty acid amide hydrolase (FAAH). Thus, FAAH represents an important target to better understand the function of these lipid mediators in numerous cellular processes. FAAH has been extensively characterized in mammalian and plant systems, and they share a conserved Ser-Ser-Lys catalytic mechanism. Here we describe procedures and experimental conditions to assay and characterize recombinant and endogenous FAAH enzymatic activity derived from plant tissues.

Key words FAAH, NAE, Lipid mediator, Protein expression, Enzymatic assay

1 Introduction

Fatty acid amide hydrolase (FAAH) is an integral membrane-associated enzyme that hydrolyzes endogenous fatty acid amide/ester substrates including *N*-acylethanolamines (NAEs), fatty acid primary amines (e.g., oleamide), and monoacylglycerols (e.g., 2-arachidonoylglycerol) (1). FAAH has been identified and characterized in many species of both animals and plants. In vertebrates, FAAH is known to regulate the endocannabinoid signaling pathway that influences various neurobehavioral processes (2, 3). Because many of the FAAH substrates are known to have biological activities implicated in neurological disorders, FAAH is currently viewed as a possible therapeutic target for the treatment of diseases, and thus efforts are being made to develop specific inhibitors for this enzyme (1, 2). In plants, FAAH appears to be involved in signaling pathways that regulate seedling growth and phytohormone

Teun Munnik and Ingo Heilmann (eds.), *Plant Lipid Signaling Protocols*, Methods in Molecular Biology, vol. 1009,
DOI 10.1007/978-1-62703-401-2_12, © Springer Science+Business Media, LLC 2013

responses, in part, by modulating endogenous levels of the NAE substrates (4–7). Thus, plant FAAH may represent an attractive tool to manipulate lipid metabolism important to plant growth, stress responses, and productivity.

Expression and enzymatic assay of FAAH are necessary to functionally characterize this enzyme for practical applications in plants. However, its membrane association (8) and low level of expression (4) make it somewhat difficult to solubilize the enzyme and purify its activity directly from plant tissues. Here we present a highly sensitive and reproducible method to detect the hydrolytic activity of plant FAAH toward NAE substrates, which has been optimized for recombinant *Arabidopsis* FAAH (AtFAAH) expressed in *Escherichia coli*. The assay as developed can be used for AtFAAH proteins that are solubilized from *E. coli* cells transformed with *AtFAAH* cDNA, or directly from *Arabidopsis* tissues (or other plants). FAAH is solubilized from membranes in the presence of dodecylmaltoside (DDM) (8). The crude extracts are then reacted with either radiolabeled or unlabeled NAE substrates and the rate of free fatty acid formation is measured. Total lipids are extracted from the reaction mixture by a modified Bligh and Dyer method (9) and separated by thin layer chromatography (TLC), followed by either radiometric scanning or iodine vapor exposure of the TLC plate for detection of the lipids. The synthesis of radiolabeled NAE substrates and data analysis for quantitative measurements will also be discussed.

2 Materials

Prepare all solutions using ultrapure water (18.2 MΩ cm at 25 °C). Prepare and store all reagents at room temperature unless indicated otherwise. Radioisotopes were purchased from PerkinElmer.

2.1 Buffers and Media

1. Murashige and Skoog (MS) medium for liquid culture of *Arabidopsis* plants: 5 % (v/v) MS Macro 10×, 5 % (v/v) MS Micro 10×, 0.5 μg/mL pyridoxine HCl, 0.5 μg/mL nicotinic acid, 1 μg/mL thiamine, 0.1 mg/mL Myo-inositol, 0.5 mg/mL 2-(*N*-morphonilo) ethanesulfonic acid (MES), and 10 mg/mL sucrose. Adjust the pH to 5.7 with KOH (*see* **Note 1**). Sterilize by autoclave at 120 °C for 20 min and store at 4 °C.
2. Luria-Bertani (LB) medium for liquid culture of *E. coli*: 5 g/L yeast extract, 10 g/L tryptone, and 10 g/L NaCl. Sterilize by autoclave at 120 °C for 20 min.
3. Homogenization buffer for crude extraction of *Arabidopsis* proteins: 10 mM KCl, 1 mM EDTA, 1 mM EGTA, 1 mM $MgCl_2$, 400 mM sucrose, and 100 mM potassium phosphate. Adjust the pH to 7.2 and store at 4 °C.

4. Lysis buffer for crude extraction of *E. coli* proteins: 50 mM Tris–HCl, 100 mM NaCl, and 0.2 mM DDM. Adjust the pH to 8.0 and store at 4 °C (*see* **Note 2**).

2.2 Preparation of Reagents

1. 1 M isopropyl β-D-1-thiogalactopyranoside (IPTG): dissolve 2.38 g of IPTG powder in water and bring the final volume to 10 mL. Sterilize by filtering with a 0.2-μm disk filter and store at −20 °C (*see* **Note 3**).
2. 50 mM ethylenediaminetetraacetic acid (EDTA): dissolve 1.86 g of EDTA powder in water and bring the final volume to 100 mL (*see* **Note 4**).
3. 50 mM ethyleneglycoltetraacetic acid (EGTA): dissolve 1.91 g of EGTA powder in water and bring the final volume to 100 mL (*see* **Note 4**).
4. 100 mM *n*-dodecyl β-D-maltoside (DDM): dissolve 0.51 g of DDM powder in water and bring the final volume to 10 mL. Store at −20 °C (*see* **Note 5**).
5. 1 M KCl: dissolve 74.55 g of KCl powder in water and bring the final volume to 1 L.
6. 50 mM Bis–tris propane (BTP): dissolve 1.41 g of BTP powder in 100 mL of water. Adjust the pH to 9.0 with HCl and store at 4 °C.

2.3 Other Chemicals and Required Equipment

1. Radiolabeled (1-^{14}C) free fatty acids (PerkinElmer) for synthesis of radiolabeled NAE substrates.
2. Unlabeled NAE substrates (Cayman Chemical) for dilution of radiolabeled NAE substrates.
3. *E. coli* TOP10 cells cloned with *AtFAAH* cDNA in an appropriate expression vector (e.g., pTrcHis, Invitrogen) as described elsewhere (8).
4. Desiccated seeds of wild-type (Col-0) *Arabidopsis thaliana*.
5. Organic solvents: chloroform, hexane, ethyl acetate, methanol, and isopropanol, all with highest purity grade (e.g., HPLC grade).
6. Glass-backed TLC plates: silica gel G (60 Å)-coated glass plate (10 × 20 cm or 20 × 20 cm, 0.25 mm thickness).
7. Glass TLC chambers with an air-tight lid.
8. AR-2000 Imaging Scanner (Bioscan) connected with P-10 gas mixture (90 % argon and 10 % methane).
9. Dichloromethane (99.6 %), dimethylformamide, oxalylchloride, and ethanolamine.
10. Mortar and pestle.
11. Liquid nitrogen and nitrogen gas.
12. Iodine crystals.

13. Bath sonicator.
14. 15 mL Corex centrifuge tubes.
15. Buchner funnel.
16. 15 mL (16 × 125 mm) glass tubes.
17. Water bath.
18. Heat block.
19. 4 mL glass vials with Teflon-coated caps for lipid storage.

3 Methods

3.1 Precautions

Carry out all procedures at room temperature unless otherwise specified. When handling radioactive materials, follow the appropriate regulations and practices for the safe handling and disposal of radioisotopes. Wear protective lab coats, gloves, and safety glasses at all times. Use only glassware and aluminum foil or Teflon-lined closures when working with lipid materials and organic solvents. Work under a fume hood with adequate exhaust to reduce exposure to volatile solvents. Rinse all washed glassware thoroughly with methanol to remove any detergent residue and then with deionized water prior to use.

3.2 Synthesis and Purification of Radiolabeled NAE Substrate

Radiolabeled NAE substrates are synthesized from corresponding radiolabeled (1-^{14}C) free fatty acids (FFA) by first producing the fatty acyl chloride and then with ethanolamine to convert the acyl chloride to the corresponding NAE (10, 11).

1. Transfer ~0.1 μmol of radiolabeled (1-^{14}C) FFA to a 4-mL glass vial and evaporate the solvent under gentle stream of N_2 gas.
2. Add 0.8 mL of dichloromethane and sonicate the mixture in a bath-type sonicator for 30 s.
3. Add 5 μL of dimethylformamide and 15 μL of oxalylchloride, mix well by vortex, and incubate on ice for 1 h, with the vial filled with N_2 gas and tightly capped.
4. Add 0.6 mL of ethanolamine, mix well by vortex, incubate on ice for 3 h, with the vial filled with N_2 gas and tightly capped, and transfer the resulting solution to a 15-mL (16 × 125 mm) test tube.
5. Add 1 mL of deionized water and 0.25 mL of dichloromethane and mix well by vortex.
6. Centrifuge at 1,000 × g for 5 min and discard aqueous upper phase by aspiration.
7. Repeat steps 5 and 6 twice without dichloromethane.

8. Evaporate the remaining organic lower phase with gentle stream of N_2 gas and resuspend the lipids in 100 μL of chloroform.
9. Separate the entire 100 μL of lipids (FFA remained and NAE produce) by performing TLC and scan the TLC plate by radiometric scanning, as described in Subheading 3.4, steps 8–11.
10. Based on the resulting radiochromatogram, mark the region corresponding to NAE (3–4 cm from the bottom of the plate) with a pencil, scrape the silica within the marked region using a razor blade, and carefully transfer the silica powder to a 4-mL glass vial (*see* **Note 6**).
11. Add 2 mL of a mixture of chloroform and methanol (1:2, v/v), mix well by vortex, and sonicate in a bath-type sonicator for 1 min.
12. Incubate the silica suspension at room temperature for 1 h, with the vial filled with N_2 gas and tightly capped.
13. Transfer the supernatant to a fresh vial and repeat steps 11 and 12 twice with the remaining silica to elute any residual NAE, combining the supernatant resulted from each extraction with the one from previous step.
14. Evaporate the resulting supernatant (~4 mL) under gentle stream of N_2 gas, and dissolve the lipids in 100 μL of methanol. Store at −20 °C until use (*see* **Note 7**).
15. The molar concentration of the synthesized NAE solution can be calculated based on its radioactivity (cpm/μL) determined by liquid scintillation counting (corrected for counting efficiency) and the specific radioactivity (Ci/mol) of the FFA used for the NAE synthesis (note, 1 μCi = 2.22×10^6 dpm). For example, if the radioactivity of the solution was measured and corrected to be 25,000 dpm/μL and the specific radioactivity of the FFA was 50 μCi/μmol, the concentration of the NAE solution is then ((25,000 dpm/μL ÷ (2.22×10^6) dpm/μCi) ÷ 50 μCi/μmol) = 0.2252 mM.
16. The solution of radiolabeled NAE synthesized is then diluted with an appropriate volume of unlabeled NAE solution to reduce its radioactivity (cpm/μL) and to achieve a final concentration desired. For example, if 100 μL of 0.2 mM radiolabeled NAE solution was combined with 900 μL of a 10 mM unlabeled NAE solution, its radioactivity is reduced by 1/10 and the final concentration of the mixture is 9.02 mM.

3.3 Preparation of Enzyme Sources

1. Bacterial lysates: grow *E. coli* Top10 cells (or other suitable strain) with *AtFAAH* cDNA under control of the lacZ promoter in 5 mL of LB medium containing appropriate antibiotic at 37 °C with shaking at 250 rpm to an A_{600} of ~0.6. Add

1 mM IPTG and incubate for 4 h to allow for induction of protein expression. Harvest the cells by centrifugation at 2,000 × *g* for 10 min at 4 °C. Break open the cells in 0.4 mL of the lysis buffer by sonication (6 cycles of 10-s burst and 10-s cooling on ice). Remove unbroken cells and cell debris by centrifugation at 1,000 × *g* for 10 min at 4 °C. Use the supernatant as an enzyme source.

2. *Arabidopsis* tissue homogenate: add ~10 mg of surface-sterilized seeds (*see* **Note 8** for sterilization) to 75 mL of sterile MS medium in a sterile 250-mL flask with aluminum foil sealed on top. Stratify the seeds for 48 h at 4 °C (by incubation in dark refrigerator). Grow the plants for 8–10 days with shaking at ~50 rpm in a controlled environment room with a 16-h light and 8-h dark cycle at 20 °C. Collect the seedlings in a Buchner funnel and briefly rinse them with deionized water. Thoroughly grind the seedlings in a mortar and pestle in the presence of liquid N_2 and immediately pour into a 15-mL Corex centrifuge tube. Mix the resulting fine powder of seedlings with 2 mL of the homogenization buffer containing 0.2 mM DDM. Incubate for 30 min on ice with intermittent vortexing. Clear the homogenate by centrifugation at 600 × *g* for 20 min. Use the supernatant as an enzyme source.
3. Measure the concentration of total proteins in the enzyme source by a common protein assay technique, such as the Bradford method (12), with bovine serum albumin (BSA) as a calibration standard.

3.4 FAAH Activity Assay

Overall procedure for the FAAH activity assay and the catalytic reaction occurring in reaction mixture are diagramed in Fig. 1.

1. Add 400 μL of 50 mM BTP buffer (pH 9.0), NAE substrate (~10,000 cpm) to a final concentration of 100 μM (less than 5 μL), and 5–50 μL of an enzyme source in a 15-mL (16 × 125 mm) test tube (*see* **Note 9**). Mix well by vortex.
2. Incubate the reaction mixture in a water bath at 30 °C for 30 min with shaking at 120 rpm.
3. Add 2 mL of boiling isopropanol preheated to 70 °C to the mixture and incubate in a heat block at 70 °C for 30 min to stop the reaction.
4. Cool to room temperature and then add 1 mL of chloroform to the mixture, mix well by vortex to ensure a monophasic mixture, and store either at 4 °C overnight or at room temperature for 2 h in a fume hood for lipid extraction.
5. Add 1 mL of chloroform and 2 mL of 1 M KCl to the mixture to induce phase separation, mix well by vortex, centrifuge at

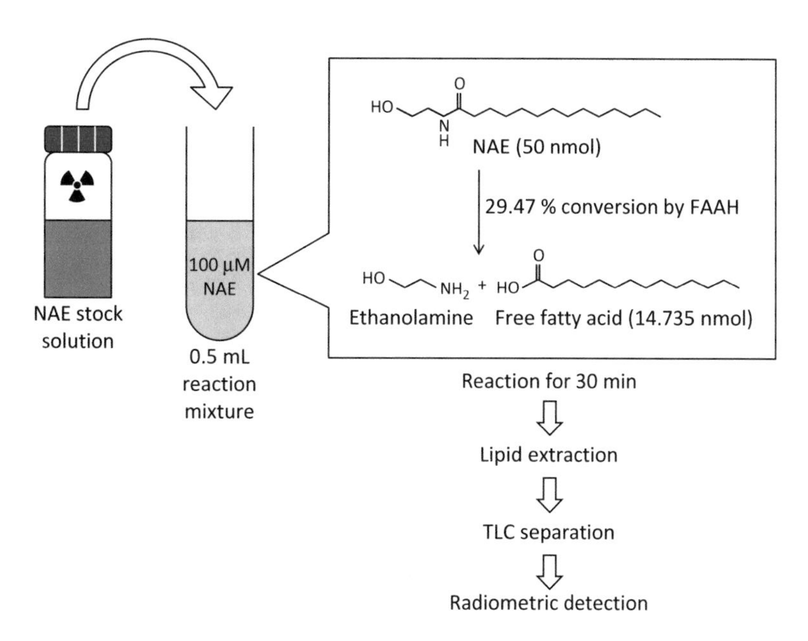

Fig. 1 Schematic diagram of FAAH activity assay. FAAH is reacted with radiolabeled NAE substrate, in which NAE is hydrolyzed into free fatty acid (FFA) and ethanolamine. Total lipids are then extracted from the reaction mixture and separated by TLC, followed by radiometric detection of the lipids. The catalytic activity of FAAH is determined as the amount of radioactive FFA produced in the reaction mixture, which is calculated based on percentage of radioactive NAE converted to FFA and the amount of NAE initially present in the reaction mixture. A numerical example of conversion of NAE to FFA is shown in *parenthesis* (also see text)

1,000 × *g* for 5 min at room temperature, and discard the aqueous upper phase by aspiration (including any precipitated material at the interface).

6. Repeat step 5 twice with 2 mL of 1 M KCl and then with 2 mL of deionized water to purify the lipids.
7. Transfer the remaining organic phase (~2 mL) to a 4-mL glass vial, evaporate the solvent with gentle stream of N_2 gas, and dissolve the lipids in 40 µL of chloroform (*see* **Note 10**).
8. Apply 20–40 µL (in small increments) of the sample at 2 cm from the bottom of a silica gel-coated TLC plate. Use a gentle stream of nitrogen gas to evaporate solvent from applied lipid sample in between applications (*see* **Note 11**). Also include a lane with appropriate substrate (NAE) and product (FFA) standards to mark their chromatographic mobility.
9. Develop the TLC plate in the TLC chamber containing a solvent mixture of hexane, ethyl acetate, and methanol (60:40:5, v/v/v) until the solvent front reaches near the top of the plate (30–40 min) (*see* **Note 12**).

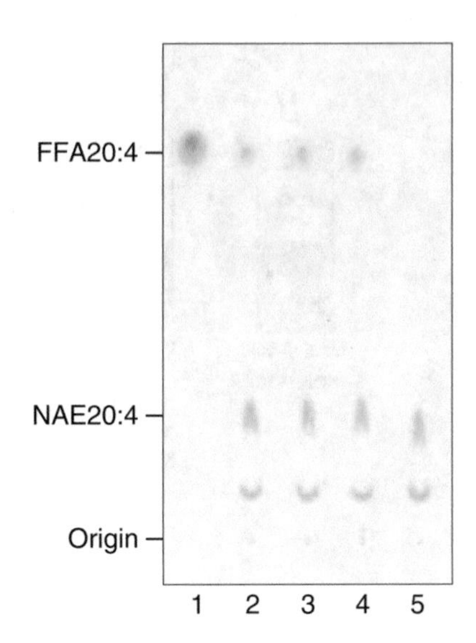

Fig. 2 Image of TLC separation of lipids followed by iodine vapor exposure. *Escherichia coli* lysates (2–20 μg proteins) expressing rat (*2*), *Arabidopsis* (*3*) and rice (*4*) FAAH proteins and empty vector (*5*) were used as the enzyme source. 100 μM *N*-arachidonoyl ethanolamine (NAE20:4) was used as the substrate. Enzyme reaction and lipid extraction and separation were performed as described in Subheading 3. Lipids were visualized by exposing the TLC plate to iodine vapors. Arachidonic acid was also loaded on the TLC plate as standard for position comparison (*1*). Positions of origin, substrate (NAE20:4), and product (FFA20:4) are indicated on the *left*

10. Remove the plate and evaporate the solvent in a fume hood for at least 10 min.
11. Scan the plate by radiometric scanning using the AR-2000 Imaging Scanner according to the manufacturer's instruction for the detection of radiolabeled lipids (*see* **Note 13**). As an alternative to radiometric scanning, the position of the standards can be marked by brief iodine exposure and the plate can be exposed to X-ray film to visualize radiolabeled lipids (1,500 dpm of ^{14}C should be detectable after a few days), or the silica can be scraped and radioactivity quantified by liquid scintillation counting.
12. For unlabeled lipids, place the plate in an empty TLC chamber with iodine crystals at the bottom. The iodine vapor will reveal lipid spots and this staining is reversible under a fume hood (*see* **Note 14** and Fig. 2).

3.5 Data Analysis for Radiometric Scanning

1. Method setup: With the WinScan Software Application Version 3.12 (Nov. 22, 2004) set up a new method to quantify the radiolabeled lipids present on the TLC plate using the Bioscan

AR-2000. Under the "general" tab, use the channel 256, set the processing option for 1 min and the collimator at 10 mm with the high efficiency type, and select the electronic resolution with the normal mode. For the data display, select Autorange Y under the "correction" tabs. In the "integration" tab, select Peak search for the integration technique and set the limit of the integration with the start at 0 mm and the end at 200 mm. The origin of the solvent must be set at 0 mm for the origin and 200 mm for the front. Under the "peak search" tab, set the slope at 0.5 mm, the minimum % total at 0.5, the minimum width at 0.1 mm, and the background region width at 5.0 mm.

2. Interpretation of results: Once a lane of the plate has been scanned, results for the lane are represented in the form of radiochromatogram (see Fig. 3). The *Y*-axis represents the number of counts while the *X*-axis represents the position (mm) of each radiation signal on the plate. The total values of the radioactivity, or Region Counts, for each signal is automatically obtained by the integration of the area of signal. Then the CPM (count per minute), or Region CPM, is calculated by dividing the Region Counts by the time taken to read the lane (*see* **Note 15**). For example, if the Region Counts value is 2,170 count and the set time for the lecture of the sample is 3 min, the CPM value will be 723.3 (2,170/3) (Fig. 3, Rgn 3). The total Count is also obtained by the Bioscan and corresponds to the total radioactivity detected for each lane: peaks + background. The percentage of each signal is calculated using either this total Count or using the total Region Count (amount of each signal only) and is designated as % of Total and % of ROI (region of interest), respectively.
3. Determination of FAAH activity: The catalytic activity of FAAH is measured as the amount (mole) of FFA molecules accumulated through breakdown of the amide bond in NAE molecules for a unit time (h). 1:1 stoichiometry between NAE and FFA enables FFA to be quantified as percentage of NAE hydrolyzed into FFA. Since NAE substrate used for the reaction is a mixture of both radiolabeled and unlabeled NAEs, the percentage of total (radiolabeled + unlabeled) NAE converted into FFA is essentially equal to the percentage of radiolabeled NAE converted into FFA regardless of the molar ratio between radiolabeled and unlabeled NAEs present in the NAE substrate. The percentage of radiolabeled NAE converted into FFA is traced by radiometric scanning of the TLC plate and reported as "% of total" of the region corresponding to FFA on the data report table (Fig. 3). Thus, the amount of total FFA produced by the reaction is simply calculated as the

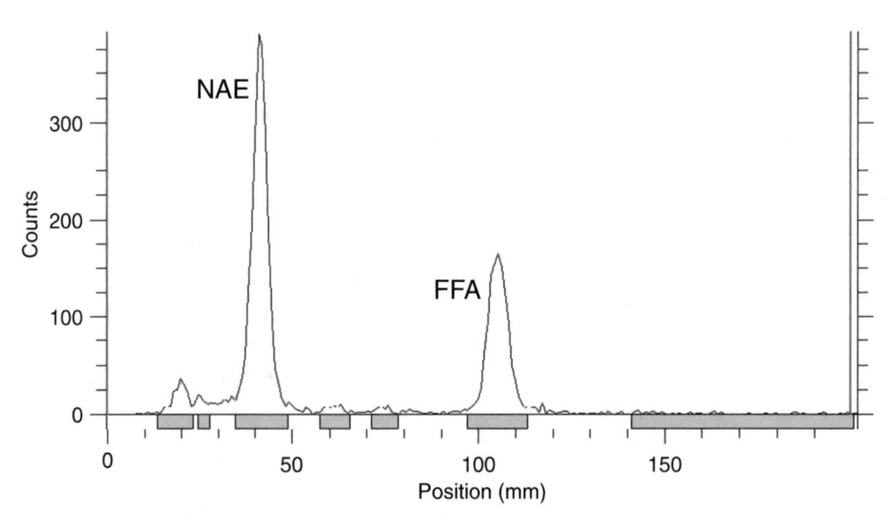

Reg	(mm) Start	(mm) Stop	(mm) Centroid	RF	Region Counts	Region CPM	% of Total	% of ROI
Rgn 1	13.4	23.5	19.4	0.097	170.0	56.7	4.29	4.60
Rgn 2	24.4	28.1	25.6	0.128	60.0	20.0	1.51	1.62
Rgn 3	34.5	49.2	41.3	0.206	2170.0	723.3	54.76	58.66
Rgn 4	57.5	65.8	61.1	0.306	51.0	17.0	1.29	1.38
Rgn 5	71.3	78.6	74.3	0.372	37.0	12.3	0.93	1.00
Rgn 6	97.0	113.6	105.2	0.526	1168.0	389.3	29.47	31.58
Rgn 7	141.1	200.9	163.2	0.816	43.0	14.3	1.09	1.16
7 Peaks					3699.0	1233.0	93.34	100.00

Total Counts: 3963.0 (1321.0 CPM)   Print   Close

Fig. 3 A representative radiochromatogram and its data report table produced by the WinScan Application Version 3.12 software. *X*- and *Y*-axes represent the position (mm from the bottom of TLC plate) of radiation signals and their radioactivity (counts), respectively. *Green bars* on the *X*-axis indicate the regions of picks that have been automatically detected to be above a threshold specified. Picks for substrate (NAE) remained and product (FFA; free fatty acid) formed are indicated. The table reports, for each region, start, stop, and center (centroid) positions, retention factor (RF), radioactivity (counts), CPM value, % of total, % of ROI, and total counts (CPM) at the bottom. Regions 3 and 6 correspond to NAE and FFA, respectively, based on their positions on the radiochromatogram

amount of NAE substrate added to the reaction mixture × "% of total" of FFA on the data report table/100. For example, if a 30-min reaction with a NAE concentration of 100 μM (as in Subheading 3.4, step 1) and a reaction volume of 0.5 mL produced a result of radiochromatogram as in Fig. 3, the amount of total FFA produced by the reaction is 50 nmol × 29.47 %/100 = 14.735 nmol (Fig. 1), and thus the enzyme activity is 14.735 nmol/0.5 h = 29.47 nmol/h.

4 Notes

1. Use solutions of KOH to adjust the pH unless noted otherwise.
2. Do not add DDM until all other components are completely dissolved to avoid excess foaming.
3. Dispense the IPTG solution into 1-mL aliquots in microcentrifuge tubes for storage. Repeated cycles of freeze and thaw affect IPTG stability and reduce the ability of the reagent to induce protein expression.
4. Use the pellet form of NaOH to adjust the pH to 8.0 to completely dissolve EDTA or EGTA.
5. Be careful to mix gently to avoid the formation of excess foam.
6. It is helpful to, with the TLC plate placed upright on a sheet of paper, flick the back side of the plate with a rod to collect any silica residue onto the paper. Then, carefully pour the powder along a corner of the paper into the glass vial.
7. Like all polyunsaturated lipids, polyunsaturated NAEs (e.g., linoleoyl (18:2) ethanolamine, linolenoyl (18:3) ethanolamine, or arachidonoyl (20:4) ethanolamine) are susceptible to oxidation. Keep these NAEs covered with solvent and an inert gas (e.g., N_2) to avoid the oxidation.
8. Immerse the seeds in 1 mL of 20 % (v/v) bleach solution and shake for 5 min. Discard the supernatant, add 1 mL of 75 % (v/v) ethanol, and shake for 5 min. Discard the supernatant and rinse the seeds at least five times with sterile water.
9. The amount of unlabeled substrates should be approximately 100 μg to ensure the visualization of lipid spots on the TLC plate. Polyunsaturated NAEs are much easier to visualize and require less lipid than completely saturated species. For this reason, iodine staining cannot be considered quantitative, so resist the temptation to quantify by densitometric scanning. The enzyme source may be a bacterial lysate expressing AtFAAH protein, *Arabidopsis* tissue homogenate, or purified AtFAAH protein. Total protein amount added may be adjusted depending on the nature of enzyme source. Due to low expression level of AtFAAH protein, total protein amount must be no less than 100 μg when *Arabidopsis* tissue homogenate is used. Other plant sources are assayed in the same manner. The enzyme is enriched in microsomal fractions (100,000 × *g* 60 min pellet of a 10,000 × *g* 30 min supernatant), so if not detectable in homogenates, prepare membrane fractions for analysis.
10. Be careful not to transfer the residual upper phase that may remain.
11. In order to make the spot as compact as possible, load a small volume (2–5 μL) of the sample and immediately evaporate the solvent and repeat this until the entire volume of the sample is

loaded. Multiple samples must be loaded in a minimum of 1.5 cm intervals to avoid cross-contamination of lanes when counting.

12. Keep solvent level well below the level of the applied lipids (usually ~100 mL) and mix combined solvent by measuring volumes separately and adding them one at a time to a flask. Mix thoroughly before placing into tank. Place 1 mm filter paper around the inside walls of the chamber to equilibrate chamber throughout, but be careful not to touch the silica surface. Any scrapes will disrupt the chromatography and lipid separation.
13. This instrument is a gas-phase proportional counter and relies on electrical signals generated by radioactivity-induced ionization in a gas-filled chamber. The radioactive materials on the plate then generate a chromatogram of radioactivity distributed along the lane.
14. The plate may be removed from the chamber and imaged by digital camera or scanner before lipid spots fade away. Since iodine reacts with the double bond of lipid molecules, iodine vapor exposure is not recommended for the detection of saturated lipids. The plate may be sprayed with sulfuric acid and heated at 90 °C for ~20 min for visualization of lipids, but this is not recommended for radiolabeled lipids. Alternatively, the lipids can be extracted and purified from the silica gel and precisely quantified by gas chromatography (GC) (providing quantitative standards have been included) as described elsewhere (13), but be careful not to expose these lipids to iodine as this sometimes interferes with recovery. Fatty acid methyl esters can be generated from FFA and analyzed by GC-FID or GC-MS directly after extraction without TLC separation to quantify FFA production by FAAH, but this is less reproducible and sensitive than the radioactivity-based assays.
15. The time used to detect radioactivity in each lane can be increased to increase the limits of detection. Generally a first scan of each lane is performed with a selected time of 1 min, and if product is not detected above background, a longer time of 10 min or 30 min per lane can be used. This is one advantage of the gas-phase proportional counters as they record accumulated counts over time.

References

1. McKinney MK, Cravatt BF (2005) Structure and function of fatty acid amide hydrolase. Annu Rev Biochem 74:411–432
2. Fowler CJ (2006) The endocannabinoid system and its pharmacological manipulation- a review with emphasis upon the uptake and hydrolysis of anandamide. Fundam Clin Pharmacol 20:549–562
3. Fezza F et al (2008) Fatty acid amide hydrolase: a gate-keeper of the endocannabinoid system. Subcell Biochem 49:101–132
4. Wang YS et al (2006) Manipulation of *Arabidopsis* fatty acid amide hydrolase expression modifies plant growth and sensitivity to *N*-acylethanolamines. Proc Natl Acad Sci USA 103:12197–12202

5. Teaster ND et al (2007) *N*-Acylethanolamine metabolism interacts with abscisic acid signaling in *Arabidopsis thaliana* seedlings. Plant Cell 19:2454–2469
6. Kang L et al (2008) Overexpression of a fatty acid amide hydrolase compromises innate immunity in *Arabidopsis*. Plant J 56:336–349
7. Kim S-C et al (2009) Mutations in *Arabidopsis* fatty acid amide hydrolase reveal that catalytic activity influences growth but not sensitivity to abscisic acid or pathogens. J Biol Chem 284:34065–34074
8. Shrestha R, Dixon RA, Chapman KD (2003) Molecular identification of a functional homologue of the mammalian fatty acid amide hydrolase in *Arabidopsis thaliana*. J Biol Chem 278:34990–34997
9. Bligh EG, Dyer WJ (1959) A rapid method for total lipid extraction and purification. Can J Biochem Physiol 37:911–917
10. Shrestha R et al (2002) *N*-Acylethanolamines are metabolized by lipoxygenase and amidohydrolase in competing pathways during cottonseed imbibitions. Plant Physiol 130: 391–401
11. Hillard CJ, Edgemond WS, Campbell WB (1995) Characterization of ligand binding to the endocannabinoid receptor of rat brain membranes using a novel method: application to anandamide. J Neurochem 64: 677–683
12. Bradford MM (1976) A rapid and sensitive method for the quantification of microgram quantities of protein utilizing the principle of protein-dye binding. Anal Biochem 72:248–254
13. Chapman KD et al (1999) *N*-Acylethanolamines in seeds: quantification of molecular species and their degradation upon imbibitions. Plant Physiol 120:1157–1164

Chapter 13

Ionization Behavior of Polyphosphoinositides Determined via the Preparation of pH Titration Curves Using Solid-State ^{31}P NMR

Zachary T. Graber and Edgar E. Kooijman

Abstract

Detailed knowledge of the degree of ionization of lipid titratable groups is important for the evaluation of protein–lipid and lipid–lipid interactions. The degree of ionization is commonly evaluated by acid–base titration, but for lipids localized in a multicomponent membrane interface this is not a suitable technique. For phosphomonoester-containing lipids such as the polyphosphoinositides, phosphatidic acid, and ceramide-1-phosphate, this is more conveniently accomplished by ^{31}P NMR. Here, we describe a solid-state ^{31}P NMR procedure to construct pH titration curves to determine the degree of ionization of phosphomonoester groups in polyphosphoinositides. This procedure can also be used, with suitable sample preparation conditions, for other important signaling lipids. Access to a solid-state, i.e., magic angle spinning, capable NMR spectrometer is assumed. The procedures described here are valid for a Bruker instrument, but can be adapted for other spectrometers as needed.

Key words Solid-state ^{31}P NMR, pH titration curves, Lipid interactions, Ionization, Phosphomonoester groups, Electrostatic-hydrogen bond switch

1 Introduction

Biological membranes contain a multitude of lipids and proteins that interact with each other via hydrophobic and electrostatic forces. Many membrane lipids are zwitterionic, i.e., contain both a positive and negative charge, but have no net charge under cellular conditions. These lipids, such as phosphatidylcholine and phosphatidylethanolamine, form the matrix of the biological membrane, whereas anionic lipids often have specific signaling functions (1–5). These anionic signaling lipids can interact with positively charged protein domains and thereby activate or inactivate the target protein. In this way, signaling lipids form crucial cofactors for many membrane protein functions.

Teun Munnik and Ingo Heilmann (eds.), *Plant Lipid Signaling Protocols*, Methods in Molecular Biology, vol. 1009,
DOI 10.1007/978-1-62703-401-2_13,

Importantly, the p*K*a of many of the ionizable groups, e.g., phosphomonoester moieties, of these lipids falls in the physiologically relevant range of $5<pH<8$ (6–8). Additionally, the exact charge is mediated by the exact lipid composition of the resident membrane (6, 8). Signaling lipids such as the polyphosphoinositides, but also phosphatidic acid and related anionic lipids, are formed transiently in space and time in a cell and the charge that they carry may influence the interaction with protein targets. Therefore, it is important to investigate the ionization behavior of these lipids in detail as a function of membrane composition. Here we provide a general protocol, optimized for the polyphosphoinositides, to determine the ionization behavior of anionic lipids carrying one or multiple phosphomonoester groups.

Previously, ^{31}P NMR was used to determine the ionization behavior of phosphomonoester-containing lipids in micellar or small unilamellar vesicle (SUV) dispersions (9, 10). In these particular cases ^{31}P NMR results in isotropic chemical shifts due to the rapid reorientation of the lipids with respect to the external magnetic field. pH titration curves can readily be prepared for such systems. The problem with these experiments is that they do not represent the native packing environment of a biological membrane. Micelles and SUVs are systems in which the lipids experience a high degree of membrane curvature (+ for the micelles, and both + and − for the SUV). A better model system is the multilamellar vesicle (MLV) dispersion where the lipids essentially reside in a flat membrane. However, ^{31}P NMR of MLV dispersions results in a chemical shift anisotropy (CSA) that is representative of the organization and orientation of the lipids but yields little information on the ionization of the individual lipids making up the lipid membrane. An example of the CSA for a mixture of phosphatidylcholine (PC) and lysophosphatidic acid (LPA) is shown in Fig. 1a. Note the different CSA profiles for PC and LPA representing differences in their motion and orientation in the membrane. Chemical shift values observed in a solid-state magic angle spinning (MAS) experiment, where we average out most orientation-dependent interactions, enable us to study lipid ionization properties (6, 11). These MAS experiments form the foundation of the method to construct pH titration curves described here. Figure 1b shows the MAS spectrum for the same mixture of PC/LPA as shown in Fig. 1a. The chemical shift of LPA (phosphomonoester) is very sensitive to pH and thus allows a pH titration curve to be determined by preparing many samples with different pH values (6, 12, 13). An example of such a pH titration curve for brain phosphatidylinositol 4,5-bisphosphate (bPI(4,5)P_2) in dioleoyl-phosphatidylcholine is shown in Fig. 2 together with the raw NMR spectra from which the data points for the curves are taken (data taken from ref. (7)).

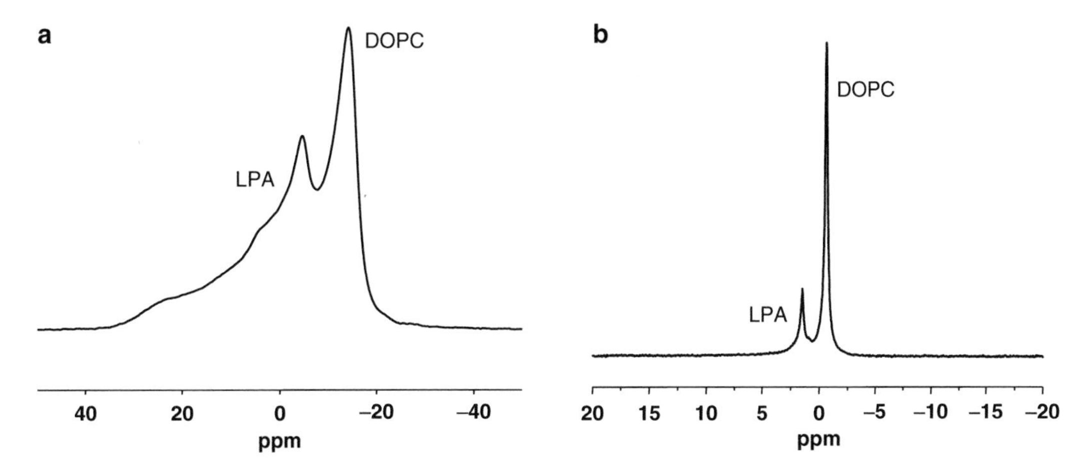

Fig. 1 Examples for NMR spectra for a mixture of 20 mol% LPA and 80 mol% DOPC. (**a**) Static ^{31}P NMR spectrum. The CSA for DOPC is ~41 ppm, and the CSA for LPA is ~10 ppm. (**b**) MAS ^{31}P NMR spectrum of the same lipid mixture. Peaks are relative to an external 85 wt% H_3PO_4 standard

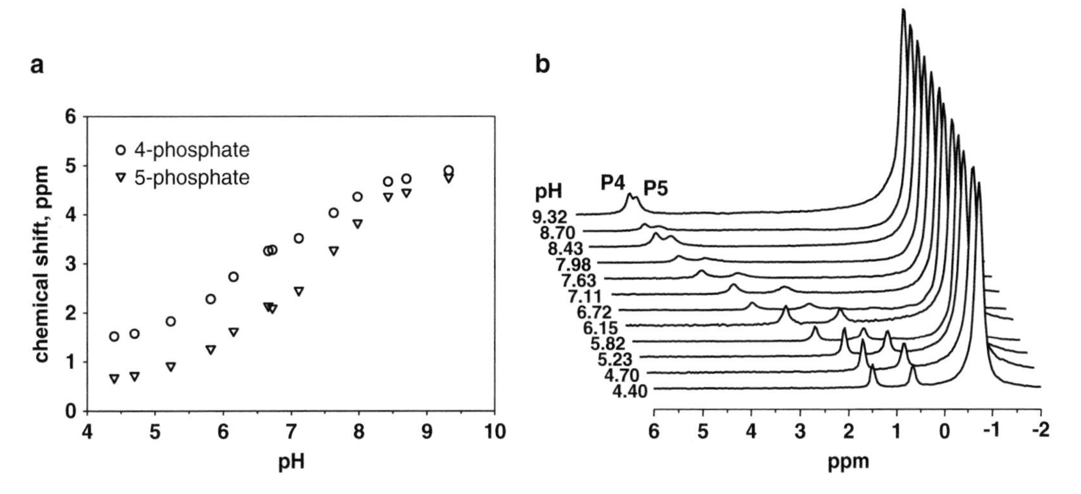

Fig. 2 Ionization properties of bPI(4,5)P_2. (**a**) pH titration curves for the 4, and 5-phosphate of bPI(4,5)P_2. (**b**) ^{31}P MAS NMR data for the data points plotted in (**a**), the 4, and 5-phosphate are indicated. Phosphorus chemical shift is relative to an 85 wt% H_3PO_4 standard. Reprinted with permission from Kooijman et al. (7). Copyright 2011 American Chemical Society

2 Materials

Water used in the preparation of buffers and final cleaning of glassware and NMR rotors should have a resistivity of at least 18.2 MΩ cm, e.g., MilliQ water or HPLC-quality water. All buffer components should be of analytical grade and organic solvents of HPLC grade. All stock solutions (buffer and lipid) and final buffers for the pH titration curve should be stored at −20 °C or lower.

2.1 Lipid Film Materials

1. 1,2-Dioleoyl-*sn*-glycero-3-phosphocholine (DOPC); L-α-phosphatidylinositol-4,5-bisphosphate (Brain, Porcine) (ammonium salt) (bPI(4,5)P_2) from Avanti Lipids (Birmingham, AL).
2. Lipid stock solutions: Prepare lipid stocks from lipid powder (*see* **Note 1**). Dissolve polar lipids in a 2:1 $CHCl_3$/methanol solution which is prepared by mixing 20 ml $CHCl_3$ with 10 ml methanol. Carefully weigh the lipid on a semimicro balance and dissolve the lipid in an exact amount of organic solvent using a volumetric flask. For highly charged lipids such as bPI(4,5)P_2, use a mixture of $CHCl_3$:methanol:water (20:9:1, v/v/v) to dissolve the lipid.
3. $CHCl_3$, HPLC grade.
4. Methanol, HPLC grade.
5. Laboratory parafilm.
6. 15 mm borosilicate glass test tubes with rotary evaporator joint fitting mouth piece (these were home-made with a 12/30 joint to fit the rotary evaporator).
7. Rotary evaporator (e.g., Buchi).
8. Vacuum oven or desiccator.

2.2 Multilamellar Vesicle Materials

1. Hydration buffers: 100 mM NaCl, 2 mM EDTA, 50 mM buffer component. We have used the following buffers for the indicated pH ranges (*see* **Note 2**): 20 mM citric acid/30 mM MES for pH 4–6.5, 50 mM HEPES for pH 6.5–8.5, and 50 mM glycine for pH 8.5–10. Use concentrated HCl and NaOH to adjust the buffer pH to the desired pH values. Store buffers in the freezer −20 °C or below) to prolong shelf life.
2. Sentron Intelli CupFET pH probe (Sentron, Roden, The Netherlands) (*see* **Note 3**).
3. pH meter.
4. Vortexer.
5. Small Dewar (~1 l).
6. Denatured ethanol.
7. Dry ice.
8. N_2 gas (or other inert gas, e.g., Argon).
9. 1.5 ml microcentrifuge tubes (able to withstand 22,000 rcf).
10. Tabletop microcentrifuge with 2 ml centrifuge tube rotor, such as Eppendorf centrifuge 5424.
11. 4 mm zirconium MAS NMR sample tubes with Kel-F cap (Bruker, Karlsruhe, Germany).

2.3 NMR Experimental Components

1. 85 % H_3PO_4: Required for ^{31}P NMR spectral reference.
2. AVANCE™ III 400 MHz WB console and spectrometer (Bruker, Karlsruhe, Germany).
3. Burker BioSpin 4 mm cross-polarization (CP) MAS probe.
4. MAS control unit.

3 Methods

Experiments can be carried out at room temperature, but it is possible to measure ionization behavior at other temperatures, e.g., 37 °C by using a pH probe calibrated at 37 °C and adjusting the *T* in the MAS NMR probe to 37 °C (*see* **Note 4**).

3.1 Preparation of Lipid Films

1. Remove lipid stocks from freezer and allow them to come to room temperature (~20 °C).
2. Make sure that the entire lipid is dissolved in the lipid stock solution. If lipid precipitate is visible (some of the bPI(4,5)P_2 can come out of solution), warm the lipid stock in a water bath to help the lipid to dissolve. The lipid stock can also be sonicated briefly to promote dissolution.
3. Lipid films should be prepared in 15 mm borosilicate glass tubes with a joint fitting enabling the tubes to be placed on the rotary evaporator (*see* **Note 5**).
4. Mix appropriate amounts of lipid stock in test tubes to form an organic lipid solution (*see* **Note 6**). The total lipid should be around 4–10 μmol (*see* **Notes 7** and **8**). Add 400–600 μl $CHCl_3$ to increase total solution volume to 800–1,200 μl (*see* **Note 9**).
5. Flush lipid tubes with gaseous N_2 and cap to prevent oxidation.
6. Use a rotary evaporator to remove the organic solvent and produce a dried lipid film (takes 3–10 min). Set the rotary evaporator water bath temp at 40 °C.
7. Refill tubes with gaseous N_2 after completing the rotary evaporation.
8. Place the dry lipid films under vacuum (in a vacuum oven, or a desiccator) overnight to remove any residual organic solvent. The vacuum should be 25 in Hg (~64 cm Hg) or higher, and the temperature should be set to no higher than 40 °C (*see* **Note 10**).
9. When the lipid films have completely dried, fill the vacuum oven with gaseous N_2, cap the test tubes, and seal the cap with parafilm. Store the dried lipid films in the freezer at −20 °C for no longer than 1 year.

3.2 Preparation of MLV Dispersions

1. Remove a lipid film and buffer from the freezer to thaw (placing the buffer in warm water will help it thaw faster).
2. Calibrate the Sentron Intelli pH probe (*see* **Note 11**).
3. Combine ethanol and dry ice in a small (~1 l) tabletop Dewar. Fill the Dewar with denatured ethanol and slowly add small chunks of dry ice until the ethanol has cooled. The ethanol level should be low enough so that the test tube with the lipid MLV dispersion may be placed upright in the Dewar without the ethanol reaching the opening of the test tube (*see* **Note 12**).
4. Measure the pH of the buffer solution to make sure that the buffer pH has not changed during storage. If the pH of the buffer shifted by more than ±0.1 pH unit a new buffer solution should be prepared.
5. Add 2 ml of the buffer to the dried lipid film. Flush with gaseous N_2 to prevent lipid oxidation and close the test tube. Vortex the lipid/buffer solution for 30 s or more to create an MLV dispersion (the solution will be milky white at this point).
6. Flash freeze the lipid solution in the ethanol/dry ice mixture, and then gently thaw in warm water while occasionally vortexing the sample. This "freeze–thaw" technique helps to remove metastable lipid phases and provides a more homogeneous size distribution of the MLVs. Repeat the freeze–thaw cycle a second time (*see* **Note 13**).
7. Measure the pH of the lipid suspension. This is the pH used to construct the pH titration curve.
8. Divide the lipid solution evenly between two centrifuge tubes (*see* **Note 14**). Sediment the MLVs for 45–60 min at the highest possible speed in a tabletop microfuge-type centrifuge using a 2 ml centrifuge tube-compatible rotor.
9. Remove supernatant from centrifuge tubes and save it. Combine the lipid pellets and transfer the lipid into a 4 mm zirconium MAS NMR tube (*see* **Note 15**). Fill the NMR tube, leaving just enough room for the Kel-F cap (*see* **Note 16**). If necessary, use some of the supernatant to fill the tube (*see* **Note 17**). Make sure not to overfill the tube; otherwise the cap may pop off during the MAS experiment (*see* **Note 18**).

3.3 NMR Experimental Procedure

1. Fill a 4 mm zirconium MAS NMR tube with 85 % (w/w) H_3PO_4 standard and cap it (*see* **Note 19**). Insert the H_3PO_4 standard into the NMR spectrometer. Use the wobble command (in Bruker's TopSpin software) to ensure that the tube has properly inserted (this is generally how we check whether the MAS rotor has inserted properly in the probe). Set the spin rate at 1,000 Hz. Once the spin stabilizes, run several scans. Use a peak picking function to determine the location of the single H_3PO_4 peak. Set this peak to zero and record the

spectral reference value. Stop the spin and eject the standard (*see* **Note 20**).

2. Create a new experimental file (e.g., copy the standard experiment file to a new experiment number in TopSpin) (*see* **Note 21**).
3. Insert the lipid sample into the NMR spectrometer; follow proper insertion using the wobble command. Set the MAS speed at 5,000 Hz (*see* **Note 22**). When the spinning stabilizes, use the wobble function to tune the spectrometer to the ^{31}P resonance (161.97 MHz). Set the spectral reference to the value that was determined from the H_3PO_4 standard.
4. Start the single 90° ^{31}P pulse MAS experiment. The experimental parameters are as follows:
   (a) Time domain (TD) is 16,384.
   (b) Acquisition time (AQ) is 0.2523636 s.
   (c) Dwell time (DW) is 15.400 μs, and DE is 10.00 μs.
   (d) Time in between pulses (D1) is 1.0000000 s.
   (e) 90° ^{31}P pulse length (P1) is 5.25 μs.
   (f) PL1 = 8.00 dB, and PL1W = 62.50282669 W.
   (g) SI = 16,384 (no zero filling used).
   (h) Free induction decay smoothening function: WDW = EM, and LB = 2.00 Hz.
5. Allow the MAS experiment to run until the peaks are sufficiently resolved (we generally use a 2 Hz exponential line broadening function as indicated above (WDW, and LB parameters)). For 0.2 μmol of target lipid (in our case used for bPI(4,5)P_2), it may take up to 50,000 scans or 17–18 h to finish. Appendix 1 shows the pulse program used for this experiment.
6. After running the MAS experiment, a static ^{31}P NMR experiment can be used to examine the phase of the lipid solution. We use a spinal64 pulse program (CPDPRG2) proton decoupling. Tune the ^{31}P NMR frequency, switch channels and tune the ^{1}H NMR frequency, switch back and recheck the ^{31}P NMR frequency, and then start the experiment.
7. The experimental parameters are as follows:
   (a) Time domain is 8,192.
   (b) Acquisition time is 0.0502943 s.
   (c) Dwell time is 6.133 μs, and DE is 10.00 μs.
   (d) Time in between pulses D1 is 1.00000000 s, and D11 is 0.03000000 s.
   (e) Phosphorus 90° as above (step 4e, f).
   (f) 1H 90° pulse length (PCPD2) is 22.25 μs.
   (g) PL12 = 14 dB, and PL12W = 2.95000005 W.

(h) SI = 8,192 (no zero filling).

(i) Free induction decay smoothening function: WDW = EM, and LB = 50.00 Hz.

8. The static experiment is fairly noisy, and takes more scans to show a well-resolved spectrum. Twenty thousand scans will give a rough spectrum and can indicate the primary phase, but 100,000 or more scans may be required to create a well-resolved, smooth, spectrum (after 50 Hz exponential line broadening as indicated under step 7i). Appendix 2 shows the pulse program.

9. Once the experiment is complete, remove the lipid solution from the NMR tube and combine with the previously stored supernatant. If desired, the lipid may be stored in the freezer in case it becomes necessary to rerun the experiment. Otherwise, the lipid may be discarded. The NMR sample tube should be cleaned with HPLC-grade water and denatured ethanol for reuse.

3.4 Data Analysis

1. Once enough data points have been collected, a titration curve may be established by plotting the chemical shifts of the phosphate peaks vs. the pH of the samples (*see* Fig. 2a for an example for 5 mol% bPI(4,5)P_2 in 95 mol% DOPC). These peaks are picked by the TopSpin software provided with the NMR spectrometer. Higher chemical shift values for the phosphomonoester peak indicate deshielding of the ^{31}P nucleus and a corresponding increase in deprotonation. The degree of protonation can be calculated from the chemical shift according to

$$f_{i,\mathrm{p}} = \frac{\delta_i^{\mathrm{obs}} - \delta_{i,\mathrm{d}}}{\delta_{i,\mathrm{p}} - \delta_{i,\mathrm{d}}}, \qquad (1)$$

where $f_{i,\mathrm{p}}$ is the degree of protonation for phosphomonoester group i, δ_i^{obs} is the pH-dependent chemical shift such as shown in Fig. 2a, and $\delta_{i,\mathrm{p}}$ and $\delta_{i,\mathrm{d}}$ are the chemical shifts of the singly protonated and completely deprotonated form of phosphomonoester group i (*see* **Note 23**). Figure 3 shows the result of this calculation which is used to determine the charge on each of the phosphomonoesters of bPI(4,5)P_2.

2. Note that we use the assumption that the chemical shift at any pH can be considered as a weighted average of the concentration of the protonation states multiplied by the chemical shift for that state:

$$\delta = \frac{[\mathrm{A}]\delta_\mathrm{A} + [\mathrm{B}]\delta_\mathrm{B}}{[\mathrm{A}] + [\mathrm{B}]}, \qquad (2)$$

where δ is the chemical shift, (A) is the concentration of the protonated form, and (B) is the concentration of the deprotonated form.

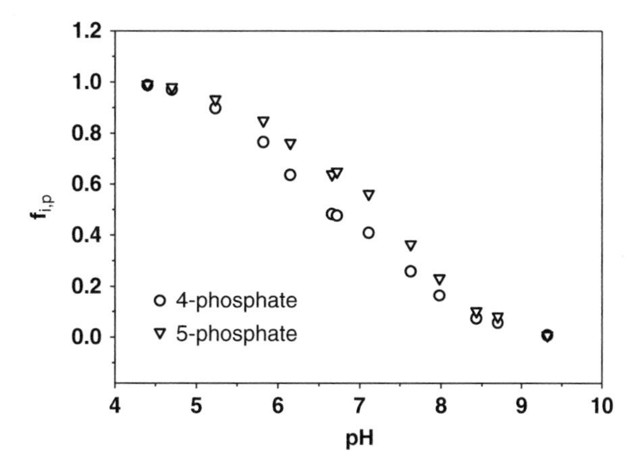

Fig. 3 Degree of protonation of bPI(4,5)P_2. The degree of protonation for the 4, and 5-phosphate of bPI(4,5)P_2 was calculated using Eq. 1 from the data shown in Fig. 2a. Reprinted with permission from Kooijman et al. (7). Copyright 2011 American Chemical Society

δ_A and δ_B are the chemical shifts of the protonated and deprotonated forms, respectively, which are determined from the data shown in Fig. 2a.

4 Notes

1. Unsaturated lipids should be purchased in powder form as the shelf life of prepared stock solutions in organic solvent is considerably less than the powder form. Both lipid powder and prepared stock solution should be stored under a N_2 atmosphere at or below −20 °C.
2. Depending on the application additional buffers can be used as needed. Note that Tris is not compatible with the suggested pH probe (*see* **Note 3**), and PBS is unsuitable for ^{31}P NMR experiments as it will cause a large background peak that might interfere with data acquisition and analysis.
3. The pH needs to be measured in concentrated liposome dispersions; a regular glass electrode is incompatible with this application. We therefore use a Sentron Intelli CupFET probe to measure the pH of liposomal dispersions. We have found this to provide reliable measurements. The Intelli probes are stand-alone probes that can be used with most commercially available pH meters.
4. Elevated temperatures may be required to ensure that the lipids in the liposomes are in the fluid phase. In order to select for lipids that are in the fluid phase we have to date carried out most of our pH titration curves with lipids carrying oleic acid (18:1 Δ^9) chains where possible. In one case (ceramide-1-phosphate)

we used a higher temperature, namely, 37 °C, to ensure that the saturated Cer-1-P was in the fluid phase (fluid phase judged from the NMR spectra) (8). Note that it is crucial to ensure that lipids are more or less acyl-chain matched as chain length critically influences the main phase transition temperature, T_m. A large difference in T_m may cause phase separation of lipids in the lipid film with the highest T_m lipid precipitating first from organic solution. When working with a mixed lipid system the film should be prepared and hydrated at a temperature above the highest T_m.

5. Previous work using bPI(4,5)P_2 showed that a special sample prep procedure was required to avoid the creation of metastable presumably non-bilayer phases (*see* supplemental material of ref. (7)). The rotovap/test tube procedure described herein was found to yield the best results. Furthermore, freeze/thawing cycles are recommended to promote the formation of bilayer phase.

6. We normally use volumetric pipettes to dispense the lipid stock solutions. Because of the high volatility of the organic solvent, there is inaccuracy in the exact volume that is dispensed and there is also the problem of dripping of the solvent from the pipette tip. To eliminate the error due to dripping, we load and discharge the pipette tip several times before dispensing the lipid stock. This saturates the pipette with organic vapor and allows the organic solvent to be dispensed. Note that this is still not a very accurate procedure. However, we are not so much concerned with an exact amount of lipid in our samples, but rather are concerned about the molar ratio of the lipids making up our liposomes and this has been shown to be a reliable procedure.

7. Although the exact number of micromoles of lipid is not important (*see* **Note 6**), the ratio of the lipids is very important. To reduce error in the lipid ratios between lipid films, we make all of the films for a single titration curve at one time. This reduces the amount of variance between the lipid films.

8. Our target lipid, bPI(4,5)P_2, is very expensive and therefore we reduced the total amount (number of μmoles) of lipid in our films for titration curves of the polyphosphoinositides. We showed that for our 400 MHz Avance III spectrometer the use of 0.2 μmol of PIPx was acceptable. This resulted in a total amount of lipid in our films of 4–10 μmol depending on the mol% incorporated. This does require long experimental times for data acquisition. If a target lipid is readily available, then larger amounts of lipid can be used, and the experimental time can be significantly reduced (a doubling of the number of micromoles of lipid results in a four times decrease in experimental time). Therefore, and if practical, the amount of total

lipid should be increased to reduce the experimental time. However, when increasing the total lipid significantly above 10 μmol the lipid may not all fit into the NMR tube (total volume of NMR rotor is ~80 μl).

9. We observed that this eliminated the appearance of metastable phases.
10. The vacuum oven temperature should be kept well below 60 °C as elevated temperatures may result in the breakdown of the polyunsaturated fatty acids of bPI(4,5)P_2.
11. When preparing a pH titration curve, accuracy in the pH readings is extremely important. Thus, we calibrate the pH meter before making each sample. The recorded pH of the MLV dispersion is used as the bulk sample pH to be plotted on the *x*-axis of the titration curve. We use a sensitive Sentron pH probe to measure the pH of the lipid solutions (also *see* **Note 3**).
12. Alternatively one may use liquid N_2. However, care should be taken as liquid N_2 is considerably colder than the dry ice-cooled ethanol and may cause glassware to break and thus lead to loss of sample.
13. Freeze–thaw cycles may be repeated as necessary to remove metastable phases. For some lipids it may be necessary to perform more than two cycles. However, care should be taken when freeze-thawing MLV dispersions containing anionic lipids as this will create small enough vesicles to interfere with the MAS NMR experiment (Brownian motion of small vesicles opposes the averaging accomplished by the MAS NMR technique). We therefore never repeat this procedure more than twice when using lipid mixtures containing bPI(4,5)P_2.
14. We split our 2 ml samples in two equal 1 ml aliquots to balance the rotor.
15. Transfer of the lipid pellet can be accomplished by using a borosilicate Pasteur pipette. Carefully remove the hydrated lipid pellet from both centrifuge tubes (*see* **Note 13**) (ensuring that it does not "shoot up" into the pipette but stays in the narrowest section of the pipette) and place the end of the Pasteur pipette in the 4 mm MAS NMR tube almost at the bottom. Dispense the sample in the NMR tube while slowly moving the pipette up. Try to prevent the formation of air bubbles while transferring the pellet (*see* **Note 17**).
16. The correct filling volume can be determined by the special tool provided with the 4 mm zirconium rotors. However, we generally estimate the filling level based on experience. If too little volume is used the rotor will not spin stably and additional lipid/supernatant will need to be added (also *see* **Notes 17** and **18**).

17. When filling the NMR tube, air bubbles can easily form in the viscous lipid dispersion. These bubbles must be eliminated; otherwise the NMR rotor may not spin properly. The air bubbles can be removed by "stirring" the solution with a thin stirring rod to pop any bubbles that may be present. We use the end of a syringe needle of which the opening has been mechanically closed.
18. It is extremely important not to overfill the NMR tube. If the tube is too full, the cap will not stay on, and may come off during the NMR experiment, leading to loss of sample and a potential mess in the MAS probe. To prevent this, we always make sure to put the Kel-F cap on tightly, and check the cap before inserting any sample into the NMR. If any lipid is seen leaking out from under the cap, then the NMR tube is too full. In this case the amount of lipid should be adjusted, and cap and top of sample tube should be cleaned to remove lipid solution that may act as lubricant. It has to be mentioned that filling the NMR rotor to the exact level takes quite a lot of skill and loss of a first sample is probably inevitable. There are O-ring sample caps available but to date we have not used these and cannot comment on their utility.
19. H_3PO_4 is a viscous acid, and requires caution. If the Kel-F cap happens to pop off this standard it can be very damaging to the instrument. Because of this, instead of filling a new MAS NMR tube with H_3PO_4 every time we run this standard, we fill the tube once and reuse it for each new standard scan without changing the solution. This reduces wear on the cap, and lowers the possibility of the cap popping off.
20. We run the standard each time before running the MAS experiment. It is important to keep this particular sequence as there will be some spectral shift during the course of the generally long NMR recording times. If a sample needs to be rerun, the standard needs to be run again.
21. Always save the standard data for future reference.
22. As stated above, it is important to avoid any possibility of the NMR cap popping off during an experiment. If there is not enough liquid in the NMR tube, it may have some difficulty spinning. If the pressure on the NMR cap is high, and it is not able to spin easily, this may lead to the cap popping off. To prevent this, we start the MAS experiment at 2,000 Hz. If the sample spins easily, we then increase the spin rate to 5,000 Hz. If the sample has difficulty reaching 2,000 Hz, we remove the sample from the spectrometer, remove the cap, adjust the solution level, and check for potential air bubbles.
23. The chemical shift for the singly deprotonated and doubly deprotonated states can be estimated from the low pH (~4) and high pH (~10) data. Alternatively the data can be fit with a Henderson–

Hasselbalch-type equation detailed in ref. (6). The chemical shift values for these two states are variables in this equation and can thus be determined exactly. This works well for sigmoidal titration curves as observed for PA, Cer-1-P, and other lipids containing a single phosphomonoester. For bPI(4,5)P_2 the titration behavior over the pH range of 4<pH<10 is considerably more complex due to interactions between the two phosphates (7). A satisfactory theory explaining the non-sigmoidal titration behavior of bPI(4,5)P_2 is currently under development.

Acknowledgments

The authors would like to acknowledge Dr. Mahinda Gangoda for his assistance and technical expertise related to the NMR experiments. This work was supported by a Farris Family Fellowship to E.E.K., Kent State University, and NSF CHE-1058719.

Appendix 1: ^{31}P MAS NMR Pulse Program

Pulse Program:

```
;zg
;avance-version (06/11/09)
;1D sequence

#include <Avance.incl>

"acqt0=-p1*2/3.1416"

1 ze
2 30m
  d1
  p1 ph1
  go=2 ph31
  30m mc #0 to 2 F0(zd)
exit

ph1=0 2 2 0 1 3 3 1
ph31=0 2 2 0 1 3 3 1

;pl1: f1 channel - power level for pulse (default)
;p1 : f1 channel - high power pulse
;d1 : relaxation delay; 1-5 * T1
;NS : 1 * n, total number of scans: NS * TD0
```

Appendix 2: Static ^{31}P NMR Pulse Program

Pulse Program:

```
;zgig
;avance-version (07/04/03)
;1D sequence with inverse gated decoupling
```

```
#include <Avance.incl>

"d11=30m"

"acqt0=-p1*2/3.1416"

1 ze
  d11 pl12:f2
2 30m do:f2
  d1
  p1 ph1
  go=2 ph31 cpd2:f2
  30m do:f2 mc #0 to 2 F0(zd)
exit

ph1=0 2 2 0 1 3 3 1
ph31=0 2 2 0 1 3 3 1

;pl1 : f1 channel - power level for pulse (default)
;pl12: f2 channel - power level for CPD/BB decoupling
;p1 : f1 channel - high power pulse
;d1 : relaxation delay; 1-5 * T1
;d11: delay for disk I/O  (30 msec)
;NS: 1 * n, total number of scans: NS * TD0
;cpd2: decoupling according to sequence defined by
cpdprg2
;pcpd2: f2 channel - 90 degree pulse for decoupling
sequence
```

References

1. Testerink C, Munnik T (2005) Phosphatidic acid: a multifunctional stress signaling lipid in plants. Trends Plant Sci 10:368–375
2. Stace CL, Ktistakis NT (2006) Phosphatidic acid- and phosphatidylserine-binding proteins. Biochim Biophys Acta 1761:913–926
3. Raghu P, Manifava M, Coadwell J, Ktistakis NT (2009) Emerging findings from studies of phospholipase D in model organisms (and a short update on phosphatidic acid effectors). Biochim Biophys Acta 1791:889–897
4. Lemmon MA (2008) Membrane recognition by phospholipid-binding domains. Nat Rev Mol Cell Biol 9:99–111
5. Stahelin RV, Subramanian P, Vora M, Cho W, Chalfant CE (2007) Ceramide-1-phosphate binds group IVA cytosolic phospholipase a2 via a novel site in the C2 domain. J Biol Chem 282:20467–20474
6. Kooijman EE, Carter KM, van Laar EG, Chupin V, Burger KN, de Kruijff B (2005) What makes the bioactive lipids phosphatidic acid and lysophosphatidic acid so special? Biochemistry 44:17007–17015
7. Kooijman EE, King KE, Gangoda M, Gericke A (2009) Ionization properties of phosphatidylinositol polyphosphates in mixed model membranes. Biochemistry 48:9360–9371
8. Kooijman EE, Sot J, Montes LR, Alonso A, Gericke A, De Kruijff B, Kumar S, Goni FM (2008) Membrane organization and ionization properties of the minor but crucial lipid ceramide-1-phosphate. Biophys J 94: 4320–4330
9. Hauser H (1989) Mechanism of spontaneous vesiculation. Proc Natl Acad Sci USA 86: 5351–5355
10. Swairjo MA, Seaton BA, Roberts MF (1994) Effect of vesicle composition and curvature on the dissociation of phosphatidic acid in small unilamellar vesicles – a ^{31}P-NMR study. Biochim Biophys Acta 1191:354–361
11. Watts A (1998) Solid-state NMR approaches for studying the interaction of peptides and proteins with membranes. Biochim Biophys Acta 1376:297–318
12. Kooijman EE, Burger KN (2009) Biophysics and function of phosphatidic acid: a molecular perspective. Biochim Biophys Acta 1791: 881–888
13. Kooijman EE, Tieleman DP, Testerink C, Munnik T, Rijkers DT, Burger KN, de Kruijff B (2007) An electrostatic/hydrogen bond switch as the basis for the specific interaction of phosphatidic acid with proteins. J Biol Chem 282:11356–11364

Part II

Enzyme Activities

Chapter 14

Phosphatidylinositol Synthase Activity from Plant Membrane Fractions and from *E. coli*- Expressed Recombinant Peptides

Sylvie Collin and Françoise Cochet

Abstract

Phosphatidylinositol (PtdIns) synthase is a lipid-synthesizing enzyme responsible for the synthesis of the phospholipid, PtdIns. Its enzymatic properties have been studied in in vitro assays using either membrane-enriched fractions or the purified protein in reconstituted lipid vesicles as a source of enzyme. More recently the specificities in terms of substrate preferences have also been studied using the recombinant protein expressed in *Escherichia coli*. This chapter deals with the purification of membranes as a source of PtdIns synthase before focusing on the in vitro assays of the enzymatic activities of the protein and, briefly, on the analysis of the product.

Key words Phosphatidylinositol synthase, Recombinant protein, Enzymatic activity, Phosphatidylinositol, Membrane purification, De novo synthesis, Exchange activity

1 Introduction

Phosphatidylinositol (PtdIns) synthase (EC 2.7.8.11, PIS) is a lipid-synthesizing enzyme responsible for the de novo synthesis of PtdIns. Also known as cytidine 5′-diphospho-1,2-diacyl-*sn*-glycerol:*myo*-inositol 3-phosphatidyltransferase, PIS catalyzes the transfer of *myo*-inositol in place of the CMP molecule present on the phosphatidic acid skeleton of CDP-DAG. Studies of PIS activity were initiated in plants in the 1970s with the aim of characterizing the regulation of phospholipid synthesis in various plant species. The optimum pH (between 8 and 9 depending on the enzyme source), the dependency of the activity on Mn^{2+} ions, with a lower preference for Mg^{2+}, and the apparent affinity constants for its two substrates, CDP-DAG and *myo*-inositol, have been determined (1–7). Our laboratory (8) compared the selectivity of PIS from maize coleoptiles, testing the activity either in microsomal fractions, where the enzyme is naturally present, or after solubilization in CHAPS or prepurification on a DEAE trisacryl column.

Teun Munnik and Ingo Heilmann (eds.), *Plant Lipid Signaling Protocols*, Methods in Molecular Biology, vol. 1009, DOI 10.1007/978-1-62703-401-2_14,

The authors were able to demonstrate that PIS reconstituted in detergent and lipid micelles had lost the selectivity it shows towards naturally occurring CDP-DAG molecules in the endoplasmic reticulum. They suggested that care has to be taken in analyzing the substrate specificities of PIS in reconstituted membrane environments.

In parallel to the description of the de novo PtdIns synthesis by PIS, several publications showed that the membrane fractions containing PIS activity also possessed a *myo*-inositol:PtdIns exchange activity. In 1981 Sexton and Moore showed for the first time that this activity, previously identified in animals, also exists in plants (9). The exchange reaction is defined as the ability to catalyze the production of PtdIns in a CDP-DAG-independent manner. This activity presumably results from the addition of *myo*-inositol on a PA backbone but it has been shown experimentally that the almost exclusive donor of this backbone is PtdIns itself. This exchange reaction therefore consists of a *myo*-inositol turnover on PtdIns in a seemingly pointless reaction whose occurrence in vivo has never been established. The exchange activity has now been shown to be a property of PIS (4, 7). In castor bean endosperm PIS is ER-localized, depends upon Mn^{2+} or magnesium with a preference for Mn^{2+}, and leads to the synthesis of PtdIns in a CDP-DAG-independent manner with a high affinity for *myo*-inositol. Because the preferred lipid substrate for the exchange is PtdIns, it is crucial, when studying PtdIns de novo synthesis activity, especially from a recombinant enzyme, to work with a membrane system devoid of endogenous PtdIns. For this reason *E. coli*, naturally devoid of PtdIns synthase and thus PtdIns, is the ideal system for expressing PIS-encoding cDNAs.

The current interest in PIS in plants is based on the fact that this enzyme plays a key role in producing PtdIns molecular species that will be directed towards signalling, after phosphorylation by PtdIns kinases, for production of PtdIns phosphates that have been shown to be crucial in many cell functions. In Arabidopsis, where two PIS encoded by two different genes exist, recent work showed that each is responsible for the synthesis of different PtdIns pools (10). PtdIns is also a substrate of inositolphosphorylceramide synthase and is involved in the formation of GPI anchors.

2 Materials

All solutions are prepared using ultrapure water (Millipore Synthesis A10 system, using Quantum™ Ex Ultrapure Organex cartridges, Millipore, Molsheim, France) and analytical grade reagents (Normapur for solvents).

2.1 Expression of PIS in Bacterial Cells

1. Luria and Bertani (LB) growth medium: For 1 L, weigh 25 g of LB solid medium (Becton Dickinson, Le Pont-de-Claix, France). Carefully add to a beaker containing 800 mL water. Dissolve using a magnetic stirrer. When the powder is dissolved, transfer to a 1-L measuring cylinder, and adjust to 1 L with water. Mix well by inverting several times. Aliquot by 100-mLvolumes and sterilize by autoclaving. If making LB agar, add 3 g of bactoagar (BDH Prolabo, Fontenay-sous-Bois, France) to a bottle containing 200 mL of LB medium. Sterilize by autoclaving. Store at room temperature (RT).
2. M9 minimal medium: Prepare 200 mL 5× M9 salts by dissolving 12.8 g Na_2HPO_4, $7H_2O$, 3.0 g KH_2PO_4, 0.5 g NaCl, and 1.0 g NH_4Cl in 150 mL ultrapure water. Add 2 g casein acid hydrolysate, vitamin-free. In a measuring cylinder, complete to 200 mL with water. Sterilize by autoclaving. Keep at RT. Proceed to items 3 and 4.
3. Autoclave 100 mL of ultrapure water. Keep at RT.
4. 20 % glucose (w/v): Dissolve 20 g glucose in 100 mL ultrapure water. Sterilize by filtration on a 0.22 μm Millipore filter into a sterile bottle. Keep at 4 °C.
5. In sterile conditions, prepare 100 mL M9 culture medium by mixing 20 mL tepid sterile 5× M9 salts supplemented with casein acid hydrolysate, 2 mL 20 % (w/w) glucose, and 78 mL sterile water.
6. Ampicillin: Make a 100 mg/mL stock solution by dissolving 1 g of ampicillin in 10 mL water in a 15-mL plastic sterile tube. Mix well. Sterilize by filtration through a 0.22 μm Millipore filter into sterile 2-mL polypropylene reaction tubes. Store at −20 °C.
7. Chloramphenicol: Make a 34 mg/mL stock solution by dissolving 340 mg in 10 mL absolute ethanol. Aliquot in 1-mL volume in sterile polypropylene reaction tubes. Store at −20 °C.
8. 100 mM D-*Myo*-inositol: Dissolve 180 mg D-*myo*-inositol in 10 mL sterile water. Sterilize by filtration through a 0.22 μm Millipore filtering unit. Store at −20 °C.
9. Spectrophotometer for visible light.
10. 400 mM Isopropyl thio-β-D-galactopyranoside (IPTG) solution: Dissolve 960 mg IPTG in 10 mL water. Sterilize by filtration (0.22 μm Millipore filtering unit). Aliquot by 1-mL volume in sterile polypropylene reaction tubes. Store at −20 °C.
11. 50-mL plastic tubes.

12. Washing buffer for cell pellet: 50 mM Tris–HCl, pH 8.0, 2 mM EDTA. To make 100 mL, mix 5 mL of 1 M Tris–HCl, pH 8.0 (12.1 g are dissolved in 90 mL water before the pH is adjusted to 8.0 with fuming HCl and the final volume adjusted to 100 mL) with 95 mL water and 0.4 mL 0.5 M EDTA, pH 8.0, in a measuring cylinder. Mix well before autoclaving. Keep at RT.

2.2 Isolation of Bacterial Membranes

1. Sonicator (Branson Sonifier 250).
2. Sonication buffer: 50 mM Tris–HCl pH 8.0. To make 100 mL see Subheading 2.1, item 11. Store at RT.
3. Membrane pellet resuspension buffer: 50 mM Tris–HCl pH 8.0, 1 mM EDTA, 20 % (v/v) glycerol (*see* **Note 1**). First prepare 10 mL Tris–HCl/EDTA by mixing 0.5 mL 1 M Tris–HCl, pH 8.0, with 9.5 mL water and 20 μL 0.5 M EDTA, pH 8.0. Separately, prepare 10 mL of an 80 % (v/v) solution of glycerol in the same buffer (8 mL 98 % (v/v) glycerol mixed with 0.5 mL 1 M Tris–HCl, pH 8.0, 1.5 mL water, and 20 μL 0.5 M EDTA, pH 8.0).

2.3 The Manipulation of Tritium

1. Specific lab space dedicated to the manipulation of ^{3}H (*see* **Note 2**).
2. Fume-hood equipped for the work with solvents (methanol and $CHCl_3$).
3. Shaking water-bath on the bench outside the fume-hood.
4. Refrigerated tabletop centrifuge.
5. 30-mL (100×26 mm) Pyrex screw-capped tubes. The cap is sealed on the inside with a removable Teflon disk (*see* **Note 3**).
6. Adaptors for the centrifuge tubes.
7. Laboratory pipettes dedicated to the manipulation of ^{3}H-labeled molecules.
8. Maximum Recovery™ pipette tips.
9. Long-sleeved latex gloves.
10. 10-mL glass tubes resistant to boiling, with screw cap.
11. Heating plate and container (Pyrex beaker) filled with water to heat the 10-mL tubes.
12. Decon®90. Prepare 2 L of 2 % (v/v) Decon by adding 40 mL of detergent to 2 L of water. Keep some in a squeeze bottle to clean surfaces.
13. A large plastic container with lid to be used as a decontaminating tank. Fill with Decon 2 % so that contaminated Pyrex tubes will soak below the surface of the liquid.
14. A bottle of gaseous N_2, equipped with a manometer.

15. Pasteur pipettes.
16. A scintillation counter and plastic vials.
17. Scintillation mix for samples containing solvents.

2.4 De Novo Phosphatidylinositol Synthase Activity (see Note 4)

1. CDP-DAG 18:1 (1,2-dioleoyl-*sn*-glycero-3-cytidine diphosphate, ammonium salt) solution: Dissolve to a final concentration of 5 mg/mL in 50 mM Tris–HCl, pH 8.0 (*see* **Notes 5** and 6).
2. Triton X-100 (*see* **Note 7**).
3. Reaction buffer: 50 mM Tris–HCl, pH 8.0, 2 mM *myo*-inositol (4×), 10 mM Triton X-100 (4×) (*see* **Note 8**). First prepare 100 mL 50 mM Tris–HCl, pH 8.0 (see Subheading 2.1, item 7). To 10 mL in a 15-mL plastic tube, add 3.6 mg *myo*-inositol and 60 mgTriton X-100 (*see* **Note 9**). Alternatively, mix 9 mL of 50 mM Tris–HCl, pH 8.0, with 1 mL *myo*-inositol 20 mM in 50 mM Tris–HCl, pH 8.0 (36 mg in 10 mL) and 60 mg Triton X-100 (*see* **Notes 10** and **11**).
4. (2-^{3}H)*myo*-inositol solution: Stored at 4 °C. Dilute the stock solution as indicated in Table 1 according to the number of reactions you wish to carry out. Dilutions are not stored.
5. EDTA (tetrasodium salt) solution: Use the 0.5 M pH 8.0 master solution to prepare a 100-mL 50 mM EDTA solution in 50 mM Tris–HCl, pH 8.0, by mixing 10 mL of EDTA stock solution with 85 mL water and 5 mL 1 M Tris–HCl, pH 8.0. Store at RT.
6. 150 mM $MnCl_2 \cdot 4H_2O$ solution: Weigh 29.6 mg in a polypropylene reaction tube and reserve until just before use. Add 1 mL 50 mM Tris–HCl, pH 8.0, and mix vigorously before adding to each tube (*see* **Note 12**).

Table 1
Dilutions of the (2-^{3}H) *myo*-inositol stock for use in the measurement of PIS de novo activity in vitro

Number of samples to be tested	Volume of (2-^{3}H) *myo*-inositol (μL)	Volume of Tris–HCl 50 mM pH 8.0 (μL)
4	6.8	93.2
7	11.9	163.1
9	13.6	186.4
11	18.7	256.1
13	19.6	269.4
15	22.6	310.9

2.5 Myo-Inositol Exchange Activity

1. Same reaction buffer as for the de novo synthesis activity.
2. 2.5 mM $MnCl_2 \cdot 4H_2O$ solution: Weigh 9.9 mg in a polypropylene reaction tube and reserve until just before use. Add 1 mL 50 mM Tris–HCl, pH 8.0 (*see* **Note 12**).
3. L-α-Phosphatidylinositol (isolated from soybean, 10 mg/mL).
4. CMP solution: Prepare a 4 μM solution of CMP by diluting 5 μL 3 mM CMP (44 mg in 2 mL 50 mM Tris–HCl, pH 8.0) in 3.995 mL 50 mM Tris–HCl, pH 8.0. Aliquot. Store at −20 °C.

2.6 Lipid Extraction

1. Methanol.
2. $CHCl_3$, ice-cold.
3. 0.9 % (w/v) NaCl solution in water: Weigh 0.9 g and add to a measuring cylinder containing 100 mL ultrapure water. Mix well. Keep at 4 °C.

2.7 Lipid Separation by Thin-Layer Chromatography

1. Lab space equipped with a fume-hood.
2. Methanol.
3. $CHCl_3$.
4. Acetic acid.
5. Thin-layer chromatography (TLC) glass chamber (20 × 20 cm, 5 spaces).
6. Saturation filter paper sheets.
7. Silica-coated TLC plates (20 × 20 cm, SIL-G200).
8. L-α-Phosphatidylinositol (isolated from soybean, 10 mg/mL).
9. L-α-Phosphatidylethanolamine.
10. Aerosol disperser.
11. 0.01 % (v/v) Primuline, in 80 % (v/v) acetone in water.
12. UV lamp (365–254 nm).

3 Methods

3.1 Expressing PIS in E. coli

The following protocol is based on the expression of a recombinant PIS in *E. coli*, expressed from a vector from the pET series. The expressing strain is BL21(DE3)pLysE (*see* **Note 13**). The PIS peptide may or may not be tagged (*see* **Note 14**).

1. Streak the chosen expression clone (*see* **Note 15**) on a plate containing 25 mL LB agar supplemented with ampicillin at a final concentration of 100 μg/mL and chloramphenicol at a final concentration of 34 μg/mL (25μL ampicillin at

100 mg/mL in 25 mL melted LB agar; 34 μL chloramphenicol at 34 mg/mL). Incubate overnight at 37 °C.

2. The next morning, remove the plate from the incubator (*see* **Note 16**). In late afternoon, inoculate a single colony into 5 mL of liquid LB medium containing 100 μg/mL ampicillin and 34 μg/mL chloramphenicol. Incubate overnight at 37 °C shaking at a speed of 180 rpm.
3. Under sterile conditions, remove 1 mL of overnight culture. Sediment cells by centrifuging at 5,000 × *g* for 5 min in a sterile polypropylene reaction tube. Discard the supernatant. Resuspend the cell pellet in 1 mL of fresh LB medium and use to inoculate 50 mL of LB supplemented with both antibiotics. Incubate at 37 °C shaking at 180 rpm.
4. When the optical density at 600 nm has reached a value of ca. 0.6, induce the expression of the recombinant protein using IPTG at a final concentration of 0.4 mM (50 μL of 400 mM IPTG in 50 mL LB medium). Grow the cells for a further 3 h at 37 °C.
5. Weigh a plastic 50-mL centrifugation tube. Transfer the cells and harvest them by centrifugation at 5,000 × *g* for 10 min at 20 °C.
6. Discard the supernatant. Resuspend the cells in 10 mL washing buffer.
7. Sediment at 5,000 × *g* for 10 min at 20 °C. Discard the supernatant. Carefully remove traces of buffer (*see* **Note 17**).
8. Weigh the tube and keep note of the fresh cell weight (FW).
9. Store the pellet at −80 °C without freezing in liquid N_2 first.

3.2 Identification of a Recombinant Peptide-Associated PIS Activity

This protocol is used when establishing the identity of a recombinant peptide as a PIS by evaluation of the capacity of the protein to produce PtdIns during growth.

1. Follow the protocol described in Subheading 3.1 up to step 3. Include a strain carrying the empty expression vector.
2. For each strain, set up three 50-mL cultures containing LB medium, 100 μg/mL ampicillin, and 34 μg/mL chloramphenicol and *myo*-inositol at the following mM concentrations: 0, 1, and 10.
3. Induce expression and then collect cell pellets as described above.
4. Remove the pellet from the −80 °C freezer. Keep on ice for 5 min.
5. Resuspend the pellets in ice-cold Tris–HCl 50 mM, pH 8.0, at a density of 1 mg cells FW per mL. Keep on ice.

6. Lipids are extracted from 2 mL of cell suspension. Transfer the cells in a plastic round-bottomed 30-mL tube. Disrupt the cells by sonication using a Branson sonifier (*see* **Note 18**). Keep the tube containing the resuspended cells in a beaker on ice. Sonicate on ice, output 1, for 2 min (duty cycle 20 %), 2 min (duty cycle constant), and 1 min (duty cycle 20 %) (*see* **Note 19**).
7. Sediment the cell debris and unbroken cells for 10 min at 4 °C, 5,000 × *g*.
8. Collect supernatant and keep on ice in a 30-mL Pyrex Teflon-capped tube. Collect 2 samples of 25 μL for protein quantitation if expression of the amount of PtdIns synthesized by each strain is to be expressed per mg protein instead of per mg FW. Lipids are extracted on the rest of the supernatant.
9. All further steps are to be carried out using gloved hands. Add 7 mL of ice-cold methanol, followed by 3.5 mL ice-cold $CHCl_3$. Vortex for 10 s. A single phase must be observed. Add 3.5 mL ice-cold $CHCl_3$. Vortex for 10 s. Add 4 mL ice-cold 0.9 % (w/v) NaCl in water. Vortex for 10 s (*see* **Note 20**).
10. Separate phases by centrifuging at 120 × *g* at 10 °C for 5 min (*see* **Note 21**).
11. Harvest the lower $CHCl_3$ phase using a Pasteur pipette (*see* **Note 22**). Reserve in a fresh tube.
12. Evaporate under streaming N_2 before resuspending the lipids in 500 μL $CHCl_3$. Aliquot twice 25 μL for dosage of total fatty acids (*see* **Note 23**). The samples can then be stored at −20 °C. If storage is to last for more than 1 week, resuspend the lipids in toluene:ethanol (4:1, v/v) instead of $CHCl_3$.
13. The day before proceeding to lipid separation by TLC, prepare the chromatographic chamber. Line the sides with two layers of saturating filter paper. Prepare the mobile phase by mixing in an Erlenmeyer flask 65 mL $CHCl_3$ with 25 mL methanol and 10 mL glacial acetic acid (11) (*see* **Note 24**). Gently pour in the chamber. Secure the lid to allow saturation.
14. Dry a TLC plate in an oven at 80 °C for 1 h. At a distance of 2 cm from the bottom side and 2 cm from each side, use a lead pencil to softly mark the loading point of each sample. Samples must be separated by at least 1.5 cm. Do not forget to include spaces for the lipid standards, PtdIns, and phosphatidylethanolamine.
15. Load samples one by one in a fume-hood using a stream of N_2 to dry the spots after each loading (*see* **Note 25**). Use 10 μL 2 mg/mL PtdIns and 10 μL 2 mg/mL PtdEtn as standards. Load 190 μL per sample, i.e., the equivalent of 80 mg of cells FW.

16. Working quickly, insert the plate into the chromatographic chamber between the pieces of saturating paper. Secure the lid. Leave the migration to proceed for ca. 1 h 15 min (*see* **Note 26**).
17. Stop the migration when the front is ca. 5 cm from the top of the plate. Dry the plate in the fume-hood for 45 min.
18. Position the plate vertically in the fume-hood against a background of aluminum paper. Spray uniformly with primuline 0.01 % over the whole surface. Leave to dry for 5 min.
19. Visualize the lipids under UV light before taking a photograph.

3.3 Purification of Membrane Samples from E. coli

The following sections are useful if one wishes to characterize the enzymatic parameters of a recombinant PIS activity.

1. Remove the pellet from the –80 °C freezer. Keep on ice for 5 min.
2. Follow the protocol described in Subheading 3.2 up to step 8.
3. Transfer the supernatant to a clean tube. Pool supernatants if necessary, i.e., if there were several sonicated fractions. Sediment at 100,000 × *g* for 1 h at 4 °C to collect membranes.
4. Resuspend the pellet in 200 μL 50 mM Tris–HCl, pH 8.0.
5. Determine the protein concentration of the homogenized sample according to Lowry (12) (*see* **Note 27**). Adjust to 20 % glycerol by adding 1/3 volume of glycerol 80 % (v/v) in 50 mM Tris–HCl, pH 8.0 (*see* **Note 28**). Homogenize gently but thoroughly. Aliquot the sample in 0.5-mL polypropylene reaction tubes in 100-μg quantities. Store at –80 °C (*see* **Note 29**) or proceed to first characterization steps.

3.4 Purification of Membrane Samples from Plant Material

Several publications describe the isolation of plant membranes (see ref. (7) for some of them). We refer the reader to these papers for the protocols used in each case, and specific to the membrane whose activity is studied, i.e., microsomes (a mixed endomembrane fraction rich in endoplasmic reticulum), endoplasmic reticulum, Golgi apparatus, mitochondria and mitochondrial membranes, or plasma membrane. In the case of our work, when a plant endomembrane fraction was to be used to study PIS activity, the final pellet was resuspended in 50 mM Tris–HCl, pH 8.0, before protein determination and storage at –80 °C in 20 % (v/v) glycerol as described for bacterial membranes (7).

3.5 De Novo Activity Assay

The whole manipulation is to be carried out in a laboratory assigned to the manipulation of tritium.

1. Set the shaking water-bath to 30 °C (*see* **Note 30**).

2. For each enzymatic sample, prepare a 10-mL glass tube containing 2.6 mL methanol. Keep at room temperature. Keep methanol, $CHCl_3$, and NaCl solution on ice.
3. Each determination of enzymatic activity is carried out in a 30-mL Pyrex tube. To each reaction add in the following order: 11.5 μL CDP-DAG, 50 μL of reaction buffer, 10 μL $MnCl_2$, 10 μL 50 mM EDTA, and 20 μL diluted tritiated inositol (see Table 1). Complete with the appropriate volume of 50 mM Tris–HCl so that only the volume of protein sample is missing to reach 200 μL. Start each reaction by adding 25–50 μg protein (*see* **Note 31**). Briefly homogenize and immediately place the tube in the water-bath and switch on agitation at 40 rpm. Switch on timer. Proceed to the next tube. Allow 1 min between each reaction (*see* **Note 32**).
4. 90 s before each enzymatic reaction is to be stopped, set a tube containing 2.6 mL methanol to boil.
5. After the desired incubation time has elapsed, remove the enzymatic reaction tube from the water-bath. Quickly add the boiling methanol and incubate the mixture in the boiling water-bath for 90 s (*see* **Note 33**).
6. Remove the tube and keep on ice (*see* **Note 34**).
7. To each tube, add 1 mL $CHCl_3$, and vortex for 4–5 s. Repeat this step. Add 1.6 mL ice-cold 0.9 % NaCl. Vortex for 10 s.
8. Centrifuge at 120 × *g* for 5 min at 15 °C.
9. Remove the lower $CHCl_3$ phase using a Pasteur pipette. Transfer to a new tube. Add 2 mL methanol. Vortex. Add 2 mL 0.9 % NaCl. Vortex. Repeat centrifugation as in step 7.
10. Collect lower phase as described in step 8. Transfer to a small glass tube.
11. Evaporate under a stream of N_2. Resuspend in 500 μL $CHCl_3$.
12. Proceed to counting by adding 100 μL of sample to 7 mL scintillation liquid in a vial. In a separate vial, include a count of the total number of dpms per reaction by adding 5 μL of the diluted tritiated inositol to 7 mL of scintillation liquid.
13. Express the data in nmols PtdIns synthesized/mg protein/min considering that 1 nmol inositol will allow the synthesis of 1 nmol PtdIns (*see* **Note 35**).

3.6 Exchange Activity Assay

1. Prepare cell pellets as in Subheading 3.1, from step 2, but use M9 medium prepared as described above instead of LB medium (*see* **Note 36**).
2. Prepare cell membranes as described in Subheading 3.3.
3. When ready to set up the enzymatic reactions, switch on the water-bath at 30 °C.

4. Prepare PtdIns liposomes. From a PtdIns solution at 10 mg/mL, retrieve 58 μL and keep in a 2-mL polypropylene reaction tube (*see* **Note 37**). Add a few drops of $CHCl_3$ and evaporate under an N_2 stream. Add 500 μL 50 mM Tris–HCl, pH 8.0, 2 mM *myo*-inositol, 0.6 % (w/v) Triton (10 mM), and then 500 μL of 50 mM Tris–HCl, pH 8.0.
5. Scrape the lipids from the side of the tube using the tip of a sealed Pasteur pipette (*see* **Note 38**).
6. Mark the level of buffer using a permanent pen.
7. Vortex vigorously without allowing the liquid to touch the tube cap.
8. Sonicate on melting ice, for 2 min at the bottom of the tube, followed by 1 min near the surface (output 1, duty cycle constant).
9. Carefully rinse the probe using a mixture of the two Tris–HCl buffers mentioned in step 7.
10. Adjust the volume of buffer to the mark if necessary.
11. To each reaction tube of the experiment, add in the following order: 100 μL of PtdIns liposomes, 20 μL tritiated *myo*-inositol (Table 1), 10 μL $MnCl_2$ prepared just before use, and 10 μL 4 μM CMP or 10 μL 50 mM Tris–HCl, pH 8.0, if you wish to compare the exchange reactions with and without CMP to show the typical activating effect of CMP. Complete with 50 mM Tris–HCl, pH 8.0, up to the final volume (200 μL) minus that of the protein sample.
12. Start the enzymatic reactions by adding the protein sample (200 μg). Allow 1 min between each point to give yourself time to stop the reaction in between samples. Incubate for 20–30 min in the same conditions as those described in Subheading 3.5, step 3.
13. At the desired time-points, stop the reaction by transferring the tube on ice and adding 3 mL methanol:$CHCl_3$ (2:1, v/v). Vortex. Add 1 mL $CHCl_3$. Vortex. Add 1.6 mL 0.9 % NaCl. Vortex. Spin at 1,000 rpm for 5 min. Remove the lower chloroform phase and transfer into a fresh tube.
14. Wash the $CHCl_3$ phase by adding 2 mL methanol, vortexing, adding 2 mL salted water, vortexing, and spinning as above.
15. Recover the $CHCl_3$ phase and evaporate under a stream of N_2 before resuspending the sample in 500 μL $CHCl_3$. Use 100 μL to count the radioactivity incorporated into PtdIns.

3.7 Analysis of the Product

Beyond a global quantitative analysis of PtdIns like that described in this paper, we refer the readers to several methods more appropriate to decipher the fatty acid composition of the PtdIns produced by GC (see ref. (13), for example) or the molecular species by radio-HPLC (8, 14) or mass spectrometry (15).

4 Notes

1. These concentrations are that of the membrane sample stored at −80 °C. Keep the Tris–HCl/EDTA and glycerol solutions separated (*see* **Note 28**).
2. PIS activity is measured using a radioactive substrate. All safety regulations must be followed. Your manipulations will lead to the production of solvent and water waste containing radioactive molecules.
3. Teflon is a material that is resistant to solvents. It ensures that no leak will occur during the manipulation of the tubes. At the end of your experiment, they must be decontaminated like the rest of the equipment in Decon 2 % (see step 13 in Subheading 2.3).
4. We have found that dissolving all the components of the enzymatic reaction, i.e., *myo*-inositol, Triton X-100, $MnCl_2$, EDTA, and CDP-DAG in 50 mM Tris–HCl, pH 8.0, is by far the preferred method in order to obtain reproducible results. Using a 1 M concentrated solution of Tris–HCl, pH 8.0, gives highly unreproducible results in our hands.
5. For dissolution, directly add the buffer to the dried CDP-DAG and gently shake the bottle in a beaker of hot water from the tap. Do not heat more than necessary. Check the progress of dissolution. It should take a few minutes but not more. This can be done in advance as CDP-DAG is then kept at −20 °C and thawed when necessary without the need to make aliquots.
6. Specific molecular species of CDP-DAG can be synthesized chemically if not commercially available (10).
7. Triton X-100 is a liquid detergent. Very viscous, it requires the use of cut pipetman tips when sampled pure.
8. The optimum pH for PIS activity has not always been determined in plant samples. For de novo synthesis, it is 8.5 in castor bean endosperm, the activity being 50 % lower at pH 8.0 (3). In cauliflower inflorescences, it is 9.0. No study has been carried out in Arabidopsis but the data published so far has used a Tris–HCl buffer at pH 8.0 (4, 6, 10).
9. Add the desired volume of Triton to the Tris–HCl/inositol mix and leave on a shaking table to agitate for a least 2 h to ensure complete dissolution. When not dissolved Triton appears as a transparent lump.
10. Working on Arabidopsis PISs, Löfke et al. (10) used slightly different conditions: 50 mM Tris–HCl, pH 8.0, 20 mM $MgCl_2$, 1.5 mM *myo*-inositol, 0.9 mM CDP-DAG, and 1 % (v/v)

Triton X-100. Vincent et al. (16) also used different conditions when working with leek endoplasmic reticulum vesicles.

11. Freezing the Tris–HCl/inositol/Triton buffer leads with time to enhanced activities from the same samples. We presume that the molecular organization of Triton changes over time. Buffers are to be kept for a maximum of 1 week at −20 °C. We prefer making them fresh.
12. Dissolve $MnCl_2$, $4H_2O$, seconds prior to adding to each tube. We find that oxidation of $MnCl_2$, visible by the appearance of a reddish brown color, occurs within 1 min of salt dissolution.
13. The two strains used so far for the expression of a recombinant PtdIns synthase from plants in *E. coli* are BL21 derivatives. We used BL21(DE3)pLysE for expressing AtPIS1 (4, 6, 17) and AtPIS2 (unpublished data). In this latter case we compared the level of PtdIns produced by the bacterial cultures after 3 h at 37 °C using BL21(DE3)pLysE, BL21(DE3)pLysS, and BL21(DE3)Codon+RP. BL21(DE3)pLysS produces a lower level of lysozyme when compared to the pLysE strain and is a weaker repressor of the T7 RNA polymerase which transcribed the cDNA of interest from the pET plasmid. BL21(DE3) Codon+RP expresses the eukaryotic tRNAs for arginine and proline to palliate a low translational level of the recombinant protein. In our hands, BL21(DE3)pLysE gave the best results. Löfke et al. (10) used C43, a BL21 derivative carrying an unmapped mutation leading to a larger membrane surface and potentially an increased capacity to accommodate recombinant membrane proteins, such as PISs.
14. Both untagged (4, 6, 17) and poly-His-tagged (10) PISs are functional in *E. coli*.
15. We do not make glycerol stocks of the expressing strains but work with fresh transformants each time to ensure a maximum level of expression.
16. Because ampicillin can be degraded by the resistant strains, transformations have to be done late in the afternoon and taken out of the incubator in the morning. Pre-cultures have to be washed before inoculating larger scale cultures to prevent the introduction of β-lactamase and a destruction of ampicillin in the medium.
17. Because they express lysozyme, BL21(DE3)pLysE cells are fragile. In this buffer they will tend to lyse. The cell pellet will be easily dislodged.
18. In our hands, we find that the ideal volume for lysis by sonication is 2 mL. With larger volumes it will take a longer time to complete lysis and the sample will be subjected to warm

temperatures for too long, with a risk to damage the proteins, even when sonication is carried out in ice. For similar reasons, we tend not to work with more than 6 sonicated samples in one go, not to let the first one wait for too long on ice before centrifugation.

19. A successful sonication is indicated by a transparent sample. Alternatively, you may use a French press, with three passages at 1,000 psi and a maximum volume of 6 mL in one go.
20. After adding 0.9 % NaCl in water, you should be able to distinguish two phases. If not try adding a little more salted water to enhance phase partitioning between $CHCl_3$ and methanol.
21. Below 10 °C, the two phases will lose their transparency. Removing the lower $CHCl_3$ phase will become more difficult.
22. This is not easy at first. We favor the use of a propipette for a better control. Do not worry if a small amount of methanol/water phase is removed at the same time as the $CHCl_3$ phase as you will easily identify it as an upper phase. Just make sure that you do not transfer it to the $CHCl_3$ collection tube since the methanol phase will contain large quantities of radioactive inositol that you will count as PtdIns. Water will also interfere with migration on the TLC plate.
23. The amount of PtdIns synthesized will be calculated by the proportion of fatty acids present in PtdIns with respect to the total amount of fatty acids in the strain.
24. $CHCl_3$ cannot be measured using a plastic pipette as it will dissolve it. The final mix will appear whitish whereas it has to be perfectly transparent and colorless, forming a single phase. Note that the proportions of solvents indicated here are specific for the separation of lipids from *E. coli*.
25. To manipulate and load the samples we use glass syringes which are very easy to clean. Alternatively you may use pipetmans and plastic tips.
26. If the temperature of the room is above 25 °C, carry out the lipid separation in a climatized room.
27. This protocol is more adequate for the dosage of membrane proteins rather the protocol based on the Bradford method, used for soluble proteins. In our conditions, the concentration of the protein samples at this stage of the protocol is usually 10–12 μg/μL.
28. Glycerol will interfere with protein dosage according to Lowry. We determine the protein concentration before adding the appropriate volume of glycerol 80 % and then recalculate the final concentration.

29. It is best to store the membrane samples at −80 °C. In our hands the enzymatic activity is stable for at least 3 months.
30. We use a standard water-bath set on top of an agitating table.
31. The amount of proteins necessary for each enzymatic reaction varies according to the objective of the manipulation as well as the enzyme source. In the characterization of the PIS1 protein from Arabidopsis, Collin et al. used 50 μg proteins per point (4). In their unpublished study of AtPIS2 they used 25 μg. In their study of the same enzymes, Löfke et al. used 20 μg (10). The amount of membrane proteins to be used has to be established experimentally for each protein sample, even between bacterial cultures of the same expression clones, and a fortiori between membranes from different sources such as plant subcellular compartments. The same is true for the incubation time. One has to determine for each sample the conditions in which the amount of PtdIns formed is linear with time for a given amount of proteins.
32. When working with a large number of samples you will need to work in pair.
33. Extreme care must be taken when manipulating boiling methanol for lipid extraction. We have found that aliquoting the methanol necessary for each reaction, and warming it in a beaker of boiling water for 1.5 min just before using it, is satisfactory and limits the risks as well as the evaporation of methanol. Wear protection equipment.
34. Once the sample has been boiled in methanol it can be kept on ice until you are ready to proceed with the extraction of lipids.
35. It is assumed that the enzyme will not choose between cold and radioactive *myo*-inositol, whose concentration is negligible in the assay.
36. Membrane samples containing PtdIns do not allow to carry out kinetic experiments for de novo synthesis as the exchange activity interferes with the labelling of PtdIns. This is also true in the study of the exchange activity, as the concentration of PtdIns needs to be controlled. In the first case, we usually work with membrane obtained from bacterial cultures grown on LB medium without the addition of *myo*-inositol, and carry out the enzymatic assays in the presence of 5 mM EDTA to inhibit the exchange activity that could occur on the traces of PtdIns present in the membranes. When working with plant membrane samples EDTA is always included in the reaction mix at a final concentration of 5 mM. The de novo synthesis activity could also be studied from strains grown on M9 medium. As growth is less efficient in these conditions the amount of membrane proteins is usually too low to allow a thorough study of the enzyme.

37. The volume indicated is calculated considering that PtdIns MM is 800 g/mol (diC16). The minimum quantity that can be prepared is for 10 enzymatic reactions.

38. The aim is to prepare PtdIns liposomes. For other purposes, such as preparing liposomes in order to test lipid transfer activity, phospholipids can also be mechanically detached from the side of the tube using glass beads. In the case of the PIS exchange activity, the conditions described here give liposomes that are a good substrate for the enzyme. We have not tested alternative methods.

Acknowledgments

The authors wish to thank E. Ruelland for his careful reading of the manuscript. Financial support from the Université Pierre et Marie Curie via the French Ministère de l'Enseignement Supérieur et de la Recherche, as well as from the C.N.R.S., is gratefully acknowledged.

References

1. Sumida S, Mudd JB (1970) The structure and biosynthesis of phosphatidylinositol in cauliflower inflorescence. Plant Physiol 45:712–718
2. Moore TS, Lord JM, Kagawa T et al (1973) Enzymes of phospholipid metabolism in endoplasmic reticulum of castor bean endosperm. Plant Physiol 52:50–53
3. Sexton JC, Moore TS (1978) Phosphatidylinositol synthesis in castor bean endosperm. Plant Physiol 62:978–980
4. Collin S, Justin A-M, Cantrel C et al (1999) Identification of AtPIS, a phosphatidylinositol synthase from Arabidopsis. Eur J Biochem 262:652–658
5. Blouin A, Lavezzi T, Moore TS (2003) Membrane lipid biosynthesis in *Chlamydomona sreinhardtii*. Partial characterization of CDP-diacylglycerol: *myo*-inositol 3-phosphatidyltransferase. Plant Physiol Biochem 41: 11–16
6. Justin A-M, Kader J-C, Collin S (2003) Synthetic capacity of Arabidopsis phosphatidylinositol synthase 1 expressed in *E. coli*. Biochim Biophys Acta 1634:52–60
7. Davy de Virville J, Brown S, Cochet F et al (2010) Assessment of mitochondria as a compartment for phosphatidylinositol synthesis in *Solanum tuberosum*. Plant Physiol Biochem 48:952–960
8. Justin A-M, Hmyene A, Kader J-C et al (1994) Compared selectivities of the phosphatidylinositol synthase from maize coleoptiles either in microsomal membranes or after solubilization. Biochim Biophys Acta 1255: 161–166
9. Sexton JC, Moore TS (1981) Phosphatidylinositol synthesis by a Mn-dependent exchange enzyme in castor bean endosperm. Plant Physiol 68:18–22
10. Löfke C, Ischebeck T, König S et al (2008) Alternative metabolic fates of phosphatidylinositol produced by phosphatidylinositol synthase isoforms in *Arabidopsis thaliana*. Biochem J 413:115–124
11. Ohta A, Obara T, Asami Y et al (1985) Molecular cloning of the *cls* gene responsible for cardiolipin synthesis in *E. coli* and phenotypic consequences of its amplification. J Bacteriol 163:506–514
12. Lowry OH, Rosebrough NJ, Farr AL et al (1952) Protein measurements with the Folin reagent. J Biol Chem 193:265–275
13. Vaultier MN, Cantrel C, Vergnolle C et al (2006) Desaturase mutants reveal that membrane rigidification acts as a cold perception mechanism upstream of the diacylglycerol kinase pathway in Arabidopsis cells. FEBS Lett 580:418–4223

14. Ahmed HA (1993) Analysis of Phospholipids by high performance liquid chromatography. In: Graham J, Higgins J (eds) Biomembrane Protocols: I. Isolation and Analysis, Methods in Molecular Biology, Springer Protocols, vol. 19, pp. 159–177
15. Quinn PJ, Rainteau D, Wolf C (2009) Lipidomics of the Red Cell in Diagnosis of Human Disorders. In: Amstrong D (ed) Lipidomics: vol. 1: Methods and Protocols, Methods in MolecularBiology, Springer Protocols, vol. 579, pp. 127–159
16. Vincent P, Maneta-Peyret L, Cassagne C et al (2001) Phosphatidylserine delivery to endoplasmic reticulum-derived vesicles of plant cells depends on two biosynthetic pathways. FEBS Lett 498:32–36
17. Justin A-M, Kader J-C, Collin S (2002) Phosphatidylinositol synthesis and exchange of the inositol head are catalysed by the single phosphatidylinositol synthase 1 from Arabidopsis. Eur J Biochem 269: 2347–2352

Chapter 15

Phosphatidylinositol 4-Kinase and Phosphatidylinositol 4-Phosphate 5-Kinase Assays

Yang Ju Im, Irena Brglez, Catherine Dieck, Imara Y. Perera, and Wendy F. Boss

Abstract

Inositol lipid kinases are perhaps the easiest and most straightforward enzymes in the phosphoinositide pathway to analyze. In addition to monitoring lipid kinase-specific activity, lipid kinase assays can be used to quantify the inositol lipids present in isolated membranes (Jones et al., Methods Mol Biol 462:75–88, 2009). The lipid kinase assays are based on the fact that the more negatively charged phosphorylated lipid products are readily separated from their lipid substrates by thin layer chromatography. We have summarized our current protocols and identified important considerations for working with inositol lipids including different methods for substrate delivery when using recombinant proteins.

Key words Lipid kinase, Phosphatidylinositol, Phosphatidylinositol phosphate, Cyclodextrin

1 Introduction

Early studies of plant membrane lipid kinase activities included biochemical analyses and optimization of pH, cations, ATP, lipid substrate, and detergent concentrations and reported apparent K_ms for ATP and added lipid substrate (2–7). With the sequencing of the Arabidopsis genome came an appreciation for the multiple isoforms of the lipid kinases (8) and a realization that to understand their functions one would have to understand the regulation and subcellular distribution of each enzyme. Ultimately, the regulation of the lipid kinase will determine the flux through selective downstream pathways. Unfortunately, once the genes for the lipid kinases were identified the interest in the biochemical properties of the enzymes waned. While some kinetic analyses of recombinant lipid kinases have been done (9–11), most studies have been limited to proof of function and more emphasis was placed on their subcellular localization (12).

Teun Munnik and Ingo Heilmann (eds.), *Plant Lipid Signaling Protocols*, Methods in Molecular Biology, vol. 1009, DOI 10.1007/978-1-62703-401-2_15, © Springer Science+Business Media, LLC 2013

To understand how the enzymes are regulated in vivo, it is essential to characterize substrate preference, kinetic properties and mechanisms of posttranslational regulation in vitro. One cannot predict function-based amino acid sequence alone. For example, Saavedra et al. (13) showed that although the two predicted phosphatidylinositol phosphate 5-kinases (PIP5Ks) in *Physcomitrella patens* are 85 % identical at the amino acid level, their substrate preference is quite different. The preferred substrate for PpPIPK2 is PtdIns, i.e., PpPIPK2 functions primarily as a phosphatidylinositol kinase (PIK) rather than a PIP5K in vitro. Another example of unpredictable function is the family of type II γ PI4Ks. At least two of the eight predicted type II γ PI4Ks, γ4 and γ7, are functionally protein kinases in vitro (14).

There are no data comparing the K_ms and V_{max}s for the each of the isoforms of the purified plant PI4Ks and PIP5Ks and little is known about how the plant PI4Ks and PIP5Ks are regulated. It is well accepted that PI4Ks and PIP5Ks can be phosphorylated (10); however, the specific phosphorylation sites of the plant PI4Ks and PIP5Ks and protein kinases that phosphorylate them have not been identified. Clearly, more biochemical and structural studies of the lipid kinases are needed. We encourage the readers to characterize substrate preferences, the effects of cations, pH, posttranslational modifications (e.g., phosphorylation), and lipid and protein cofactors that regulate the specific activity of each of the lipid kinase isoforms. Hopefully, some day sufficient protein can be obtained for structural studies, which will reveal insights as to how these lipid kinases function at the membrane surface in vivo. Protein structure data also will increase our understanding of how interacting proteins such as GTP-binding proteins (15–18) and as yet unidentified cellular proteins regulate the PI4Ks and PIP5Ks to both affect basal metabolism and mediate stimulus response mechanisms.

What we present here are our current protocols along with suggestions for optimizing the specific activity and product recovery for individual systems. Different membrane fractions will contain competing ATP consuming reactions and different lipids and the lipid kinases will have different K_ms for ATP and substrates. For all of these reasons, we encourage investigators to do concentration and time course reactions for each type of enzyme they are analyzing. V_{max} conditions for one enzyme can easily mask the activity of another (e.g., PI3K, (4)). Recombinant proteins present a special challenge due to their solubility and stability. Furthermore, different tags and isolation buffers will affect protein folding, solubility, protein–protein interaction, and enzyme kinetics. It is hoped that this chapter stimulates the community to maintain an active Web site to post newly optimized protocols and to establish uniform assay conditions.

2 Materials

2.1 Reagent and Consumables

1. *PI4K assay buffer*. 120 mM Tris–HCl, 30 mM $MgCl_2$, 4 mM $NaMoO_4$ pH 7.2.
2. *PI4K resuspension buffer*. 30 mM Tris–HCl, 7.5 mM $MgCl_2$ pH 7.2.
3. *PIP5K assay buffer*. 200 mM Tris–HCl, 40 mM $MgCl_2$, 4 mM $NaMoO_4$, 4 mM EGTA, pH 7.5.
4. *PIP5K resuspension buffer*. 50 mM Tris–HCl, 10 mM $MgCl_2$ pH 7.5.
5. *Nuclei resuspension buffer*. 0.2 molal D-sorbitol, 5 mM HEPES, 10 mM $MgCl_2$, 2 mM EGTA, 1 μg/mL leupeptin, 100 μM PMSF, pH 7.
6. *PtdIns stock*. 5 mg/mL PtdIns stock in 1 % (v/v) Triton X-100 (store −20 °C until used) (*see* **Note 1**).
7. *PtdInsP stock*. 1 mg/mL stock in 0.1 % (v/v) Triton. Make fresh (*see* **Note 1**).
8. *NBD-PtdOH*. 1-palmitoyl-2-(12-[(7-nitro-2-1,3-benzoxadiazol-4-yl)amino)lauroyl)-*sn*-glycero-3-phosphate in $CHCl_3$ (Avanti Polar Lipids, Inc.).
9. *ATP*. 45 or 10 mM ATP in water *[$\gamma^{32}P$]-ATP* (6,000 Ci/mmol, 1 mCi/0.1 L).
10. *Tris–HCl*. 30 mM pH 7.2.
11. *Tris–HCl*. 50 mM pH 7.2.
12. *Phosphate-buffered saline (PBS) buffer*. 140 mM NaCl, 2.7 mM KCl, 10 mM Na_2HPO_4, 1.8 mM NaH_2PO_4 pH 7.4.
13. Chloroform ($CHCl_3$).
14. 2.4 N HCl.
15. 0.1 M EDTA.
16. Ice-cold $CHCl_3$:MeOH (2:1, v/v).
17. 1 N HCl:MeOH (1:1, v/v).
18. $CHCl_3$:MeOH:H_2O (2:1:0.01, v/v/v).
19. 1 % (w/v) Potassium Oxalate.
20. 13 × 100 mm disposable glass tube.
21. Filter paper (Whatman #1, 20 × 20 cm).
22. X-ray film.

2.2 Equipment

1. Table top centrifuge capable of obtaining 1,000 × *g*.
2. Nitrogen gas tank.
3. Vacuum desiccator.

4. TLC plate (20 × 20 cm Whatman 4855-620 silica gel LK5 plates).
5. TLC glass tank (preferably 8–10 cm higher than the TLC plates) with a lid.
6. Radioisotope imaging scanner or X-ray film.

3 Methods

3.1. Keep all solutions on ice until the reaction is started for PI4K and PIP5K assays of isolated membranes or nuclei.

3.2. Prepare protein samples for PI4K and PIP5K assays.

3.2.1. Freshly isolated membranes in the membrane resuspension buffer: For isolated membranes resuspend in 30 mM Tris–HCl, 7.5 mM $MgCl_2$ pH 7.2 for the PI4K assay and in 50 mM Tris–HCl, 10 mM $MgCl_2$ pH 7.5 for the PIP5K kinase assay. Vortex gently so as not to froth the membranes and pipette the solution up and down until the suspension is uniform. Measure the protein using the BCA or Bradford (Pierce, Thermo Fisher Scientific, Inc.) mini assays according to the manufacturer's directions. If necessary dilute with the resuspension buffer until the final protein concentration is at least 1–2 μg/25 μL for plasma membranes and 15–30 μg/25 μL for microsomal membranes.

3.2.2. Freshly isolated nuclei: Resuspend freshly isolated nuclei (19) in 0.2 molal D-sorbitol, 5 mM HEPES, 2 mM EGTA, 1 μg/mL leupeptin, 100 μM PMSF, pH 7. Work quickly as the nuclei are fragile and the enzymes are not stable. Just prior to assaying for protein (BCA assay) and lipid kinase activity, vortex the nuclei briefly to break up any clumps. Use 5–10 μg of protein per reaction.

3.2.3. PI4K recombinant protein in resuspension buffer: PI4Kβ can be produced as a functional protein with a His Tag™ when expressed in *Sf9* insect cells (9, 20) as well as a histidine–maltose binding protein tag when expressed in *E. coli* [pDEST-HisMBP (plasmid # 11085 from Addgene), Im and Brglez, unpublished results) (*see* **Note 2**). His-tagged PI4Kα has only been produced as a functional protein in *Sf9* insect cells (20). The recombinant HisMBP and His-tagged proteins can be purified using an Ni affinity column according to the manufacturer's directions. The eluted protein can be diluted so that the concentration is at least ~0.5–2 μg/μL in a final buffer of 150 mM imidazole in 250 mM NaCl, 1.8 mM KCl, and 30 mM Na_2HPO_4 pH 7.9 for the His-tagged protein.

3.2.4. PIP5K recombinant protein resuspension buffer: PIP5Ks have been produced with an MBP tag (12) and GST tag (21) in *E. coli* and with a His tag in *Sf9* insect cells (11). For proteins expressed in insect cells, use a microsomal fraction for the assays and use the vector control or non-transfected insect microsomes as a control. For purified recombinant proteins, the GST-tagged proteins can be assayed on the GST beads in PBS buffer or off the beads with or without GST enzymatically cleaved from the protein. The specific activity will vary with the conditions used (*see* **Note 2**). To release the protein from the beads use 20 mM glutathione in 50 mM Tris–HCl pH 8.0. The eluted protein can be diluted so that the concentration is at least ~0.5–2 μg/μL in a final buffer of 20 mM glutathione and 50 mM Tris–HCl pH 8.0 for the GST-tagged protein or in the 0.15 M imidazole in 0.26 M NaCl, 1.7 mM KCl, and 30 mM Na_2HPO_4 pH 7.9 for the His-tagged protein.

3.3. Pipette protein (total protein per assay should be 1–2 μg for plasma membrane, 15–30 μg for microsomes, 2–5 μg for nuclei, 5–10 μg for recombinant proteins) into a 13 × 100 mm disposable glass tube and add the PI4K or PIP5K resuspension buffer to give a final volume of 25 μL. When comparing enzyme activities of different samples always use the same amount of protein per assay and assay duplicate samples. Mix the protein in the resuspension buffer with gentle vortexing prior to adding the reaction mixture.

3.4. Prepare the reaction mixture.

3.4.1. To prepare the PI4K reaction mixture without substrate mix the following solutions (volumes are for ten reactions):

125 μL of PI4K assay buffer.

50 μL of 0.1 % (v/v) Triton X-100.

10 μL of 45 mM ATP stock (except for isolated nuclei: use 11 μL of a 4.5 mM stock).

100 μCi of [γ^{32}P]-ATP (the volume will vary with the specific activity).

Add Tris–HCl 30 mM to give 250 μL total volume for ten reactions. Use 25 μL of the reaction mixture for each assay tube.

3.4.1.1. When substrate is added to the PI4K reaction: add 50 μL of PtdIns stock solution. Do not add 50 μL of 0.1 % (v/v) Triton X-100.

3.4.2. To prepare the PIP5K reaction mixture without substrate mix the following (volumes are for ten reactions):

125 μL of PIP5K buffer.

50 μL of 0.1 % (v/v) Triton X-100.

2.5 μL of ATP stock (for nuclei use 5 μL of ATP stock).

100 μCi of [γ^{32}P) ATP (the volume will vary with the specific activity).

Add 50 mM Tris–HCl pH 7.5 to give 250 μL total volume for ten reactions.

3.4.2.1. When substrate is added: add 50 μL of PtdInsP stock solution. Do not add 50 μL of 0.1 % (v/v) Triton X-100.

3.4.3. Substrate for recombinant protein reactions for both PI4K and PIP5K reactions: As an alternative to using Triton X-100, for recombinant proteins, the substrate can be suspended in a 5 % (w/v) suspension (w/v) of β-cyclodextrin. This increases specific activity several fold (11, 21) (*see* **Note 3**). Aliquot the lipid in $CHCl_3$ to a glass tube, dry under N_2 and add a 5 % (w/v) suspension of β-cyclodextrin to give a suspension of equivalent to 5 mg/mL stock solution of substrate. Add this to the reaction buffer in place of substrate in Triton X-100 as described above. For example, for PtdIns one would add 50 μL of the PtdIns β cyclodextrin solution to the reaction buffer for ten reactions.

3.5. Start the reaction by adding 25 μL of the PI4K or PIP5K reaction mixture to the 25 μL resuspended protein. This gives a total reaction volume is 50 μL. Incubate at RT (~25 °C) with gentle shaking for 5, 10, and 15 min. The PI4K and PIP5K reactions are usually linear for at least 10 min with microsomal or nuclear fractions.

3.5.1. For the PI4K assay without added substrate, the final concentration of reagents in the 50 μL reaction should be 30 mM Tris–HCl pH 7.2, 7.5 mM $MgCl_2$, 1 mM $NaMoO_4$, 0.01 % (v/v) Triton X-100, 0.9 mM ATP, and 10 μCi (γ^{32}P) ATP per reaction. When substrate is added, the final Triton X-100 concentration is 0.1 % (v/v) and PtdIns is 0.1 μg/μL (~0.12 mM). When assaying nuclei, the final ATP concentration is 0.1 mM with 10 μCi [γ^{32}P) ATP per reaction and the final Triton concentration should be 0.01 % (v/v).

3.5.2. For the PIP5K assay without added substrate, the final concentration of reagents in the 50 μL reaction mixture should be 50 mM Tris–HCl pH 7.5, 10 mM $MgCl_2$ 1 mM $NaMoO_4$, 1 mM EGTA, 0.01 % (v/v) Triton X-100, 0.05 mM ATP, and 10 μCi [γ^{32}P]-ATP per reaction. When substrate is added, the final Triton X-100 concentration is 0.1 % (v/v) and PtdInsP is 0.15 mg/mL (~0.15 mM).

3.6. Stop the reaction by adding 1.5 mL of ice-cold $CHCl_3$:MeOH (2:1, v/v) and place on ice for 5 min or store at −20 °C overnight. Once the lipid extraction is initiated do not stop until the lipids are dried and under N_2.

3.7. Extract the lipids (*see* **Note 4**): Warm the reaction tubes to RT. Work in a fume hood.

3.7.1. To the stopped reaction, add 0.5 mL of 2.4 N HCl, 0.5 mL EDTA (0.1 M), and 0.5 mL $CHCl_3$. Vortex after each addition.

3.7.2. Centrifuge in a table top centrifuge (~1,000 ×*g*) for 6 min to facilitate phase separation.

3.7.3. To remove lower $CHCl_3$ phase, use a pasture pipette with positive pressure as you go through upper phase. If there is a lot of protein, it will be at the interface. Use positive pressure to avoid this protein interface. Place the lower $CHCl_3$ phase in a clean disposable glass tube.

3.7.4. Add 0.5 mL $CHCL_3$ to the remaining upper phase and re-extract (i.e., vortex and centrifuge as above except for 3 min). Combine $CHCl_3$ lower phase with previous $CHCl_3$. Repeat the $CHCl_3$ extraction once more.

3.7.5. To the combined $CHCl_3$ lower phase, add 2 mL of a mixture of 1 N HCl:MeOH (1:1, v/v). Vortex and centrifuge for 6 min. Remove the HCl:MeOH upper phase, which contains residual [γ^{32}P]-ATP. Repeat the HCl:MeOH backwash of the $CHCl_3$ lower phase. This time avoid the upper HCl:MeOH layer. Very carefully (using a Pasteur pipette with positive pressure) remove the $CHCl_3$ lower phase and place it in a clean disposable tube. Be careful and do not remove any of the upper phase. If you do, as the lipids dry, they will be hydrolyzed and most of the ^{32}P will be at the bottom of the TLC plate.

3.7.6. To recover all the remaining lipid, add 0.5 mL $CHCl_3$ to the upper phase, recentrifuge and combine the $CHCl_3$ fractions.

3.7.7. Dry the lipids in a vacuum desiccator, blow N_2 into the dried tubes, cover with Parafilm and store in the freezer (−20 °C) until you spot the lipids on TLC plates.

3.8. TLC tank preparation and separation of the lipids.

3.8.1. Prepare the TLC tank in advance so that solvents have migrated to the top of the filter paper in the tank. This usually takes 3–5 h (*see* **Note 5**).

3.8.1.1. For analyzing PtdInsP from isolated membranes use the solvent mixture 86:76:6:16 ($CHCl_3$:MeOH:NH_3OH:H_2O, v/v/v/v). For analyzing $PtdInsP_2$ use a more polar solvent so that the $PtdInsP_2$ migrates farther from the preabsorbant region, 90:90:7:22 ($CHCl_3$:MeOH:NH_3OH:H_2O; v/v/v/v). If needed, different solvents should be used to separate PtdIns3P and PtdIns4P (22).

3.8.1.2. Add the desired solvents in order in a fume hood. Add 20 cm Whatman #1, 20 × 20 cm filter paper to each end of the tank to equilibrate the solvents throughout the tank.

3.8.1.3. Cover the tank and keep it tightly sealed until use.

3.8.2. Prepare the TLC plate.

3.8.2.1. Just prior to spotting the lipids, gently soak the plates in 1 % (w/v) potassium oxalate by carefully dipping the plate (preabsorbant end first) just long enough for the solution to wet the plate entirely (~80 s).

3.8.2.2. Hold the plate vertically to allow the excess solution to drain, place the plate on a small glass Petri dish in a microwave oven.

3.8.2.3. Heat the oven on high for 5 min. Keep the plates in the oven until you use the plate.

3.8.3. Prior to spotting the samples, warm them to RT and add 50 μL of $CHCl_3$:MeOH:H_2O (2:1:0.01, v/v/v) and mix well by vortexing.

3.8.4. Spot the lipid evenly in the preabsorbant region high enough from the bottom of the plate so that they will be above the solvent [about 2.5 cm up from the bottom of the plate (*see* **Note 6**)). For the radioactive assays, the best standards can be generated readily by using bovine brain extract (*see* **Note** 7). To monitor the solvent migration, spot fluorescent PtdOH (e.g., 1-palmitoyl-2-(12-[(7-nitro-2-1,3-benzoxadiazol-4-yl)amino)lauroyl)-*sn*-glycero-3-phosphate from Avanti Polar Lipids) in a middle lane.

3.9. Add the TLC plate to the tank with the silica gel side facing the center of the tank. You can put two plates in one tank. When the NBD-PtdOH standard has reached 15 cm (~2 h), remove the plate from the tank and air dry it in the fume hood.

3.10. Quantitation: The ^{32}P-labeled lipid products can be visualized by autoradiography and the relative amount determined by densitometry or by scraping the plates and quantifying the ^{32}P with scintillation counting (*see* **Note 8**). If available, a radioisotope scanner is extremely convenient, linear, and more reliably consistent than either densitometry or scraping plates. Report both cpm and nanomoles ^{32}P-labeled lipid recovered/mg protein min. If fluorescent substrates are used, they can be imaged on the plate and quantified by scraping the fluorescent spots from the plate, extracting the fluorescence using the TLC solvent system, and monitoring the relative fluorescence in a spectrofluorometer. HPLC analysis of the glycerophosphoinositol phosphate head groups is required to confirm the specific $PtdInsP_2$ isomer formed (23).

4 Notes

1. The inositol phospholipids, especially PtdInsP and $PtdInsP_2$ are not stable and will stick to glass. To make the PtdIns stock, PtdIns (5 mg) in $CHCl_3$ (Avanti Polar Lipids, Inc.) is dried under a gentle stream of N_2 with swirling so that the lipid coats the bottom of the glass tube. Add 1 % (v/v) Triton X-100 (for all reactions except isolated nuclei) to give a final concentration of 5 mg/mL. Vortex vigorously; store in aliquots at −20 °C. For isolated nuclei, add 0.1 % (v/v) Triton to make the 5 mg/mL stock.

   The PtdInsP stock must be made fresh. For ten reactions, use PtdInsP (Avanti Polar Lipids, Inc.) in $CHCl_3$, aliquot 75 μL, dried with N_2 in a glass tube with swirling so that the lipid coats the bottom of the tube; solubilize with 50 μL of 0.1 % (v/v) Triton by vigorous vortexing (prepare fresh).

   Silanized glass (glass coated with dimethyldichlorosilane) should be used throughout these protocols for best results. For reasons of cost and convenience, we routinely use disposable glass tubes that are not silanized and have been able to get reproducible results comparing relative radioactive products formed. Aliquots of PtdIns–Triton X-100 stock solution can be stored at −20 °C and are stable for several months. The PtdInsP is not stable when stored in Triton and should be made fresh. If you don't want to add Triton, PtdIns and PtdInsP can be resuspended with brief sonication on ice in a bath sonicator. Do not use a tip sonicator.

2. The position and type of recombinant protein tag and the protein purification methods used will affect protein folding, specific activity, and protein–protein interactions. The MBP or His-tagged enzymes seem to give the highest specific activity. We have found the GST tag itself will bind to *E. coli* DAG kinases. Even without adding DAG as a substrate, PtdOH is a major product formed when assaying recombinant GST-tagged lipid kinases. For these reasons, when comparing recombinant proteins, use the same tag and purification method. Also, do not rely on protein assays to ascertain whether you have the same amount of the desired recombinant enzyme per assay.

Use gel electrophoresis and immunoblotting to estimate the amount of the desired recombinant protein, and when comparing activities of proteins of different molecular weights, be sure to use the same molar ratio of the desired recombinant enzyme per assay.

3. β-Cyclodextrin has proven to be an excellent means for substrate delivery for recombinant proteins (24); however, when added to membranes, β-cyclodextrin preferentially extracts and binds the endogenous membrane lipids, which decreases the availability of the added substrate and decreases product formation.

4. The acidic solvents are used to force the protonated lipids into the organic phase; however, the inositol lipids are readily hydrolyzed by acid. For this reason, when extracting with the acidic solvents, mix the solutions quickly. Most importantly, when back washing, it is essential that the aqueous phase is not carried over into the final organic phase. If it is, you will recover lysolipids and inositol phosphates. The tell-tale sign of lipid hydrolysis is that the bottom of the TLC plate will have increased radioactivity.

5. We usually prepare the TLC tanks the night before use to ensure that the solvents have saturated the tank. The solvent composition will vary over time. For consistent separations, prepare and use the tank at the same time for each experiment.

6. For consistent spotting, we place a piece of paper under the 20 cm TLC plate that has a line marked at 2.5 cm from the bottom of the plate and spot each sample along the line.

7. To prepare standards, in a separate lipid kinase reaction, use 2–5 μg bovine brain extract as the enzyme and add either PtdIns or PtdInsP to obtain the desired product. We make the bovine brain extract from bovine brain acetone powder (#BO 508, Sigma-Aldrich, Co.). Grind the bovine brain powder in a ground glass homogenizer in ice-cold buffer (25 mM sucrose, 3 mM EGTA, 2 mM EDTA, 2 mM DTT, 10 mM β mercapto-ethanol 1 mM PMSF, 30 mM Tris–HCl, pH 7.5), centrifuge at $40{,}000 \times g$ at 4 °C for 1 h; remove the supernatant and resuspend the pellet in the membrane resuspension buffer at a concentration of 1 mg/mL protein, freeze aliquots at −20 °C until use. The bovine brain extract will lose activity if thawed and refrozen. Standards can be prepared in this manner using [γ^{32}P]-ATP to produce radioactive products or by using fluorescent PtdIns or PtdInsP as the substrates to form fluorescently labeled products.

8. With isolated plant membranes, the first thing you will notice is that the primary phosphorylated lipid is PtdOH (70–90 % of

the product formed). This is not because the DAG is higher in plants than animals but rather because the DAG kinase specific activity is higher (25). A discussion of DAG kinase can be found in Chapter 6.

Acknowledgments

We would like to acknowledge the funding from National Science Foundation (# MCB0718452), US Department of Agriculture (IYP #1009-65114-06019), graduate training fellowships for C.D. from the NIH/NCSU Molecular Biotechnology Training Program and the Initiative for Future Agriculture and Food Systems Grant No. 2001-52101-11507 from the USDA Cooperative State Research, Education, and Extension Service and support from the North Carolina Agricultural Research Service (W.F.B.).

References

1. Jones DR, Bultsma Y, Keune WJ, Divecha N (2009) Methods for the determination of the mass of nuclear PtdIns4P, PtdIns5P, and PtdIns(4,5)P_2. Methods Mol Biol 462:75–88
2. Sommarin M, Sandelius AS (1988) Phosphatidylinositol and phosphatidylinositol-phosphate kinases in plant plasma membranes. Biochim Biophys Acta 958:268–278
3. Sandelius AS, Sommarin M (1986) Phosphorylation of phosphatidylinositols in isolated plant membranes. FEBS Lett 201:282–286
4. Dove SK, Lloyd CW, Drøbak BK (1994) Identification of a phosphatidylinositol 3-hydroxy kinase in plant cells: association with the cytoskeleton. Biochem J 303:347–350
5. Yang W, Boss WF (1994) Regulation of the plasma membrane type III phosphatidylinositol 4- kinase by positively charged compounds. Arch Biochem Biophys 313:112–119
6. Westergren T, Ekblad L, Jergil B, Sommarin M (1999) Phosphatidylinositol 4-kinase associated with spinach plasma membranes. Isolation and characterization of two distinct forms. Plant Physiol 121:507–516
7. Stevenson-Paulik J, Love J, Boss WF (2003) Differential regulation of two Arabidopsis type III phosphatidylinositol 4-kinase isoforms. A regulatory role for the pleckstrin homology domain. Plant Physiol 132:1053–1064
8. Mueller-Roeber B, Pical C (2002) Inositol phospholipid metabolism in Arabidopsis. Characterized and putative isoforms of inositol phospholipid kinase and phosphoinositide-specific phospholipase C. Plant Physiol 130:22–46
9. Xue HW, Pical C, Brearley C, Elge S, Muller-Rober B (1999) A plant 126-kDa phosphatidylinositol 4-kinase with a novel repeat structure. Cloning and functional expression in baculovirus-infected insect cells. J Biol Chem 274:5738–5745
10. Westergren T, Dove SK, Sommarin M, Pical C (2001) AtPIP5K1, an Arabidopsis thaliana phosphatidylinositol phosphate kinase, synthesizes PtdIns(3,4)P_2 and PtdIns(4,5)P_2 in vitro and is inhibited by phosphorylation. Biochem J 359:583–589
11. Perera IY, Davis AJ, Galanopoulou D, Im YJ, Boss WF (2005) Characterization and comparative analysis of Arabidopsis phosphatidylinositol phosphate 5-kinase 10 reveals differences in Arabidopsis and human phosphatidylinositol phosphate kinases. FEBS Lett 579:3427–3432
12. Stenzel I, Ischebeck T, Konig S, Holubowska A, Sporysz M, Hause B, Heilmann I (2008) The type B phosphatidylinositol-4-phosphate 5-kinase 3 is essential for root hair formation in *Arabidopsis thaliana*. Plant Cell 20: 124–141
13. Saavedra L, Balbi V, Dove SK, Hiwatashi Y, Mikami K, Sommarin M (2009) Characterization of phosphatidylinositol phosphate kinases from the moss *Physcomitrella patens*: PpPIPK1 and PpPIPK2. Plant Cell Physiol 50:595–609
14. Galvão RM, Kota U, Soderblom EJ, Goshe MB, Boss WF (2008) Characterization of a

new family of protein kinases from Arabidopsis containing phosphoinositide 3/4-kinase and ubiquitin-like domains. Biochem J 409: 117–127

15. Yang W, Burkhart W, Cavallius J, Merrick WC, Boss WF (1993) Purification and characterization of a phosphatidylinositol 4-kinase activator in carrot cells. J Biol Chem 268: 392–398
16. Kost B, Lemichez E, Spielhofer P, Hong Y, Tolias K, Carpenter C, Chua NH (1999) Rac homologues and compartmentalized phosphatidylinositol 4, 5- bisphosphate act in a common pathway to regulate polar pollen tube growth. J Cell Biol 145:317–330
17. Preuss ML, Schmitz AJ, Thole JM, Bonner HKS, Otegui MS, Nielsen E (2006) A role for the RabA4b effector protein, PI-4Kβ1, in polarized expansion of root hair cells in Arabidopsis. J Cell Biol 172:991–998
18. Zhao Y, Yan A, Feijo JA, Furutani M, Takenawa T, Hwang I, Fu Y, Yang Z (2010) Phosphoinositides regulate clathrin-dependent endocytosis at the tip of pollen tubes in Arabidopsis and tobacco. Plant Cell 22: 4031–4044
19. Hendrix W, Assefa H, Boss WF (1989) The polyphosphoinositides, phosphatidylinositol monophosphate and phosphatidylinositol bisphosphate are present in nuclei isolated from carrot protoplasts. Protoplasma 151:62–72
20. Stevenson JM, Perera IY, Boss WF (1998) A phosphatidylinositol 4-kinase pleckstrin homology domain that binds phosphatidylinositol 4-monophosphate. J Biol Chem 273: 22761–22767
21. Im YJ, Davis AJ, Perera IY, Johannes E, Allen NS, Boss WF (2007) The N-terminal membrane occupation and recognition nexus domain of Arabidopsis phosphatidylinositol phosphate kinase 1 regulates enzyme activity. J Biol Chem 282:5443–5452
22. Walsh JP, Caldwell KK, Majerus PW (1991) Formation of phosphatidylinositol 3-phosphate by isomerization from phosphatidylinositol 4-phosphate. Proc Natl Acad Sci USA 88: 9184–9187
23. Hama H, Torebinejad J, Prestwich G, DeWald D (2004) Measurement and immunofluorescence of cellular phosphoinositides. In: Dickson R (ed) Methods in molecular biology: signal transduction protocols. Humana, Clifton, pp 243–258
24. Davis AJ, Perera IY, Boss WF (2004) Cyclodextrins enhance recombinant phosphatidylinositol phosphate kinase activity. J Lipid Res 45:1783–1789
25. Heilmann I, Perera IY, Gross W, Boss WF (1999) Changes in phosphoinositide metabolism with days in culture affect signal transduction pathways in *Galdieria sulphuraria*. Plant Physiol 229:1331–1339

Chapter 16

Assaying Inositol and Phosphoinositide Phosphatase Enzymes

Janet L. Donahue, Mustafa Ercetin, and Glenda E. Gillaspy

Abstract

One critical aspect of phosphoinositide signaling is the turnover of signaling molecules in the pathway. These signaling molecules include the phosphatidylinositol phosphates (PtdInsPs) and inositol phosphates (InsPs). The enzymes that catalyze the breakdown of these molecules are thus important potential regulators of signaling, and in many cases the activity of such enzymes needs to be measured and compared to other enzymes. PtdInsPs and InsPs are broken down by sequential dephosphorylation reactions which are catalyzed by a set of specific phosphatases. Many of the phosphatases can act on both PtdInsP and InsP substrates. The protocols described in this chapter detail activity assays that allow for the measurement of PtdInsP and InsP phosphatase activities in vitro starting with native or recombinant enzymes. Three different assays are described that have different equipment requirements and allow one to test a range of PtdInsP and InsP phosphatases that act on different substrates.

Key words Inositol, Inositol phosphate, Phosphatidylinositol, Phosphatase, HPLC

1 Introduction

The enzymes that break down phosphatidylinositol phosphates (PtdInsPs) and inositol phosphates (InsPs) are important regulators of signaling (1). To examine the activity of PtdInsP and InsP phosphatases we describe here three methods that can measure the activity of native or recombinant enzymes incubated with PtdInsP or InsP substrates. In each method purified, commercial substrates are presented to purified native or recombinant enzymes and during catalysis inorganic phosphate and dephosphorylated PtdInsP or InsP are produced. Each method entails the measurement of either the phosphate produced or the remaining substrate and PtdInsP or InsP products. Each method has different sensitivity, limitations, and equipment requirements, which allows investigators flexibility in choosing an approach.

The first method utilizes unlabelled substrates followed by measurement of phosphate release (2). The advantage of this first

Teun Munnik and Ingo Heilmann (eds.), *Plant Lipid Signaling Protocols*, Methods in Molecular Biology, vol. 1009,
DOI 10.1007/978-1-62703-401-2_16, © Springer Science+Business Media, LLC 2013

method is that radioisotopes are not required, and no specialized equipment is required. While phosphate release is the least sensitive of the three protocols we describe, it is very amenable to performing multiple time points and replicates, a necessity for kinetic analyses of enzymes (3). In addition, since only a few radiolabelled PtdInsPs and InsPs are available commercially, measuring phosphate release of unlabelled substrates may be the only approach possible in some instances. One disadvantage of this assay is that the assay measures only one of the products of catalysis, the inorganic phosphate; thus a definitive identification of all reaction products is not possible with this approach. This could be problematic if one is characterizing an enzyme that removes sequential and multiple phosphates from substrates, like one of the yeast inositol 5-phosphatase enzymes (4).

A second method for examining InsP phosphatase activity utilizes radioactive substrates followed by separation of products by high-performance liquid chromatography (HPLC). This method is very sensitive, but requires both a source of radiolabelled substrate and an HPLC equipped with a radioisotope detector (5). One advantage of this method is that both substrates and products are measured; thus one gains more information about the products of catalysis as compared to the phosphate release method.

A third method is also described that can be used to examine PtdInsP phosphatase activity without specialized equipment. In this assay commercially available fluorescent PtdInsP substrates are incubated with enzyme and substrates and products are measured utilizing thin-layer chromatography (TLC) separation. This method provides high sensitivity, but may be limited by the availability of fluorescent substrates. In our hands, this assay is the most straightforward and reproducible way to analyze the activity of enzymes that act on PtdInsPs. Use of fluorescent substrates and TLC avoids the complicated process of deacylation required for the analysis of PtdInsPs by HPLC. We modified our protocol from a previously described phosphatase assay (6).

Together, these methods allow investigators to address multiple facets of PtdInsP and InsP phosphatase activity with different technologies of separation. The caveats of each approach are discussed to allow for a critical interpretation of data.

2 Materials

All solutions should be prepared with ultrapure water and analytical grade reagents, and can be stored at room temperature unless otherwise indicated.

2.1 Materials for Use with Unlabelled Substrates and Phosphate Release Assays

1. Reaction buffer 1:50 mM Tris–HCl, pH 7.0, 5 mM $MgCl_2$.
2. Unlabelled InsP and PtdInsP substrates can be purchased from Sigma-Aldrich, Echelon Research Laboratories, Cayman Chemical, or Avanti Polar Lipids. We have used d-*myo*-Ins(1,4,5)P_3 potassium salt (Sigma-Aldrich), d-*myo*-Ins(4,5) P_2sodium salt (Cayman), d-*myo*-Ins(1,5)P_2sodium salt (Cayman), d-*myo*-Ins(1,3,4,5)P_4 (Echelon), and PtdIns(4,5) P_2 (Echelon) (3, 7). All InsPs and PtdInsPs should be stored at −20 °C unless indicated otherwise by the supplier.
3. Malachite green solution: Three volumes 0.045 % (w/v) malachite green oxalate in H_2O, one volume 4.2 % (w/v) ammonium molybdate in 4 N HCl. Prepare on the day of use. Let sit at room temperature for 20 min before use. Filter if necessary. The 0.045 % malachite green oxalate is not stable for more than a month at room temperature. If a fresh solution is prepared, a new standard curve should be made. Malachite green is a hazardous substance. Use with caution and discard as hazardous waste. The malachite green solution contains acid that can corrode metals. Clean up thoroughly, including the inside of pipettors.
4. 34 % (w/v) sodium citrate, prepared in water.
5. 0.5 mM K_2HPO_4, using salt that has been dried several hours at 100 °C, for a standard curve.

2.2 Materials for Use with Labelled Substrates and HPLC Separation of Products

1. Reaction buffer 2:50 mM Tris, pH 7.5, 5 mM $MgCl_2$, 250 mM KCl.
2. Sources for labeled substrates include Perkin Elmer, Sigma-Aldrich, and American Radiolabelled Chemicals. We have used [^{3}H]d-*myo*-inositol (1,4,5)P3 (10 μCi/ml, 22 Ci/mmol, Perkin Elmer) and [^{3}H] D-*myo*-inositol (1,3,4,5)P4 (10 μCi/ml, 22 Ci/mmol, Perkin Elmer) (7–10). Since suppliers often change the products they carry, we suggest checking other suppliers and new companies for products.
3. For HPLC we use a Waters Spherisorb S5 SAX 4 × 125-mm analytical column and a 1 cm guard column (Waters).
4. 4.10 mM ammonium phosphate (AP) buffer at pH 3.8 is made by adding $NH_4H_2PO_4$ to water, followed by pH adjustment to 3.8 with phosphoric acid. We also use a 1.7 M $NH_4H_2PO_4$, pH 3.8 buffer prepared the same way. Both buffers are passed through a 0.2 μm nylon filtration apparatus before use. Note that the gradients are provided by two reservoirs attached to the pumps which contain each buffer.
5. For in-line radioisotope counting we use Ready Flow III (Beckman Instruments, Fullerton, CA).

2.3 Materials for Use with Fluorescent Substrates and Thin-Layer Chromatography

1. Reaction buffer 3:50 mM HEPES, pH 7.5, 5 mM $MgCl_2$, and 250 mM KCl.
2. The source for fluorescent substrates is Echelon Research Laboratories, Salt Lake City. Store reconstituted stocks in reaction buffer 3 at −20 °C. We have used fluorescent di-C6-NBD6-phosphatidylinositol (4,5) bisphosphate, di-C6-NBD6-phosphatidylinositol (3,4,5) trisphosphate, di-C6-NBD6-phosphatidylinositol (5) phosphate, and di-C6-NBD6-phosphatidylinositol (3,5) bisphosphate (9). The availability of these products may be limited, so a new group of BODIPY-labelled molecules by Echelon carrying an amide linkage could be substituted (i.e., BODIPY FL Phosphatidylinositol(5) Phosphate, etc.). These molecules also carry a C6 group and are water soluble. Note that similar C16 compounds are not water soluble and must be sonicated before use. We have used C16 substrates and found that catalysis with these does not differ for the At5PTase enzymes; however, migration of substrates and products are slower, and the resulting bands are often distorted and hard to quantify (9).
3. TLC plates are silica gel 60. Plates are pretreated in 1.2 % potassium oxalate in methanol:water (60:40, v/v) made by adding methanol and water together in a beaker, and then adding the potassium oxalate salt. TLC plates are developed in 180 ml of $CHCl_3$:methanol:acetone:glacial acetic acid:water (70:50:20:20:20, v/v/v/v/v) made by mixing water and then the other solvents in a flask or a graduated cylinder.

3 Methods

3.1 Analyzing Activity with Unlabelled Substrates and Phosphate Release Assays

1. Immunoprecipitated or purified enzymes (50–500 ng, *see* **Note 1**) are added to reaction buffer 1 on ice. Set up a control reaction without enzyme.
2. Reactions are started with the addition of unlabelled inositol phosphate substrates (i.e., Ins(1,4,5)P_3, Ins(1,3,4,5)P_4, etc.) to a final concentration of 200 μM. PtdInsP substrates must first be dissolved in reaction buffer with the aid of sonication before addition (*see* **Note 2**).
3. Reactions are incubated for 15–30 min at 22 °C (*see* **Note 3**).
4. Reactions are terminated by removing 50 μl from enzyme reactions and adding to 800 μl of malachite green solution at room temperature.
5. The phosphate in the reaction is stabilized by the immediate addition of 100 μl 34 % (w/v) sodium citrate. Mix well.
6. Make the blank for the spectrophotometer with the malachite green solution, water, and sodium citrate in appropriate proportions.

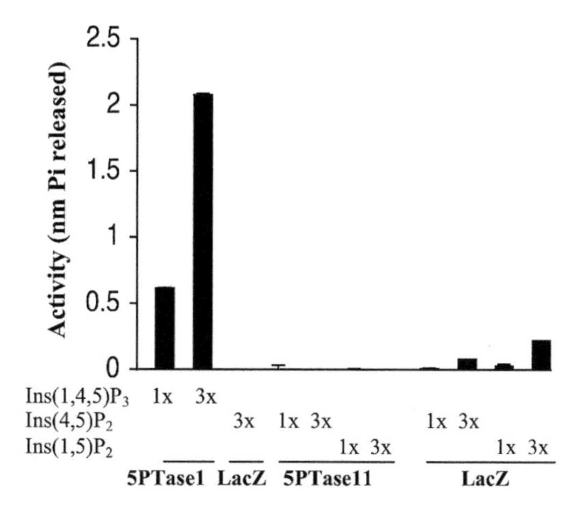

Fig. 1 Sample results using unlabelled substrates and phosphate release assays. Recombinant V5-tagged proteins were expressed in yeast and purified with anti-V5 antibody linked to agarose and then added to in vitro assays. Phosphate release from 0.5 mM (1×) and 1.5 mM (3×) of the indicated InsP was measured after incubation with the indicated protein. 5PTase1 hydrolysis of Ins(1,4,5)P_3 was used as a positive control, and LacZ was used with each substrate as a negative control. Note that using different substrate concentrations is helpful in situations in which kinetic constants are unknown

7. Allow the color to develop for 15–30 min, before observing the absorbance at 660 nm. Color is stable for 4 h.
8. Make a standard curve to use for calculating the phosphate released in enzyme assays. Use 0.5 mM K_2HPO_4 to make up samples of 1–10 nmols phosphate in a volume of 50 μl with H_2O. Add 800 μl malachite green solution and 50 μl 34 % sodium citrate. Allow color to develop, record the A_{660}, and plot a standard curve (*see* **Note 4**).
9. The absorbance of each enzyme reaction sample is compared to the standard curve of inorganic phosphate and the total phosphate released is calculated (*see* **Note 5**).
10. Interpreting data: In Fig. 1 we present data that addresses whether the At5PTase11 enzyme can hydrolyze three different InsP substrates (11). In this experiment we performed phosphate release assays with two different concentrations of each substrate. This approach is helpful in cases in which kinetic constants are unknown and it is difficult to predict the optimal substrate concentration to be used. The nmol of phosphate released in each reaction was determined and compared. In this example we used a negative control protein (LacZ) instead of assays containing no added enzyme. We present the values for the negative control as well as positive control (At5PTase1). The use of proper controls brings confidence that the At5PTase11 enzyme does not hydrolyze the substrates used in this experiment.

4 Analyzing Activity with Labelled Substrates and HPLC Separation of Products

1. Immunoprecipitated or purified enzymes (50–500 ng; *see* **Note 1**) are added to reaction buffer 2 on ice in 300 μl total volume. Set up control reactions without enzyme.
2. Reactions are started with the addition of 30 nCi of ^{3}H-labelled inositol phosphate substrates.
3. Reactions are incubated for 15–60 min at 22 °C (*see* **Note 2**).
4. Reactions are terminated by freezing at −20 °C; reactions can be stored this way until analysis by HPLC.
5. For HPLC separation of inositol phosphates we use a Beckman Gold System and a Waters Spherisorb S5 SAX 4 × 125-mm analytical column. The column is first equilibrated with 10 mM ammonium phosphate buffer at pH 3.8 (*see* **Note 6**). A thorough protocol of HPLC use for separating inositol phosphates has been described by others (5; see Chapter 2).
6. 1–50 μl samples are applied to the column using an autosampler, and products are eluted with a linear gradient from 10 to 340 mM AP over 30 min; 340–1.02 M AP over 15 min; and constant 1.02 M AP over 5 min. We use an in-line IN/US radioisotope detector which is set to mix 3:1 scintillant:eluant and provides a determination of the radioactivity in the column eluate. Beckman software (32 Karat software; Beckman Coulter, Fullerton, CA) provides a plot of time versus radioactivity which can be converted and analyzed by Microsoft Excel (Microsoft, Seattle). Alternative means for collection of eluate fractions followed by scintillation counting may be used (*see* **Note 7**).
7. The amount of product formed is quantified by calculating the area under each peak for each chromatogram and using the specific activity of the substrate to convert this number to moles of product formed. Enzyme activity can then be presented as the amount of product formed per time per unit protein (*see* **Note 8**).
8. Interpreting results: Figure 2 shows sample chromatograms of standards (a) and enzyme reactions (b). To compare different enzymes, product formed per minute per mg enzyme used was calculated for three different enzymes (c) (At5PTase1, At5PTase2, and At5PTase3). These results indicate that all three enzymes hydrolyze Ins(1,4,5)P_3, and At5PTase3 has a higher specific activity. The strength of this experimental approach is that intact product, Ins(1,4)P_2, has been measured as opposed to the indirect method of measuring phosphate released from Ins(1,4,5)P_3. This can be especially important when dealing with enzymes that can catalyze removal of multiple phosphates from InsP and PtdInsP substrates.

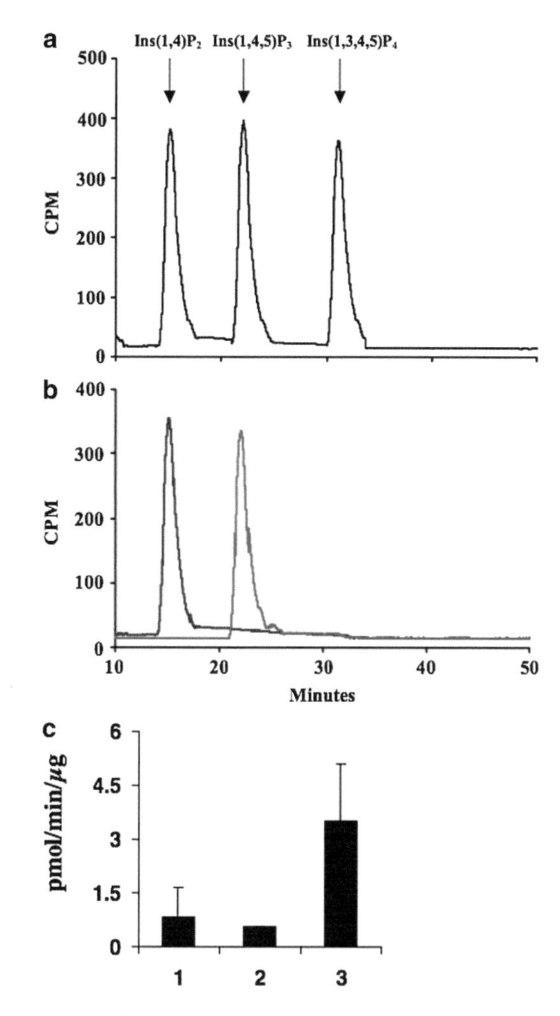

Fig. 2 Sample results from activity assays using radiolabelled substrate and HPLC separation. (**a**) A mixture of [^{3}H]-Ins(1,4)P_2, [^{3}H]-Ins(1,4,5)P_3, and [^{3}H]-Ins(1,3,4,5)P_4 standards or (**b**) reaction products from immunoprecipitated At5PTase1-V5 incubated with [^{3}H]-Ins(1,4,5)P_3 (*first peak*) or [^{3}H]-Ins(1,3,4,5)P_4 (*second peak*) were separated via HPLC. Note that co-migration with standards on HPLC is critical for assessing the reaction products. (**c**) Comparison of activities of 200 ng of At5PTase1, At5PTase2, and At5PTase3 with [^{3}H]-Ins(1,4,5)P_3. Note that the amount of radioactivity in product formed can be converted to picomoles product formed to compare specific activity of different enzymes. The mean and standard error for independent experiments is presented

4.1 Analyzing Activity with Fluorescent Substrates and Thin-Layer Chromatography

1. Immunoprecipitated or purified enzymes (50–500 ng; *see* **Note 1**) are added to reaction buffer 3 on ice. Set up control reactions without enzyme.
2. Reactions are started with the addition of 1.5 µg of fluorescent substrate (i.e., di-C6-NBD6-phosphatidylinositol (4,5) bisphosphate, di-C6-NBD6-phosphatidylinositol 3,4,5-trisphosphate, or di-C6-NBD6-phosphatidylinositol (5) monophosphate) (*see* **Note 9**).

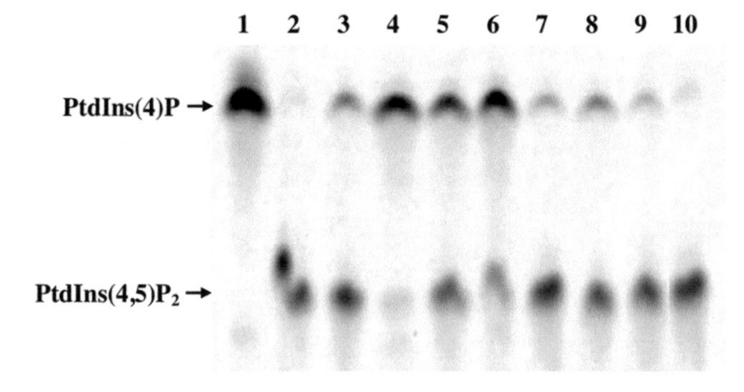

Fig. 3 Sample results using fluorescent substrates and TLC. Phosphatase reactions (*Lanes* 3–10) containing 1.5 μg of fluorescent PtdIns(4,5)P2 and 80 ng of immunoprecipitated (IP) recombinant At5PTase enzymes produced in Drosophila S2 cells were incubated for 1 h at room temperature. *Lane 1*: PtdIns(4)P incubated with reaction buffer only. *Lane 2*: PtdIns(4,5)P2 incubated with reaction buffer only. Other lanes contain reactions with the following IP proteins: *Lane 3*: At5PTase11; *Lane 4*: At5PTase1; *Lane 5*: At5PTase2; *Lane 6*: At5PTase3; *Lane 7*: At5PTase4; *Lane 8*: At5PTase7; *Lane 9*: At5PTase9; *Lane 10*: Control IP from mock-transfected cells. The migration of fluorescent standards is indicated

3. Reactions are incubated at 22 °C for 15–60 min. Agitate or vortex reactions every 15 min to maintain enzyme access to substrate.
4. Reactions are stopped by addition of 100 μl acetone and are dried in a speedvac under low-heat vacuum (*see* **Note 10**).
5. Silica gel 60 TLC plates are pretreated in 1.2 % potassium oxalate in methanol:water (60:40, v/v) and dried at 100 °C for 30 min (*see* **Note 11**).
6. Dried reactions are dissolved in 10 μl of methanol:2-propanol:glacial acetic acid (5:5:2, v/v/v), and equal amounts are spotted onto the prepared TLC plates. To decrease spreading, 1–2 μl can be spotted sequentially on the same spot with drying intervals.
7. Standards must also be spotted to identify substrates and products (*see* **Note 12**).
8. TLC plates are developed in 180 ml of $CHCl_3$:methanol:acetone:glacial acetic acid:water (70:50:20:20:20, v/v/v/v/v) until the solvent front reaches the top of the plate.
9. Plates are dried and can be analyzed using a fluorescence or chemifluorescence detector. We use a Storm 860 Blue detector and Scanner Control Software version 4.1 (Molecular Dynamics, Piscataway, NJ). For densitometric analysis of the images produced via Storm 860, we have utilized AlphaEaseFC software version 3.1.2 (Alpha Innotech, San Leandro, CA).
10. Interpreting results: Figure 3 shows sample results using different immunoprecipitated At5PTases incubated with fluorescent PtdIns(4,5)P_2. Note that PtdIns(4)P migrates faster than PtdIns(4,5)P_2 and one may see trace contamination

of standards with other PtdInsPs. In lanes 3–9, hydrolysis of PtdIns(4,5)P_2 is evident as seen by the increase in PtdIns(4)P.

5 Notes

1. The source of enzymes is critical. In our hands eukaryotic expression systems are necessary to maintain catalytic activity of recombinant proteins. Thus we have used Drosophila S2 cells and yeast as expression systems (9). Enzymes do not need to be purified to homogeneity; however, crude eukaryotic extracts will contain significant amounts of background activity, so some type of purification is required. We have successfully used immunoprecipitated enzymes in the past. In this case, one must be able to quantify the immunoprecipitated proteins in order to compare different enzymes. We find that in general, 50–500 ng is usually enough enzyme for each protocol described here. Optimizing the amount of enzyme used is desirable and is required for kinetic analyses.
2. PtdInsPs are not soluble in aqueous solution and must be sonicated in a bath sonicator for 1 min. This treatment disperses the PtdInsPs and ensures that the enzyme has access to the substrate.
3. The length of incubation is critical when examining any enzyme activity. It is usually necessary to optimize the amount of enzyme added to a reaction, and the time course of the reaction as well. Monitoring the enzyme activity over time can best be accomplished by starting with a 500 μl reaction, and removing aliquots of 50 μl at predetermined times to quantify the total phosphate released at each timepoint. A plot of time versus product will reveal the linear range of product accumulation, and a suitable length of incubation can be chosen from this range. For kinetic analyses, reactions should be assayed when the phosphate released is less than 10 % of the substrate present.
4. Care should be taken to measure the phosphate released when it is in the linear range of the standard curve. Note that phosphate release can detect as little as 1 nmol phosphate, and is linear to 15 nmol phosphate in our hands.
5. The biggest pitfall in phosphate release assays is not accounting for background-free phosphate present in the enzyme, buffers, or substrate. The amount of phosphate released in control reactions without enzyme is thus critical to determine background phosphate levels and should be subtracted from reactions containing enzymes. Note that background levels may be higher with certain unstable inositol phosphate compounds. It is also imperative that phosphate buffers be removed from enzyme preparations, e.g., by dialysis, before testing the phosphate content of the enzyme reaction. One excellent

control to ensure that phosphate released is dependent on catalysis is to double the time of one reaction and double the amount of enzyme in another reaction, each of which should result in twice the amount of phosphate released.

6. Other HPLC columns can be used such as a Partisphere SAX by Whatman. Note that the gradients are provided by two reservoirs attached to the pumps which contain 10 mM $NH_4H_2PO_4$, pH 3.8 and 1.7 M $NH_4H_2PO_4$, pH 3.8.
7. Non-in-line detection can also be used, if desired. In this case 0.3 ml fractions can be collected every 20 s and mixed with 2 ml of water-miscible scintillation cocktail, and counted in a liquid scintillation counter.
8. Identifying peaks with standards: Note that purified standards are a necessity and allow one to determine which substrates and products are present. The area under these curves can be calculated and converted to picomoles of product formed using the specific activity of the substrate. Consideration should be given to which standards are required and the limits of identification. For example, migration of Ins(1,4)P_2 and Ins(4,5)P_2 could be similar, so using one standard or the other could aid identification. A discussion of elution times of InsPs under different conditions has been described (5).
9. The C6 fluorescent substrates are mostly water soluble which means that one does not need to sonicate these substrates (12). However, it is useful to vortex reactions frequently to increase substrate access to enzymes.
10. An alternative to speed vacuum drying of reaction products is to use a stream of N_2 gas.
11. Pretreated, dried TLC plates can be stored for several weeks without loss of sensitivity or separation.
12. Migration of standards is critical for identifying substrates and products. In general one should use both the substrate and expected product in each experiment. This method is very sensitive, and in our hands, is the most sensitive of the three techniques described here.

References

1. Astle MV, Horan KA, Ooms LM, Mitchell CA (2007) The inositol polyphosphate 5-phosphatases: traffic controllers, waistline watchers and tumour suppressors? Biochem Soc Symp 74:161–181
2. Lanzetta PA, Alvarez LJ, Reinach PS, Candia OA (1979) An improved assay for nanomole amounts of inorganic phosphate. Anal Biochem 100:95–97
3. Torabinejad J, Donahue JL, Gunesekera BN, Allen-Daniels MJ, Gillaspy GE (2009) VTC4 is a bifunctional enzyme that affects *myo*inositol and ascorbate biosynthesis in plants. Plant Physiol 150:951–961
4. Guo S, Stolz LE, Lemrow SM, York JD (1999) SAC1-like domains of yeast SAC1, INP52, and INP53 and of human synaptojanin encode polyphosphoinositide phosphatases. J Biol Chem 274:12990–12995
5. Stevenson-Paulik J, Chiou ST, Frederick JP, Dela Cruz J, Seeds AM, Otto JC, York JD (2006) Inositol phosphate metabolomics:

merging genetic perturbation with modernized radiolabeling methods. Methods 39: 112–121
6. Maehama T, Taylor GS, Slama JT, Dixon JE (2000) A sensitive assay for phosphoinositide phosphatases. Anal Biochem 279:248–250
7. Ercetin M, Torabinejad J, Robinson J, Gillaspy G (2008) A phospholipid-specific *myo*-inositol polyphosphate 5-phosphatase required for seedling growth. Plant Mol Biol 67:375–388
8. Berdy S, Kudla J, Gruissem W, Gillaspy G (2001) Molecular characterization of *At*5PTase1, an inositol phosphatase capable of terminating IP_3 signaling. Plant Physiol 126:801–810
9. Ercetin ME, Gillaspy GE (2004) Molecular characterization of an Arabidopsis gene encoding a phospholipid-specific inositol polyphosphate 5-phosphatase. Plant Physiol 135:938–946
10. Gunesekera B, Torabinejad J, Robinson J, Gillaspy GE (2007) Inositol polyphosphate 5-phosphatases 1 and 2 are required for regulating seedling growth. Plant Physiol 143: 1408–1417
11. Ercetin ME, Ananieva EA, Safaee NM, Torabinejad J, Robinson JY, Gillaspy GE (2008) A phosphatidylinositol phosphate-specific *myo*-inositol polyphosphate 5-phosphatase required for seedling growth. Plant Mol Biol 67:375–388
12. Best MD, Zhang H, Prestwich GD (2010) Inositol polyphosphates, diphosphoinositol polyphosphates and phosphatidylinositol polyphosphate lipids: structure, synthesis, and development of probes for studying biological activity. Nat Prod Rep 27: 1403–1430

Chapter 17

Determination of Phospholipase C Activity In Vitro

S.M. Teresa Hernández-Sotomayor and J. Armando Muñoz-Sanchez

Abstract

Measurement of phospholipase C (PLC) activity in vitro is a valuable biochemistry technique easily applicable in samples from different organisms. It quantifies the enzymatic activity of a key protein involved in critical developmental functions in organisms such as plants, animals, and bacteria. A protocol is described which assays the formation of two main products of the PLC hydrolysis reaction on radioactively labeled phospholipid substrates.

Key words Phospholipase C, Phosphoinositides, Lipid pathway, PLC, Phospholipids

1 Introduction

Phospholipids are now recognized as playing a role in plant signal transduction (1). Their multiple intracellular functions are regulated by a series of metabolizing enzymes including lipases (phospholipase A, phospholipase C (PLC), phospholipase D), kinases, and phosphatases. Several phosphorylated derivatives of phosphatidylinositols (PtdIns), collectively termed phosphoinositides, have been identified in eukaryotic cells in life forms ranging from yeasts to mammals (2). Although PtdIns represent only a small percentage of total cellular phospholipids, they play a crucial role in signal transduction as the precursors of several second-messenger molecules. The signal transduction cascade involving these lipid-derived compounds seems to operate in plant cells much as it does in animal cells (3). A number of recent studies have focused on this cascade and it is now known that external signals such as hormones, growth regulators, and environmental stress may have important roles in their regulation. Inositol phospholipids may also have a major role in cytoskeletal structure regulation in higher plants (4, 5).

PLC plays a key role in the signal transduction pathway in animal cells. Using amino acid sequences, the 14 different PLC isoenzymes in mammalian cells have been identified and classified into

Teun Munnik and Ingo Heilmann (eds.), *Plant Lipid Signaling Protocols*, Methods in Molecular Biology, vol. 1009,
DOI 10.1007/978-1-62703-401-2_17, © Springer Science+Business Media, LLC 2013

five types: β, γ, δ, ε, and ζ (6–8). Activation of PLC and stimulation of phosphatidylinositol 4,5-bisphosphate (PIP_2) hydrolysis in animal cells occur through agonists which bind to either G-protein-coupled receptors or tyrosine kinase receptors (6). PLC catalyzes PIP_2 hydrolysis, generating two potential intracellular second messengers: diacylglycerol (DAG) and inositol 1,4,5-trisphosphate (IP_3). IP_3 binds to an intracellular membrane receptor, usually a Ca^{2+} channel, releasing Ca^{2+} into the cytosol and elevating cytosolic Ca^{2+} levels (9, 10). Ca^{2+} is the principal second messenger in plants, and one of the mechanisms which regulate Ca^{2+} levels involves the enzyme PLC. In addition, PIP_2 regulates the activity of several enzymes, and participates in membrane trafficking regulation and in processes such as vesicle trafficking and ion transport (11). DAG can function as a substrate for DAG kinase-generating phosphatidic acid (PA), and IP_3 can be phosphorylated by inositol kinases, generating IP_6 (12), among other compounds.

PLC has been found in a broad spectrum of organisms, including bacteria, eukaryotes, plants, and animals (13, 14). Intracellular studies have identified the biochemical presence of this enzyme; specifically, PLC activity has been reported in several plant species such as oat, wheat, rice, rose periwinkle, coffee, and soybean (15–21). PLC signaling may be implicated in a variety of plant physiological processes such as germination, osmoregulation, pathogen response, stress, and cell fusion (22–25). Given its general importance, measuring PLC can form a vital part of studies in numerous organisms. This chapter describes a method for measuring PLC activity in vitro which is applicable in a broad range of organisms.

2 Materials

2.1 Protein Extraction

1. Extraction buffer: 50 mM NaCl, 1 mM EGTA, 50 mM Tris–HCl, pH 7.4, 250 mM sucrose, 10 % glycerol, 1 mM phenylmethylsulfonyl fluoride, 10 mM sodium pyrophosphate, 0.2 mM orthovanadate, 1 mM β-mercaptoethanol.
2. 74 kBq (^3H)-Phosphatidylinositol-4,5-bisphosphate (NEN Life Science Products, NET-895 (*see* **Note 1**)).
3. L-α-Phosphatidyl-D-myo-inositol-4,5-bisphohate (1 mg) from bovine brain, ammonium salt.

2.2 Stock Solutions

1. 50 mM NaH_2PO_4, 100 mM KCl, pH 6.8.
2. 2 % (w/v) deoxycholic acid.
3. 50 mM NaH_2PO_4, 100 mM KCl, 1 mM EGTA, pH 6.8.
4. 10 mM $CaCl_2$.
5. 1 % (w/v) bovine serum albumin (BSA).
6. 10 % (w/v) trichloroacetic acid (TCA).

7. Scintillation liquid.
8. General equipment: Probe sonicator, water bath, scintillation counter.

3 Methods

3.1 Protein Sample Extraction

1. Freeze cells (0.5–1.0 g) quickly with liquid nitrogen and pulverize them with a mortar and pestle. The cells can be kept frozen during pulverization by adding liquid nitrogen.
2. Transfer the powdered cells to a glass vial containing 1.25–2.5 mL (1:2.5; 1 g of cells in 2.5 mL) of extraction buffer.
3. Homogenize the cells for 1 min at 4 °C with a polytron homogenizer.
4. Filter the extract through gauze or cheese cloth.
5. Transfer the homogenate to a 1.5 mL polypropylene reaction tube and centrifuge at 20,000 × *g* in a microcentrifuge for 30 min at 4 °C.
6. Recover the supernatant and use it as total protein extract (the concentration of protein is usually 3.5–5.0 mg/mL). Store the protein sample at −70 °C.
7. Measure protein concentration following the BCA protein assay relative to BSA standard.

3.2 Preparation of PLC Substrate (^{3}H)PIP$_2$/PIP$_2$ Stock: 2.65 mM

1. Transfer 1 mg of PIP$_2$ ammonium salt to a cryogenic vial. Add 74 kBq of (^{3}H)PIP$_2$ (approximately 200 μL) and dissolve, powered by vortexing. Dry the solvent with a direct nitrogen stream over the sample for 1 min or until attaining approximately 10 μL.
2. Add 180 μL of 50 mM $NaHPO_4$, and 100 mM KCl, pH 6.8. Mix and sonicate for 2 min.
3. Add 180 μL of 2 % (w/v) deoxycholic acid. Mix and sonicate for 2 min.
4. Transfer 4 μL of solution to 5 mL of scintillation liquid in a counting vial. Mix and quantify by liquid scintillation counting. At this point, total counts must be approximately 18,000–20,000 cpm (*see* **Note 2**).
5. Store under refrigeration (4 °C) until use.

3.3 Reaction Mix Preparation

1. Add the following components to a 1.5-mL microcentrifuge tube on ice: 22 μL 50 mM $NaHPO_4$; 100 mM KCl; 1 mM EGTA, pH 6.8; 4 μL 2.65 mM (^{3}H)PIP$_2$, PIP$_2$; and 4 μL 10 mM $CaCl_2$. Total reaction mix volume per sample should be 30 μL.
2. If desired, a Master Mix can be prepared for multiple reactions. Preparation of this Master Mix eliminates the need to repeatedly pipette volumes, resulting in increased consistency between

samples. When preparing the Master Mix, include two additional samples to produce a reaction mix background and calculate total counts. Mix contents of tube and sonicate for 1 min.

3. If a Master Mix has been prepared, place 30 μL aliquots in 1.5 mL microcentrifuge tubes immediately before initiating the reaction. Each aliquot corresponds to one reaction mix for one sample.

3.4 PLC Activity Measurement

1. Initiate reaction by adding 20 μL protein sample (*see* **Note 3**). If desired, small protein sample volumes can be used but distilled H_2O must be added to make up the 20 μL total sample volume. Total reaction volume must be 50 μL (30 μL reaction mix + 20 μL protein sample). Briefly vortex the mixture.
2. Incubate tubes in a water bath at 30 °C for 10 min (*see* **Note 4**).
3. Stop reaction by adding 100 μL 1 % (w/v) BSA and 250 μL TCA. Mix contents by vortexing.
4. Centrifuge tubes in an Eppendorf microcentrifuge at 20,000 × *g* for 10 min.
5. Remove all the supernatant, carefully avoiding pellet remains, and transfer into 5 mL scintillation liquid in a counting vial. Mix and quantify by liquid scintillation, counting for 5 min.
6. Total counts for the 30 μL reaction mix are done by first transferring into 5 mL scintillation liquid in a counting vial. Mix and quantify by liquid scintillation, counting for 5 min.
7. Use the following formulas to calculate specific activity in pmol/min/mg^{-1} protein:

$$\text{pmol per count} = \frac{10{,}000\,\text{pmol}}{\text{Total cpm in } 30\mu\text{L reaction mix}}.$$

$$\text{pmol/min sample} = \frac{(\text{cpm sample} - \text{cpm background}) \times \text{pmol per count}}{10}.$$

$$\text{pmol/min mg/protein} = \frac{(\text{pmol/min}^{-1})\,\text{of sample}}{\text{mg protein in } 20\mu\text{L sample}}.$$

4 Notes

1. The (^{3}H) PIP_2 is shipped dissolved in CH_2Cl_2:ethanol:H_2O (20:10:1, v/v/v), in a sealed glass ampoule under inert gas. Open the ampoule (snap off the neck) and separate 200 μL aliquots into cryogenic vials. Once opened, the radiochemical is no longer under optimum conditions and decomposition rate may increase. If possible, replace the air in the cryogenic vials with nitrogen gas before resealing. Store the cryogenic vials at −70 °C to prevent solvent loss through evaporation.

2. If a low number of counts is to be done, sonicate for an additional minute. Alternatively, add 30–50 μL (^{3}H)PIP$_2$, mix, sonicate for 2 min, and verify count number by liquid scintillation.
3. Protein sample concentration can vary. For an enzyme assay, it must first be demonstrated that assay response is linearly proportional to protein concentration. Assay a concentration range (1, 5, 10, 15, and 20 μL protein sample) to identify the range within which this proportionality holds true and choose a concentration in this range.
4. Another key consideration in any enzyme assay is to identify reaction conditions that allow the full working linear range to be used during the standard assay. In the present case, the key conditions are protein concentration, temperature, and time at which the reaction maintains a linear range. Perform the assay at different time intervals (2, 4, 6, 8, and 10 min) and temperatures (26, 30, and 37 °C).

References

1. van der Luit AH, Piatti T, van Doorn A, Musgrave A, Felix G, Boller T, Munnik T (2000) Elicitation of suspension-cultured tomato cells triggers the formation of phosphatidic acid and diacylglycerol pyrophosphate. Plant Physiol 123:1507–1515
2. Fruman DA, Meyers RE, Cantley LC (1998) Phosphoinositide kinases. Annu Rev Biochem 67:481–507
3. Gawer M, Norberg D, Chervin D, Guern N, Yaniv Z, Mazliak P, Kader JC (1999) Phosphoinositides and stress-induced changes in lipid metabolism of tobacco cells. Plant Sci 141:117–127
4. Gross W, Boss W (1993) Inositol phospholipids and signal transduction. In: Verma DPS (ed) Control of plants gene expression. CRC, Boca Raton, FL, pp 17–32
5. Wang X (2004) Lipid signaling. Curr Opin Plant Biol 7:329–336
6. Rhee SG (2001) Regulation of phosphoinositides-specific phospholipase C. Annu Rev Biochem 70:281–312
7. Saunders CM, Larman MG, Parrington J et al (2002) PLCζ: a sperm-specific trigger of Ca^{2+} oscillation in eggs and embryo development. Development 129:3533–3544
8. Harden TK, Sondek J (2006) Regulation of phospholipase C isozymes by Ras superfamily GTPases. Annu Rev Pharmacol Toxicol 76: 355–379
9. Suh PG, Park JI, Manzoli L, Cocco L, Peak JC, Katan M, Fukami K, Kataoka T, Yun S, Ryu SH (2008) Multiple roles of phosphoinositide-specific phospholipase C isozymes. BMB Rep 41:415–434
10. Singer W, Brown A, Sternweis P (1997) Regulation of eukaryotic phosphatidylinositol-specific phospholipase C and phospholipase D. Annu Rev Biochem 66:475–509
11. Stevenson JM, Perera IY, Heilmann I, Persson S, Boss WF (2000) Inositol signaling and plant growth. Trends Plant Sci 5:252–258
12. Meijer HJG, Munnik T (2003) Phospholipids-based signaling in plants. Annu Rev Plant Biol 54:265–306
13. Katan M (1998) Families of phosphoinositide-specific phospholipase C: structure and function. Biochim Biophys Acta 1436:5–17
14. Mueller-Roeber B, Pical C (2002) Inositol phospholipid metabolism in Arabidopsis. Characterized and putative isoforms of inositol phospholipid kinase and phosphoinositide-specific phospholipase C. Plant Physiol 130: 22–46
15. Melin P, Pical C, Jergil B, Sommarin M (1992) Polyphosphoinositides phospholipase C in wheat root plasma membranes. Partial purification and characterization. Biochim Biophys Acta 1123:163–169
16. Yotsushima K, Nakamura K, Mitsui T, Igane I (1992) Purification and characterization of

phosphatidylinositol-specific phospholipase C in suspension-cultured cells of rice (Oryza sativa L.). Biosci Biotechnol Biochem 56:1247–1251

17. Martínez-Estévez M, Racagni-Di PG, Muñoz-Sanchez JA, Brito-Argáez L, Loyola-Vargas VM, Hernandez-Sotomayor SMT (2003) Aluminium differentially modifies lipid metabolism from the phosphoinositide pathway in *Coffea arabica* cells. J Plant Physiol 160: 1297–1303
18. Park SK, Lee JR, Lee SS et al (2002) Partial purification and properties of a phosphatidylinositol 4,5-bisphosphate hydrolyzing phospholipase C from the soluble fraction of soybean sprouts. Mol Cells 13:377–384
19. Huang C-H, Tate BF, Drain RC, Coté GG (1995) Multiple phosphoinositide-specific phospholipase C in oat roots: characterization and partial purification. Plant J 8: 257–267
20. McMurray WC, Irvine RF (1988) Polyphosphatidylinositol 4,5-bisphosphate phosphodiesterase in higher plants. Biochem J 249:877–881
21. Piña-Chable ML, Hernandez-Sotomayor SMT (2001) Phospholipase C activity from *Catharanthus roseus* transformed roots: aluminum effects. Prostaglandins Other Lipid Mediat 56:19–31
22. Cote GG, Crain RC (1993) Biochemistry of phosphoinositides. Annu Rev Plant Physiol Plant Mol Biol 44:333–356
23. Chapman KD (1998) Phospholipase activity during plant growth and development and in response to environmental stress. Trends Plant Sci 3:419–426
24. Munnik T, Irvine RF, Musgrave A (1998) Phospholipid signaling in plants. Biochim Biophys Acta 1389:222–272
25. Kashem MA, Itoh K, Iwabuchi S, Hori H, Mitsui T (2000) Possible involvements of phosphoinositide-Ca^{2+} signaling in the regulation of α-amylase expression and germination of rice seed (*Oryza sativa* L.). Plant Cell Physiol 41:399–407

Chapter 18

Assaying Nonspecific Phospholipase C Activity

Přemysl Pejchar, Günther F.E. Scherer, and Jan Martinec

Abstract

Plant nonspecific phospholipase C (NPC) is a recently described enzyme which plays a role in membrane rearrangement during phosphate starvation. It is also involved in responses of plants to brassinolide, abscisic acid (ABA), elicitors, and salt. The NPC activity is decreased in cells treated with aluminum. In the case of salt stress, the molecular mechanism of NPC action is based on accumulation of diacylglycerol (DAG) by hydrolysis of phospholipids and conversion of DAG, the product of NPC activity, to phosphatidic acid (PA) that participates in ABA signaling pathways.

Here we describe a step-by-step protocol, which can be used to determine in situ or in vitro NPC activity. Determination is based on quantification of fluorescently labeled DAG as a product of cleavage of the fluorescently labeled substrate lipid, phosphatidylcholine. High-performance thin-layer chromatography is used for separation of fluorescent DAG. The spot is visualized with a laser scanner and the relative amounts of fluorescent DAG are quantified using imaging software.

Key words Bodipy, Diacylglycerol, Nonspecific phospholipase C, Phosphatidylcholine, Thin-layer chromatography

1 Introduction

Nonspecific phospholipase C (NPC; also called phosphatidylcholine-hydrolyzing phospholipase C, PC-PLC) catalyzes the hydrolysis of phospholipids to diacylglycerol (DAG) and the polar head group. Rapid down regulation of NPC activity in elicitor-treated plant cells was described by Scherer et al. (1). Molecular characterization of NPC in Arabidopsis was published by Nakamura et al. (2). It was shown that NPC is greatly induced in response to phosphate deprivation (2, 3). NPC plays a role in root development and brassinolide signaling in Arabidopsis (4). Aluminum ions inhibit the formation of DAG generated by NPC activity in tobacco cells (5). Also, the molecular mechanism of Arabidopsis responses to salt stress includes NPC action (6, 7). During the response to salt, DAG as a product of NPC activity is converted by DAG kinase to

Teun Munnik and Ingo Heilmann (eds.), *Plant Lipid Signaling Protocols*, Methods in Molecular Biology, vol. 1009,
DOI 10.1007/978-1-62703-401-2_18, © Springer Science+Business Media, LLC 2013

phosphatidic acid (PA) that participates in signaling events linked to abscisic acid (ABA) (6). The Arabidopsis genome contains six NPC genes (*NPC1–NPC6*).

Determination of NPC activity is based on the use of fluorescently labeled phosphatidylcholine (PC) as a substrate and the quantification of labeled DAG. The fluorescent marker is covalently linked to fatty acid residues. There are several such substrates that differ in fluorescent marker, number of labeled fatty acid residues, and position of labeled fatty acid residues. We have obtained the best results using bodipy-PC labeled on the *sn*-1 position. Note that in this substrate, the bond on *sn*-1 position is not an ester bond but an ether bond. Therefore this substrate is not cleaved by phospholipase A_1 activity. It must be considered when interpreting results that bodipy-PC is an artificial substrate, the properties of which may differ from natural PC. We confirmed our published inhibition of NPC activity in aluminum-treated tobacco cells (5) by using radiolabeled PC (unpublished results); however, results should be understood in the view of this characteristic.

Another complication is the fact that generation and conversion or utilization of DAG are part of lipid metabolism (5, 8). For example, DAG can originate from phospholipase D activity which generates PA and subsequent dephosphorylation of PA by PA phosphatase. Therefore, possible alternative pathways should be considered and checked. As a tool, DAG synthesis inhibitors might be used.

2 Materials

Use gradient-grade organic solvents, analytical grade reagents, and purified deionized water (resistance 18 MΩ at 25 °C) for preparing all aqueous solutions. Strictly follow instructions for disposing waste materials.

2.1 Fluorescently Labeled Substrates

1. Bodipy-phosphatidylcholine (bodipy-PC, D-3771, Invitrogen, Carlsbad, CA, USA; 2-decanoyl-1-(*O*-(11-(4,4-difluoro-5,7-dimethyl-4-bora-3a,4a-diaza-s-indacene-3-proprionyl)amino) undecyl) *sn*-glycero-3-phosphocholine). Prepare stock solution by adding 300 μL of ethanol to 1 mg of bodipy-PC. Store at −20 °C (*see* **Note 1**).

2.2 In Situ NPC Activity Assay

1. Plant material (*see* **Note 2**).
2. Orbital shaker.
3. Tissue culture filter unit: Polypropylene filter flask (Nalgene) connected to the water pump, nylon net filter (pore size 20 μm), and magnetic filter funnel (Pall/Gelman Sciences).

4. Working bodipy-PC solution: Dry out bodipy-PC stock solution under streaming N_2. Dried bodipy-PC disolve with DMSO. Final volume is ten-times higher than volume of used bodipy-PC stock solution (*see* **Note 3**). Store at −20 °C.
5. Lipid extraction solution: Methanol:$CHCl_3$ (2:1, v/v). Store at −20 °C (*see* **Note 4**).
6. 0.1 M KCl.

2.3 In Vitro NPC Activity Assay

1. N_2 cylinder.
2. Water sonicator.
3. Preparation of your protein sample (*see* **Note 5**), including determination of protein content (9).
4. PC stock solution: 7 mM dipalmitoyl-PC dissolved in toluene/ethanol (1:1, v/v), stored at −20 °C.
5. Bodipy-PC stock solution (cf. Subheading 2.1).
6. 10 % (v/v) Triton X-100 (*see* **Note 6**).
7. 100 mM Bis–Tris propane–HCl, pH 6.5.
8. 5 mM $CaCl_2$.
9. Lipid extraction solution: Methanol:$CHCl_3$ (2:1, v/v), stored at −20 °C (*see* **Note 4**).
10. 0.1 M KCl.

2.4 Preparation of Fluorescently Labeled Lipids

1. 50 mM MES–NaOH, pH 6.0.
2. 10 % (v/v) Triton X-100 (*see* **Note 6**).
3. 1 M $CaCl_2$.
4. Bodipy-PC (stock solution, cf. Subheading 2.1).
5. 400 mM SDS.
6. Bacterial PC-PLC (100 U/mL, P 6135, Sigma-Aldrich Co., St. Louis, MO, USA) (*see* **Note 7**).
7. 0.1 M KCl.

2.5 High-Performance Thin-Layer Chromatography

1. Centrifuge for glass tubes and plastic microtubes.
2. Vacuum evaporator.
3. Amber 2 mL glass vials, caps, seals (silicone/PTFE), glass inserts (250 μL) with bottom spring.
4. HP-TLC silica gel-60 plates (20 × 10 cm) (*see* **Note 8**).
5. Automatic sample applicator for TLC plates (e.g., Camag ATS4) with operating software winCATS (Camag) connected to N_2-cylinder (*see* **Note 9**).
6. Horizontal developing chamber for 20 × 10 cm TLC plates (*see* **Note 10**).

7. Mobile-phase solution: $CHCl_3$:methanol:water (65:25:4, v/v/v), stored at 4 °C.
8. Laser scanner, e.g., Fuji FLA-7000 (Fuji) with operating software (*see* **Note 11**).
9. Software for computer-assisted quantification, e.g., Multi Gauge (Fujifilm) (*see* **Note 12**).

3 Methods

Carry out all procedures at room temperature unless specified otherwise. Because you are working with fluorescent substrates, avoid exposure to direct sunlight. When possible, cover microtubes, glass tubes, developing chamber, etc. with aluminum foil or dark cloth.

3.1 In Situ NPC Activity Assay

1. Cultivate plant material (*see* **Note 13**).
2. Calculate the amount of cells and volume of culture medium (*see* **Note 14**) needed for the whole experiment. Use approximately 56 mg of cells for one reaction (final reaction volume is 1 mL).
3. Put magnetic stir bar into 25 mL beaker and position it on magnetic stirrer.
4. Harvest cells and filter them using tissue culture filter unit. Collect filtrate and transfer appropriate volume of filtered culture medium into 25 mL beaker on magnetic stirrer. Weigh required amount of cells and gently resuspend them in filtered culture medium (*see* **Note 15**).
5. Transfer (*see* **Note 16**) 1 mL of cell suspension into glass tube (*see* **Note 17**) and add 2 μL of bodipy-PC working solution (final concentration should be 0.66 μg/mL). Mix gently and incubate on an orbital shaker at 26 °C in darkness for 10 min (*see* **Note 18**).
6. Add your effective compound (*see* **Note 19**) now when assaying NPC activity under treated conditions, mix gently, put the tube back in the same position to the tube holder, and continue with incubation on orbital shaker for 1–2 h (*see* **Note 20**).
7. Add 4 mL of cold extraction solution to cell suspension and mix (*see* **Note 21**). Keep at room temperature for 30 min.
8. Add 2 mL of 0.1 M KCl, mix vigorously, and keep at 4 °C for at least 30 min (*see* **Note 22**).
9. Continue to Subheading 3.4.

3.2 In Vitro NPC Activity Assay

1. Prepare substrate for in vitro assay for the whole experiment. Calculate the number of samples including reserve for pipetting inaccuracy. Substrate for one reaction contains 0.714 μL of

7 mM PC, 0.2 μL bodipy-PC stock solution, 2.5 μL 10 % (v/v) Triton X-100, and 7.5 μL water. Prepare mixture as follows: Add PC solution into polypropylene reaction tube and evaporate under streaming N_2; add bodipy-PC and evaporate under streaming N_2; add Triton X-100 and sonicate for 1 min using water sonicator; add water and sonicate for 1 min using water sonicator.

2. Dilute your samples to appropriate protein concentrations (e.g., 15 μg/20 μL).
3. Add 10 μL of 100 mM Bis–Tris propane–HCl, pH 6.5, and 10 μL of 5 mM $CaCl_2$ to a new polypropylene reaction tube (*see* **Note 21**).
4. Add 20 μL of sample (containing, e.g., 15 μg proteins) and mix.
5. Start reaction by adding 10 μL of the substrate for in vitro assay, mix, and incubate for 30 min at 30 °C.
6. Stop reaction by adding 400 μL of cold extraction solution to the reaction mixture and mix (*see* **Note 22**). Keep at room temperature for 30 min.
7. Add 200 μL of 0.1 M KCl, mix vigorously, and keep at 4 °C for at least 30 min (*see* **Note 23**).
8. Continue to Subheading 3.4 and follow the procedure with the modification of step 3: Transfer 110 μL of lower phase into the new set of polypropylene reaction tubes.

3.3 Preparation of Fluorescently Labeled DAG Standard

1. Prepare reaction mixture for bodipy-DAG preparation by mixing 987 μL of 50 mM MES–NaOH, pH 6, 5 μL of 10 % (v/v) Triton X-100, 5 μL of 1 M $CaCl_2$, and 1 μL of bodipy-PC stock solution. Start reaction by adding 2 μL of bacterial PC-PLC (100 U/mL). Incubate for 30 min at 37 °C.
2. Stop either reaction by adding 4 mL of cold extraction solution to the reaction mixture and mix (*see* **Note 22**). Keep at room temperature for 30 min.
3. Add 2 mL of 0.1 M KCl, mix vigorously, and keep at 4 °C for at least 30 min (*see* **Note 23**).
4. Continue to Subheading 3.4: Follow steps 1 and 2; skip step 3; follow step 4; modify step 5: Redissolve sample in 100 μL of ethanol; modify step 6: Set up sequence to apply whole-sample volume (4 × 28 μL) and wider band length (4 × 36 mm); follow step 7; skip step 8; follow step 9–14; modify step 15: Identify spot corresponding to bodipy-DAG. This spot is major spot in this reaction. Using the described mobile phase, bodipy-DAG will migrate in front of the mobile phase; skip step 16.

5. Visualize fluorescently labeled spots under UV light and mark spots for bodipy-DAG on HP-TLC plate with a pencil. Protect yourself against UV light by wearing suitable eye protection.
6. Scrape out the pencil-marked areas of silica gel and transfer the silica powder into a fresh glass tube.
7. Add 3 mL of methanol and two drops of glacial acetic acid. Cover the tube with parafilm, mix, and incubate overnight at 4 °C.
8. Add 3 mL of $CHCl_3$ and 3 mL of water and mix (*see* **Note 23**).
9. Centrifuge samples at 420 × *g* at 4 °C for 15 min to separate phases.
10. Transfer whole lower phase into fresh polypropylene reaction tubes (*see* **Note 24**).
11. Evaporate samples to dryness using a vacuum evaporator (*see* **Note 25**).
12. Redissolve samples in 50 μL of DMSO store at −20 °C.
13. Use 0.5 μL for spotting on HP-TLC plate as fluorescently labeled standard (cf. Subheading 3.4, step 8).

3.4 Separation of Lipids by HP-TLC

1. Centrifuge samples at 420 × *g* at 4 °C for 15 min.
2. Transfer whole lower phase into fresh polypropylene reaction tube (*see* **Note 26**).
3. Transfer 1 mL of lower phase into a new set of polypropylene reaction tubes (*see* **Note 27**).
4. Evaporate samples to dryness using a vacuum evaporator.
5. Redissolve samples in 50 μL of ethanol. Keep at room temperature for 10 min, mix, and briefly centrifuge to sediment liquid.
6. Set up sample sequence, volume applied (40 μL), band length, starting positions, and other settings according to manufacturer's instruction when using automatic sample applicator. Set up rinsing between different samples (*see* **Note 28** for typical ATS4 settings for 20 samples and for preparation of standard).
7. Transfer sample to the glass insert, put insert into vial, and assemble vial. Teflon part of seal should be positioned towards the insert with sample.
8. Do not forget to spot also the fluorescently labeled lipid standard onto the HP-TLC plate. You can add it to the sequence or spot it manually (only small volumes, 0.5–1 μL) before developing the HP-TLC plate.
9. Place HP-TLC plate and samples to ATS4 sample applicator and run the sequence.
10. Add 20 mL of mobile-phase solution into conditioning tray of horizontal developing chamber. Place HP-TLC plate (silica gel

layer facing down) into chamber and cover it with glass cover plate. Let the HP-TLC plate saturate for 10 min.

11. Add 5 mL of mobile-phase solution into the reservoir for developing solvent. Move glass strip to start developing.
12. Stop developing when the mobile phase will reach the desired developing distance (usually 0.5–1 cm from the opposite end of HP-TLC plate).
13. Dry HP-TLC plate in a fume hood.
14. Scan HP-TLC plate by the laser scanner (*see* **Note 29** for typical FLA-7000 settings).
15. Identify individual spots based on the comparison with fluorescently labeled lipid standard (*see* **Note 30**).
16. Quantify the spot corresponding to bodipy-DAG using Multi Gauge quantification software.

4 Notes

1. Mix prior to experiment and carry out short spin to minimize evaporation of ethanol from the tube walls.
2. The tobacco cell line BY-2 is a frequently used plant model when assaying NPC activity in our laboratory (4, 5). However, we described NPC activity also using tobacco pollen tubes (5) and Arabidopsis seedlings (7). Arabidopsis suspension cultures and leave protoplasts are also routinely used (unpublished results).
3. This step is implemented to minimize pipetting inaccuracy. Final volume of bodipy-PC in one reaction would be 0.2 μL of stock solution. Therefore 2 μL of working solution are preferred. It also helps to minimize evaporation of ethanol and contaminations of stock solution.
4. Use cold extraction solution. This solution is widely used in lipid biology. Enzymatic reactions in cells are stopped and total lipids are extracted. This technique is based on the method developed by Bligh and Dyer (10). Their examination of the $CHCl_3$:methanol:water-phase diagram led to the hypothesis that "optimum lipid extraction should result when the tissue is homogenized with a mixture of $CHCl_3$ and methanol which, when mixed with the water in the tissue, would yield a monophasic solution. The resulting homogenate could then be diluted with water and/or $CHCl_3$ to produce a biphasic system, the $CHCl_3$ layer of which should contain the lipids and the methanol-water layer the nonlipids." The hypothesis was readily confirmed by experimentation and the method was proven to be very effective (11).

5. We used cellular fractions prepared according to (12).
6. Use cutaway tips when pipetting Triton X-100.
7. Prepare aliquots of enzyme solution to avoid protein damage by repeated freezing and thawing. Store at −20 °C.
8. Performing thin-layer chromatographic separation on HP-TLC layers has several advantages over those on conventional TLC: higher resolution of zones due to higher number of theoretical plates, shorter developing times, less solvent consumption, and less background noise due to narrow size distribution of particles.
9. Alternatively, you can use other types of semiautomatic sample applicators (Linomat IV or V, Camag) or spot samples manually. In addition to automatic sampling the next advantage of automatic sample applicators is that samples are sprayed onto the chromatographic layer in the form of narrow bands. This technique allows larger sample volumes to be applied than by contact transfer (manually spotting). While the solvent is almost completely evaporated in the process due to the flow of nitrogen over the spotting area, the sample is concentrated on the layer surface into a narrow band of selectable length.
10. The chamber is suitable for all kinds of solvents.
11. Any type of laser scanner can be used. Alternatively, plates can be scanned under UV light by other convenient documentation systems or using a video camera. In that case protect yourself against UV light by wearing suitable eye protection.
12. The choice of quantification software depends on the documentation system used and the types of files supported for computer-assisted quantification.
13. The subsequent steps of this section describe the use of tobacco BY-2 cells as a plant model. For details of cultivation see ref. (5).
14. Include a reserve of about 5 mL of cell suspension. This volume will cover magnetic stir bar and will help to maintain a homogenous distribution of cells in suspension.
15. Carry out this step as fast as possible. In particular, do not allow the cells to dry. Switch the stirrer on 5 s after contact of cells with medium. Be careful when setting rotation speed. It is necessary to provide homogenous distribution of cells in suspension; the cells must not settle down, neither be damaged by too high rotation speed.
16. Use cutaway tips for pipetting cell suspensions; rinse the tip twice with cell suspension before transferring samples.
17. We use glass tubes (inner diameter 13 mm) with round bottom for approximately 13 mL total volume. Check whether tubes to be used are suitable for centrifugation at $420 \times g$.

18. Place glass tubes into tube holder at an angle to provide sufficient cell mixing. Set shaker rpm to not allow cells to settle down.
19. It is critical to control the concentration of primary alcohol (methanol, ethanol, butanol) in the cell suspension if these alcohols are used as a solvent. Primary alcohols can affect PLD enzymatic activity (13). We do not exceed 0.2 % (v/v) concentration of ethanol.
20. When you need to assay NPC activity at shorter times (e.g., 10–30 min of treatment) double the amount of bodipy-PC in reaction. Prepare bodipy-PC working solution by diluting bodipy-PC stock solution five times. Add 2 μL of more concentrated bodipy-PC working solution into the reaction mix (cf. Subheading 3.1, step 5).
21. From six *Arabidopsis thaliana* NPCs only NPC4 and NPC5 have been cloned and partly characterised biochemically (14). NPC4 were found to be calcium independent. pH and concentration of Ca^{2+} used here are optimum for "tobacco BY-2 plasma membrane NPC activity" (unpublished results).
22. Rinse the tip with extraction solution at least once before adding extraction solution.
23. Two-phase system is formed. The procedure can be interrupted for longer time after this step (e.g., overnight). If so, store the samples at −20 °C.
24. Total volume of lower phase is approximately 3 mL. Transfer 1 mL into each of the three 1.5 mL polypropylene reaction tubes. Be careful to avoid contamination by upper phase.
25. Evaporate samples partially first, keeping approximately 400 μL in each tube. Pool identical samples into one tube and continue evaporating to dryness.
26. This first transfer is a basic one. It may contain limited contamination from the upper phase.
27. Prevent this transfer from contamination by upper phase. Carry out pipetting fast but accurately and reproducibly.
28. Typical ATS4 sampler settings for 20 samples (using winCATS software). Application volume: 40 μL; filling speed: 8 μL/s; dosage speed: 330 nL/s; rinsing cycles: 2; flush nozzle: on; filling cycles: 1; fill only programmed volume: on; vial adjustment: −1.5 mm (the sampler needle has to be moved upper when using glass inserts in vials; warning: wrong adjustment may destroy the needle); syringe size: 100 μL; band length: 3 mm; application mode: spray band; number of tracks: 20; first application position X: 12 mm (do not use sides on HP-TLC plate, at least 1 cm from the edge, for spotting samples); application position Y: 10 mm; and distance between tracks: 9 mm (3 mm

band plus 6 mm gap). Settings for preparation of DAG standard (only differences): application volume: 28 μL; band length: 36 mm; tracks: 4; first application position X: 35 mm; and distance between tracks: 44 mm.

29. Maximal absorption and fluorescence emission spectra for fluorescent label bodipy are 503 and 512 nm, respectively. Typical settings for FLA-7000 (using FLA-7000IR software): use fluorescence mode, laser: 473 nm, filter: O580, PMT: 500, pixel size: 50 μm, latitude: L4, mode: grid, ND filter: on, and scan mode: standard.
30. When assaying samples from plant material containing pigments, e.g., Arabidopsis seedlings, bodipy-DAG can coincide with pigments when developed using described mobile-phase solution. In that case use two-step developing. First, develop HP-TLC plate in mobile phase of acetone:$CHCl_3$ (1:1, v/v). In this mobile phase, only nonpolar lipids will move from start position. Dry plate in a fume hood and scan plate and quantify spots for bodipy-DAG. Then you can develop plate in described mobile-phase solution to detect other products of phospholipase activities.

Acknowledgments

This work was supported by the Czech Science Foundation (grants no. P501/12/1942 and no. P501/12/P950).

References

1. Scherer GFE, Paul RU, Holk A et al (2002) Down-regulation by elicitors of phosphatidylcholine-hydrolyzing phospholipase C and up-regulation of phospholipase A in plant cells. Biochem Biophys Res Commun 293:766–770
2. Nakamura Y, Awai K, Masuda T et al (2005) A novel phosphatidylcholine-hydrolyzing phospholipase C induced by phosphate starvation in *Arabidopsis*. J Biol Chem 280:7469–7476
3. Gaude N, Nakamura Y, Scheible WR et al (2008) Phospholipase C5 (NPC5) is involved in galactolipid accumulation during phosphate limitation in leaves of Arabidopsis. Plant J 56:28–39
4. Wimalasekera R, Pejchar P, Holk A et al (2010) Plant phosphatidylcholine-hydrolyzing phospholipases C NPC3 and NPC4 with roles in root development and brassinolide signalling in *Arabidopsis thaliana*. Mol Plant 3:610–625
5. Pejchar P, Potocký M, Novotná Z et al (2010) Aluminium ions inhibit the formation of diacylglycerol generated by phosphatidylcholine-hydrolysing phospholipase C in tobacco cells. New Phytol 188:150–160
6. Peters C, Li MY, Narasimhan R et al (2010) Nonspecific phospholipase C NPC4 promotes responses to abscisic acid and tolerance to hyperosmotic stress in *Arabidopsis*. Plant Cell 22:2642–2659
7. Kocourková D, Krčková Z, Pejchar P et al (2011) The phosphatidylcholine-hydrolyzing phospholipase C NPC4 plays a role in response of Arabidopsis roots to salt stress. J Exp Bot 62:3753–3763
8. Wang X (2001) Plant phospholipases. Annu Rev Plant Physiol Plant Mol Biol 52: 211–231
9. Bradford MM (1976) A rapid and sensitive method for the quantification of microgram

quantities of protein utilizing the principle of protein-dye binding. Anal Biochem 72: 248–254
10. Bligh EG, Dyer WJ (1959) A rapid method of total lipid extraction and purification. Can J Biochem Physiol 37:911–917
11. Bligh EG (1978) Citation classic—rapid method of total lipid extraction and purification, Commentary on ref. (10). http://www.garfield.library.upenn.edu/classics1978/A1978FZ82000002.pdf
12. Novotná Z, Martinec J, Profotová B et al (2003) In vitro distribution and characterization of membrane-associated PLD and PI-PLC in *Brassica napus*. J Exp Bot 54:691–698
13. Munnik T, Irvine RF, Musgrave A (1998) Phospholipid signalling in plants. Biochim Biophys Acta Lipids Lipid Metab 1389:222–272
14. Pokotylo I, Pejchar P, Potocký M et al (2013) The plant non-specific phospholipase C gene family. Novel competitors in lipid signalling. Prog Lipid Res 52:62–79

Chapter 19

Assaying Different Types of Plant Phospholipase D Activities In Vitro

Kirk L. Pappan and Xuemin Wang

Abstract

Over the past decade, tremendous progress has been made toward understanding the physiological functions of individual members of the diverse phospholipase D (PLD) family of enzymes in plants. For instance, the involvement of plant PLD members has been shown or suggested in a wide variety of the cellular and physiological processes such as regulating stomatal opening and closure; signaling plant responses to drought, salt, and other abiotic and biotic stresses; organizing microtubule and actin cytoskeletal structures; promoting pollen tube growth; cycling phosphorus; signaling nitrogen availability; regulating *N*-acylethanolamine stress signaling; and remodeling membrane phospholipids in plant responses to phosphate deprivation and during and after freezing. There are at least a dozen PLDs in *Arabidopsis* that can be separated into six classes, phospholipases Dα, Dβ, Dγ, Dδ, Dε, and Dζ, based on their molecular and enzymatic characteristics. Several of the classes have distinguishing enzymatic properties that can be used to discriminate among the various classes. Here we provide four variations of in vitro PLD activity assays using choline-labeled phosphatidylcholine to distinguish, to the extent possible, among the different PLD classes.

Key words *Arabidopsis*, Plant, Phospholipase D, PLD, Classes, Isoforms, In vitro assay, Activity assay, Protocol, Method

1 Introduction

Phospholipase D (PLD) catalyzes the hydrolysis of glycerophospholipids into phosphatidic acid (PA) and a free headgroup. For instance, PLD's hydrolysis of phosphatidylcholine (PC) generates PA and choline. There are currently 12 PLD genes known in *Arabidopsis* (1) and similar or greater diversity exists in many plants including grape and poplar (2), rice (3), and poppy (4). The complexity of the PLD family has necessitated the use of functional genomics and gene silencing techniques, in addition to biochemical assay methods, to elucidate individual PLD functions. Even with genetic techniques and the availability of isoform-specific antibodies, measurement of PLD activity remains a major approach for detecting and identifying PLD in plants. Within certain limitations,

Teun Munnik and Ingo Heilmann (eds.), *Plant Lipid Signaling Protocols*, Methods in Molecular Biology, vol. 1009,
DOI 10.1007/978-1-62703-401-2_19,

the distinct cofactor and activating conditions of different classes of PLDs offer a way to distinguish PLD classes from one another. The goal of this report is to describe key features of the major PLD classes, based on our current understanding of *Arabidopsis*, and to describe assay methods pertinent to these classes.

Phylogenic analysis places the 12 *Arabidopsis* PLDs into six classes of variable size (1), and, for the most part, these classes differ from each other by catalytic properties (see Table 1). For instance, classes vary by their dependence on polyphosphoinositides (PPI), narrowly defined here as phosphatidylinositol-4, 5-bisphosphate (PIP_2) or phosphoinositol-4-phosphate (PIP), calcium, oleic acid, and phospholipid substrates. The properties of the *Arabidopsis* PLD classes that are particularly relevant to measuring their activity by in vitro activity assay are briefly reviewed below.

PLDα: PLDα family members are involved in maintaining proper water balance, including responding to drought (5), salt (6), and freezing stresses (7), and in responding to pathogens (8, 9). The three members of PLDα comprise the iconic class of PLD that is active at 20–100 mM Ca^{2+} without detergent or down to 5 mM Ca^{2+} with detergents. At millimolar calcium, PLDα members hydrolyze PC, phosphatidylethanolamine (PE), and phosphatidylglycerol (PG) without requiring mixed phospholipid vesicles or PPI. In addition to these iconic activating conditions, PLDα members have recognized activity at more physiologically relevant conditions (10). At a moderately acidic pH, PLDα can hydrolyze PC in the presence of mixed lipid vesicles containing PE, PPI, and calcium as low as 50 μM. PLDα members have calcium-binding C2 domains but lack the full complement of acidic amino acid residues believed to coordinate Ca^{2+} binding.

PLDβ and PLDγ: Little is known about the functions of PLDβ and PLDγ family members but both classes have been implicated in early responses to pathogens and PLDγ may have a role in signaling leading to hypersensitive responses (9). These two PLD classes may also be involved in regulating PLDα since both efficiently utilize *N*-acylphosphatidylethanolamine (NAPE), generating phosphatidic acid and *N*-acylethanolamine (11), the latter of which is a potent PLDα inhibitor (12). Real-time PCR profiling indicated that *PLDγ1* is predominately expressed in roots whereas *PLDγ2* and *γ3* expression was strong in inflorescence stems but nearly absent in roots (13). Members of these two PLD classes were the first non-PLDα-class enzymes described in plants and are characterized by activity toward a broad range of phospholipid substrates in the presence of PPI, PE, and micromolar calcium (14, 15). The C2 calcium-binding domain and Lys/Arg-rich PIP_2-binding motifs of PLDβ and PLDγ members are prototypical in structure and may explain their activity at low calcium concentrations and PPI-dependence (16). In practice, it is not possible to design in vitro assays that discriminate PLDβ from PLDγ, but it is possible to distinguish them from other classes.

Table 1
Signature enzymatic properties and molecular features of *Arabidopsis* PLD classes

Arabidopsis PLD classes (members)	Known cofactor(s)	Known substrate(s)	Distinguishing molecular features	References
PLDα (1,2,3)	Ca^{2+} (mM) and detergent (pH 6.5) or Ca^{2+} (μM) and PIP_2 (pH 5.0)	PC, PE, PG	C2 domain with fewer acidic residues; one K/R-rich motif	(10, 11, 16)
PLDβ (1,2)	Ca^{2+} (μM), PIP_2 or PIP, and PE	PC, PE, PG, NAPE > PS	C2 domain; two K/R-rich motifs	(11, 15)
PLDγ (1,2,3)	Ca^{2+} (μM), PIP_2 or PIP, and PE	PE, NAPE, PG > PC, PS	C2 domain; two K/R-rich motifs	(11, 16)
PLDδ	Oleic acid, Ca^{2+}	PE > PC	One K/R-rich motif; R-399; C2 domain	(22, 23)
PLDε	Cofactors similar to those of PLDα = PLDδ > PLDβ	PC > PE > PG	Altered C2 domain lacking acidic residues; partial K/R-rich motif	(1, 24)
PLDζ (1,2)	PIP_2	PC	PH and PX domains; K/R-rich motif	(1)

PLDδ: PLDδ increases freezing tolerance (7, 17, 18), regulates cytoskeletal organization (19), helps plants cope with drought stress (20), and improves plant stress tolerance by dampening H_2O_2-induced apoptosis (21). This enzyme is optimally active in the presence of its substrate, PC, mixed with 0.5 mM oleic acid and 100 μM Ca^{2+}, whereas PLDα, PLDβ, and PLDγ1 are inactive under these conditions (22). Polyunsaturated fatty acids, linoleic acid and linolenic acid, as well as PIP_2 stimulate PLDδ about half as well as monounsaturated oleic acid (22) and the enzyme prefers PE to PC as a substrate (23).

PLDε: Prior to its biochemical characterization, PLDε was tentatively designated as PLDα4 based on its shared gene architecture with PLDα1–3 (1) but has recently been reclassified as PLDε due to its unique biochemical and sequence properties (24). Of all the PLD classes in *Arabidopsis*, PLDε has the most permissive activation conditions, showing activity under conditions appropriate for detecting PLDα (50 mM Ca^{2+}, SDS, and PC), PLDδ (100 μM Ca^{2+}, 0.6 mM oleate, and PC), and PLDβ/γ (PE/PIP_2/PC and 50 μM Ca^{2+}). However, PLDε has a clear calcium requirement and shows no PC hydrolysis activity using calcium-free PLDζ assay conditions (PC and PIP_2; no calcium). The broad calcium-dependent activity of this enzyme could stem from alterations in both its C2 calcium-binding domain and Lys/Arg-rich motif (1). Over-expression of PLDε leads to increased nitrate uptake, biomass accumulation, and root elongation whereas knockouts of PLDε display stunted growth and lessened root elongation (24).

PLDζ: PLDζ is distinct among the PLD enzymes in *Arabidopsis* by virtue of its calcium-independent activity and the presence of Phox homology (PH) and pleckstrin homology (PX) domains in lieu of the calcium-binding C2 domain found in the other plant PLD classes (1). The PH and PX domains are common to mammalian PLDs but are absent in the other *Arabidopsis* PLD classes. Some PX domains can bind PIP_2 and SH3 adaptor proteins and PH domains appear to bind PPI. Although a role in membrane targeting can be speculated, the significance of these domains in PLDζ remains to be demonstrated. PLDζ requires PIP_2, but not calcium, for activity. In fact, 1 mM Ca^{2+} inhibits nearly 75 % of PLDζ activity and 10 mM Ca^{2+} completely inhibits it, which offers an interesting contrast to PLDα's preference for millimolar calcium. PLDζs are involved in conserving and recycling phosphorus under phosphorus-limited growth conditions by initiating the remodeling pathway that converts phospholipids into galactolipids (25, 26). They also play a role in vesicular trafficking and auxin response (27).

2 Materials

2.1 Lipids and Other Materials

PC (Sigma-Aldrich, Cat. No. P3556; average MW 768 g/mol), PE (Avanti Polar Lipids, Cat. No. 840021; average MW 747 g/mol), and PIP_2 (Avanti Polar Lipids, Cat. No. 840046; average MW 1,098 g/mol) can be obtained from the indicated sources. Dipalmitoylglycero-3-*P*-[*methyl*-^{3}H]choline is available from Perkin Elmer. Other chemicals are available from vendors such as Fisher Scientific or Sigma-Aldrich. All solutions should be prepared with laboratory-grade deionized water except for lipid stock solutions that are prepared in solvents as indicated. Tritium should be used in a dedicated radioactivity area and following each use, benches, pipettes, centrifuges, and other equipment should be cleaned with alcohol or Count Off decontamination solution and the area and equipment monitored for radioactivity by swab testing.

2.2 Lipid Stock Solutions

1. Phosphatidylcholine stock solution: 100 mg/ml and 10 mg/ml 1,2-diacyl-*sn*-glycero-3-phosphocholine (i.e., egg yolk PC) in chloroform (*see* **Note 1**).
2. Phosphatidylethanolamine stock solution: 25 mg/ml L-α-phosphatidylethanolamine (i.e., egg PE) in chloroform to a stock (*see* **Note 2**).
3. PIP_2 stock solution: 5 mg/ml L-α-phosphatidylinositol 4, 5-bisphosphate (i.e., porcine brain) in 20:9:1 chloroform: methanol:water (*see* **Note 3**).
4. Oleate stock solution: 15 mg/ml oleic acid in chloroform.

2.3 Assay Reaction Buffers and Other Components

1. PLDα reaction buffer (A buffer; 10×; see Subheading 3.1 below): 1 M 4-morpholineethanesulfonic acid (MES; pH 6.5) and 0.25 M $CaCl_2$.
2. PLDβ/γ/δ reaction buffer (BGD buffer; 10×; see Subheadings 3.2 and 3.3 below): 1 M Tris(hydroxymethyl) aminomethane hydrochloride (Tris–HCl, pH 7.0), 1 mM $CaCl_2$, 5 mM $MgCl_2$, and 0.8 M KCl.
3. PLDζ reaction buffer (Z buffer; 10×; see Subheading 3.4 below): 1 M Tris(hydroxymethyl) aminomethane hydrochloride (Tris–HCl, pH 7.0) and 0.8 M KCl.
4. 5 mM sodium dodecyl sulfate (SDS).

2.4 Phospholipid Substrate Emulsions

1. PLDα substrate (10×; see Subheading 3.1 below)—For 1 ml of substrate (enough for ~100 reactions), mix 150 μl of 100 mg/ml egg yolk PC with 2 μl of 1 μCi/μl dipalmitoylglycero-3-*P*-[*methyl*-^{3}H]choline in a 1.5 ml microcentrifuge tube (*see* **Note 4**). An aliquot of this 20 mM stock PC substrate will be used to obtain a final working concentration of 2 mM PC in PLD reactions.

2. PLDβ/γ substrate (10×; *see* Subheading 3.2 below)—For 1 ml of substrate (enough for ~100 reactions), mix 104 μl of 25 mg/ml PE, 68 μl 5 mg/ml PIP_2, 16.5 μl 10 mg/ml PC, and 2 μl of 1 μCi/μl dipalmitoylglycero-3-*P*-[*methyl*-^{3}H]choline in a 1.5 ml microcentrifuge tube (*see* **Note 4**). The total phospholipid concentration of this stock PE (87.5 mol %), PIP_2 (7.5 mol %), and PC (5 mol %) substrate is 4 mM and a final concentration of 0.4 mM is used in PLD reactions.
3. PLDδ substrate (10×; *see* Subheading 3.3 below)—For 1 ml of substrate (enough for ~100 reactions), mix 115 μl of 10 mg/ml egg yolk PC, 113 μl 15 mg/ml oleic acid, and 2 μl of 1 μCi/μl dipalmitoylglycero-3-*P*-[*methyl*-^{3}H]choline in a 1.5 ml microcentrifuge tube (*see* **Note 4**). An aliquot of this stock 1.5 mM PC and 6 mM oleate substrate will be used to obtain a final working concentration of 0.15 mM PC and 0.6 mM oleate in PLD reactions.
4. PLDζ substrate (10×; *see* Subheading 3.4 below)—For 1 ml of substrate (enough for ~100 reactions), mix 57 μl 5 mg/ml PIP_2, 29 μl 100 mg/ml PC, and 2 μl of 1 μCi/μl dipalmitoylglycero-3-*P*-[*methyl*-^{3}H]choline in a 1.5 ml microcentrifuge tube (*see* **Note 4**). The total phospholipid concentration of this stock (93.5 mol %) PC and (6.5 mol %) PIP_2 is 4 mM and a final concentration of 0.4 mM is used in PLD reactions.

2.5 PLD Protein Source

Plant PLD can be prepared from plant tissues or recombinant protein expressed in *Escherichia coli* in varying degrees of purity ranging from crude preparations to highly purified forms. It is important to note that different PLDs are associated with different subcellular fractions (22, 28). For example, PLDα is associated with both soluble and membrane-particulate fractions and its allocation between soluble and membrane-associated fractions changes during leaf development and in response to stress, such as wounding (29, 30). PLDβ, γ, δ, ε, and ζ are all primarily associated with membrane, but differ some in respect to specific membranes and the strengths of their membrane associations. PLDγs are associated mostly with intracellular membranes, PLDε is predominantly associated with the plasma membrane (14, 16), whereas PLDδ is exclusively associated with the plasma membrane (22). PLDζ2 is found in tonoplasts whereas PLDζ1 is likely to be associated with the plasma membrane. A major portion of PLDα, β, and γ can be dislodged from membrane fractions by salt treatment of the membranes (22). On the other hand, solubilization of PLDδ requires detergent. Readers are referred to other reports for detailed information on purification or fractionation of PLDα (27), PLDδ (22), and PLDε (24) as well as recombinant expression and purification of PLDα, PLDβ, and PLDγ (13, 16); PLDδ (22); and PLDζ (1).

The main focus of this chapter is to describe assays capable of discriminating among the classes of plant PLD, so given below is a brief general protein extraction procedure for preparing crude-soluble and peripheral membrane PLD fractions (14). It is useful to note also that *E. coli* contain no detectable PLD activities. Given the lack of background PLD activity, the activities of various recombinant PLDs expressed in *E. coli* can be detected directly using cell lysates without any purification.

1. Grind leaf or other plant tissue in a chilled mortar and pestle at a ratio of one part tissue (mg) to three parts of a homogenization buffer (ml) containing 50 mM Tris–HCl (pH 7.5), 10 mM KCl, 1 mM EDTA, 0.5 mM phenylmethylsulfonyl fluoride, and 2 mM dithiothreitol at 4 °C.
2. Centrifuge the homogenate at 2,000 × *g* for 10 min at 4 °C to remove cellular debris. Keep supernatants.
3. Centrifuge the supernatant at 100,000 × *g* for 45 min at 4 °C. Remove the supernatant containing soluble proteins. The pellet contains microsomal proteins that can be further fractionated by salt extraction.
4. Resuspend the pellet in salt-extraction buffer consisting of 0.44 M KCl in the homogenization buffer. Incubate for 1 h at 4 °C and then collect the salt-insoluble materials by centrifugation at 100,000 × *g* for 45 min.
5. Remove the supernatant containing the peripherally membrane-associated proteins.
6. Measure the protein content in the soluble and membrane-associated protein fractions by the Bradford or other suitable protein quantification method.
7. Activity assays can be performed using 1–20 μg per reaction. Fractions can be used directly or stored in aliquots at −80 °C.

2.6 Reaction Stop and Extraction Solutions

1. 2:1 (v/v) chloroform:methanol.
2. 2 M KCl.

3 Methods

Described below are four basic in vitro assay methods that can be used to measure the known classes of PLD in *Arabidopsis*. Each assay has been given a descriptive name and lists, to the extent possible, which classes are active and/or inactive under the specific assay conditions. The four methods use PC with a radiolabeled choline headgroup for relatively rapid and moderately high-throughput screening.

3.1 Iconic High Ca^{2+}-Dependent PLD Assay

This assay can be used to detect PLDα and PLDε enzyme classes. Other classes have no detectable activity under these conditions (i.e., millimolar calcium).

1. A single reaction contains the following components.

1× (μl)	Reaction component
10	10× A buffer
10	10× PLDα substrate
10	5 mM SDS
X	PLD protein source
Y	H_2O
100	Total reaction volume

2. Determine the number of reactions (factoring in replicates and including blanks and controls) and prepare a master mix, omitting the protein source, in slight excess of that number (*see* **Note 5**).
3. Transfer the appropriate volume of master mix, to individual 1.5 ml microcentrifuge reaction tubes, such that the final reaction volume will be 100 μl when the protein sample is added (*see* **Note 6**).
4. Initiate the reaction by adding the PLD protein source, mixing with a brief vortex pulse, and incubate in a shaking water bath for 30 min at 30 °C.
5. Stop the reaction by adding 1 ml of 2:1 (v/v) chloroform:methanol and 100 μl of 2 M KCl.
6. Cap tubes tightly, vortex for ~10 s, and centrifuge at 12,000 ×*g* for 5 min.
7. Remove a 200 μl aliquot of the aqueous upper phase, mix with 3 ml of scintillation fluid, firmly cap the scintillation vial, vortex vigorously, and measure the release of [^{3}H]choline by liquid scintillation counting.

3.2 PIP$_2$- and Ca^{2+}-Dependent PLD Assay

PLDβ, PLDγ, and PLDζ classes exhibit robust activity under the described reaction conditions below, but if Ca^{2+} is omitted, only PLDζ will retain its activity. PLDα has minimal activity under these conditions (i.e., pH 7) but has considerable activity against this mixed phospholipid substrate at pH 5. PLDδ and PLDε have limited activity representing about 25 % of their maximal activity under their preferred assay conditions.

1. A single reaction contains the following components.

1× (μl)	Reaction component
10	10× BGD buffer
10	10× PLDβ/γ substrate
X	PLD protein source
Y	H_2O
100	Total reaction volume

2. Prepare a master mix for the appropriate number of reactions using the recipe above (*see* **Note 5**). The remainder of the protocol is the same as steps 3–7 from Subheading 3.1.

3.3 Oleate-Dependent PLD Assay

This assay can be used to detect PLDδ and PLDε classes. Other classes have no activity under these conditions (i.e., micromolar calcium, oleate, and no PIP_2).

1. A single reaction contains the following components.

1× (μl)	Reaction component
10	10× BGD buffer
10	10× PLDδ substrate
X	PLD protein source
Y	H_2O
100	Total reaction volume

2. Prepare a master mix for the appropriate number of reactions using the recipe above (*see* **Note 5**). The remainder of the protocol is the same as steps 3–7 from Subheading 3.1.

3.4 Ca^{2+}-Independent PLD Assay

This assay can be used to detect PLDζ enzymes. Other classes have no activity under these conditions (i.e., no calcium; *see* **Note 7**). EGTA (2 mM) may be added to reduce Ca^{2+} contamination in the assay.

1. A single reaction contains the following components.

1× (μl)	Reaction component
10	10× Z buffer
10	10× PLDζ substrate
X	PLD protein source
Y	H_2O
100	Total reaction volume

2. Prepare a master mix for the appropriate number of reactions using the recipe above (*see* **Note 5**). The remainder of the protocol is the same as steps 3–7 from Subheading 3.1.

3.5 Calculations

1. One microcurie (μCi) is equivalent to 2.22×10^6 dpm. Each reaction contains 0.02 μCi, so the maximum potential release of [^{3}H]choline is $0.02\ \mu Ci \times 2.22 \times 10^6\ dpm/\mu Ci = 44{,}400$ dp m of radioactivity. A phosphatidylcholine conversion factor is calculated to convert cpm to dpm and to account for the total

volume of aqueous phase following the two-phase partitioning with chloroform and methanol (*see* **Note 8**):

$$\text{PC conversion factor} = (X\,\text{cpm} \times \text{counting efficiency} \times \text{volume correction})/44,400\ \text{dpm} = (X\,\text{cpm} \times 2\ \text{dpm/cpm} \times 2.67)/44,400\ \text{dpm}.$$

2. PLD-specific activity (nmol/mg protein/min) is calculated by multiplying the PC conversion factor by the amount of non-radiolabeled PC (in nanomoles) and dividing by the amount of protein added (in milligrams) and duration of the assay (30 min).
3. The amount of PC per reaction, which differs in each of the four assay methods due to the differences in substrate composition, is 200, 2, 15, and 37 nmol for the assays described in Subheadings 3.1–3.4, respectively.
4. Example: If a sample gave a measurement of 4,000 cpm, its PC conversion factor would be

$$(4,000\ \text{cpm} \times 2\ \text{dpm/cpm} \times 2.67)/44,400\ \text{dpm} = 0.481.$$

If this result came from the oleate-dependent PLD assay (see Subheading 3.3) using 10 μg of protein, its specific activity would be

$$\text{Oleate-dependent PLD}(\text{nmol/mg/min}) = (\text{PC conversion factor})(\text{nmol PC})/(\text{mg protein})(30\,\text{min}) = (0.481)(15\ \text{nmol})/(0.01\ \text{mg})(30\,\text{min}) = 24\ \text{nmol/mg/min}.$$

4 Notes

1. It is convenient to purchase 100 mg of lyophilized egg yolk PC, add 1 ml of chloroform, remove an aliquot for substrate preparation, and store the rest at –20 °C in the original brown glass vial. To prevent oxidation of lipids during storage, wrap the rubber septum with parafilm and purge the headspace with N_2 delivered through a syringe needle with a second needle allowing purged air to escape the vial.
2. It is convenient to purchase 25 mg of egg PE powder, add 1 ml of chloroform, remove an aliquot for substrate preparation, and store the rest at –20 °C after purging with N_2.
3. It is convenient to purchase 5 mg of brain PIP_2 powder, add 1 ml of 20:9:1 chloroform:methanol:water, remove an aliquot for substrate preparation, and store the rest at –20 °C after purging with N_2.
4. Thoroughly remove the solvent under a gentle stream of N_2. After the solvent is completely evaporated, add 1 ml of deionized water. Sonicate just prior to use to emulsify the substrate

as follows: using a low-to-moderate-intensity sonicator setting, pulse (~1 s/pulse) the solution with the probe tip immersed in the liquid for 10 s followed by 10 s of resting on wet ice. Repeat this cycle until the milky/cloudy consistency becomes clear and no visible remnants of lipids appear stuck to the tube wall. Avoid frothing the solution during sonication (it gets easier with practice, but using a lower power setting and more cycles helps). Unused substrate can be stored at –20 °C but should be completely thawed and then sonicated before each use and a control without enzyme should be used to ensure that background degradation is low.

5. The volume of protein sample (denoted as X) can be varied to provide flexibility. We typically run triplicates of all experimental, control, and blank conditions. For the purpose of illustration, it is assumed that a protein sample volume of 15 μl will be used in all reactions. The volume of water needed per reaction (denoted as Y) in this example is therefore = 100 – 10 – 10 – 10 – 15 = 55. If 30 reactions are needed to account for all samples, blanks, and controls, it is advisable to prepare a master reaction mix for ~35 reactions by mixing the following components:

1× (μl)	35× (μl)	Reaction component
10	350	10× A buffer
10	350	10× PLDα substrate
10	350	5 mM SDS
15	–	PLD protein source
55	1,925	H_2O
100	2,975	

In this example, 85 μl (2,975 μl/35 reactions = 85 μl/reaction) of master mix should be transferred into individual reaction tubes. Assays for the other PLD classes are scaled up in an analogous fashion using the master mix recipes specified.

6. In practice, we choose to make a master mix omitting both the substrate and protein. After transferring the master mix and appropriate PLD protein samples to individual tubes, the reaction is initiated by adding the substrate. This approach works better when handling a large number of samples; allows for timely, synchronized assay initiation; and limits the chances for inadvertently mixing up samples.

7. To remove any residual calcium present in plant protein samples, one can supplement the PLDζ reaction mix with 2 mM EGTA to eliminate the possibility of background calcium-dependent PLD activity. In practice, we observe negligible levels of Ca^{2+}-dependent PLD when calcium is omitted from the reaction mix.

8. Tritium's (^{3}H) low energy makes its detection about 50 % efficient in most liquid scintillation counters, so a factor, 2 dpm/cpm, compensates for the difference between measured (cpm) and actual (dpm) radioactive decay. The aqueous volume comprises 100 μl of the reaction, 333 μl or 1/3 of 1 ml of 2:1 chloroform:methanol, and 100 μl of 2 M KCl for a total aqueous volume of 533 μl. A 200 μl aliquot is used for scintillation counting; thus a factor of 533/200 or 2.67 is needed to adjust for the radioactivity present in the total aqueous volume.

Acknowledgments

The authors thank all members and collaborators, past and present, of the Wang lab. This work was supported in part by grants from the National Science Foundation (IOS-0818740) and the US Department of Agriculture (2007-35318-18393).

References

1. Qin C, Wang X (2002) The Arabidopsis phospholipase D family. Characterization of a calcium-independent and phosphatidylcholine-selective PLDζ1 with distinct regulatory domains. Plant Physiol 128:1057–1068
2. Liu Q, Zhang C, Yang Y et al (2010) Genome-wide and molecular evolution analyses of the phospholipase D gene family in poplar and grape. BMC Plant Biol 10:117
3. Li G, Lin F, Xue H-W (2007) Genome-wide analysis of the phospholipase D family in *Oryza sativa* and functional characterization of PLDβ1 in seed germination. Cell Res 17:881–894
4. Oblozinsky M, Bezakova L, Mansfeld J et al (2011) The transphosphatidylation potential of a membrane-bound phospholipase D from poppy seedlings. Phytochemistry 72:160–165
5. Sang Y, Zheng S, Li W et al (2001) Regulation of plant water loss by manipulating the expression of phospholipase Dα. Plant J 28:135–144
6. Hong Y, Pan X, Welti R et al (2008) Phospholipase Dα3 is involved in the hyperosmotic response in Arabidopsis. Plant Cell 20:803–816
7. Li W, Wang R, Li M et al (2008) Differential degradation of extraplastidic and plastidic lipids during freezing and post-freezing recovery in *Arabidopsis thaliana*. J Biol Chem 283:461–468
8. Laxalt AM, ter Riet B, Verdonk JC et al (2001) Characterization of five tomato phospholipase D cDNAs: rapid and specific expression of *LePLDβ1* on elicitation with xylanase. Plant J 26:237–247
9. Zabela M, Fernandez-Delmond I, Niittyla T et al (2002) Differential expression of genes encoding *Arabidopsis* phospholipases after challenge with virulent or avirulent *Pseudomonas* isolates. Mol Plant Microbe Interact 15:808–816
10. Pappan K, Wang X (1999) Plant phospholipase Dα is an acidic phospholipase active at near-physiological Ca^{2+} concentrations. Arch Biochem Biophys 368:347–353
11. Pappan K, Austin-Brown S, Chapman K et al (1998) Substrate selectivities and lipid modulation of plant phospholipase Dα, -β, and -γ. Arch Biochem Biophys 353:131–140
12. Austin-Brown SL, Chapman KD (2002) Inhibition of phospholipase Dα by *N*-acylethanolamines. Plant Physiol 129:1892–1898
13. Qin C, Li M, Qin W et al (2006) Expression and characterization of Arabidopsis phospholipase Dγ2. Biochim Biophys Acta 1761: 1450–1458
14. Pappan K, Zheng S, Wang X (1997) Identification of a novel plant phospholipase D that requires polyphosphoinositides and submicromolar calcium for activity in *Arabidopsis*. J Biol Chem 272:7048–7054
15. Pappan K, Qin W, Dyer JH et al (1997) Molecular cloning and functional analysis of polyphosphoinositide-dependent phospholipase D, PLDβ, from *Arabidopsis*. J Biol Chem 272:7055–7061
16. Qin W, Pappan K, Wang X (1997) Molecular heterogeneity of phospholipase D (PLD). Cloning of PLDγ and regulation of plant

PLDγ, -β, and -α by polyphosphoinositides and calcium. J Biol Chem 272:28267–28273

17. Li W, Li M, Zhang W et al (2004) The plasma membrane-bound phospholipase Dδ enhances freezing tolerance in *Arabidopsis thaliana*. Nat Biotechnol 22:427–433
18. Chen Q-F, Xiao S, Chye M-L (2008) Overexpression of the Arabidopsis 10-kilodalton acyl-coenzyme A-binding protein ACBP6 enhances freezing tolerance. Plant Physiol 148:304–315
19. Gardiner JC, Harper JDI, Weerakoon ND et al (2001) A 90-kD phospholipase D from tobacco binds to microtubules and the plasma membrane. Plant Cell 13:2143–2158
20. Katagiri T, Takahashi S, Shinozaki K (2001) Involvement of a novel Arabidopsis phospholipase D, AtPLDδ, in dehydration-inducible accumulation of phosphatidic acid in stress signaling. Plant J 26:595–605
21. Zhang W, Wang C, Qin C et al (2003) The oleate-stimulated phospholipase D, PLDδ, and phosphatidic acid decrease H_2O_2-induced cell death in Arabidopsis. Plant Cell 15:2285–2295
22. Wang C, Wang X (2001) A novel phospholipase D of Arabidopsis that is activated by oleic acid and associated with the plasma membrane. Plant Physiol 127:1102–1112
23. Qin C, Wang C, Wang X (2002) Kinetic analysis of Arabidopsis phospholipase Dδ. Substrate preference and mechanism of activation by Ca^{2+} and phosphatidylinositol 4,5-bisphosphate. J Biol Chem 277:49685–49690
24. Hong Y, Devaiah SP, Bahn SC et al (2009) Phospholipase Dε and phosphatidic acid enhance Arabidopsis nitrogen signaling and growth. Plant J 58:376–387
25. Li M, Welti R, Wang X (2006) Quantitative profiling of Arabidopsis polar glycerolipids in response to phosphorus starvation. Roles of phospholipases Dζ1 and Dζ2 in phosphatidylcholine hydrolysis and digalactosyldiacylglycerol accumulation in phosphorus-starved plants. Plant Physiol 142:750–761
26. Cruz-Ramirez A, Oropeza-Aburto A, Razo-Hernandez F et al (2006) Phospholipase Dζ2 plays an important role in extraplastidic galactolipid biosynthesis and phosphate recycling in *Arabidopsis* roots. Proc Natl Acad Sci USA 103:6765–6770
27. Li G, Xue HW (2007) Arabidopsis PLDzeta2 regulates vesicle trafficking and is required for auxin response. Plant Cell 19:281–295
28. Fan L, Zheng S, Cui D, Wang X (1999) Subcellular distribution and tissue expression of phospholipase Dα, Dβ, and Dγ in Arabidopsis. Plant Physiol 119:1371–1378
29. Ryu SB, Wang X (1996) Activation of phospholipase D and the possible mechanism of activation in wound-induced lipid hydrolysis in castor bean leaves. Biochim Biophys Acta 1303:243–250
30. Wang X, Dyer JH, Zheng L (1993) Purification and immunological analysis of phospholipase D from castor bean endosperm. Arch Biochem Biophys 306:486–494

Chapter 20

Measuring PLD Activity In Vivo

Teun Munnik and Ana M. Laxalt

Abstract

Phospholipase D (PLD) hydrolyzes structural phospholipids like phosphatidylcholine (PC) and phosphatidylethanolamine (PE) into phosphatidic acid (PA) and free choline/ethanolamine. In plants, this activity can be stimulated by a wide variety of biotic and abiotic stresses (Li et al., Biochim Biophys Acta 1791:927–935, 2009; Testerink and Munnik, J Exp Bot 62(7):2349–2361, 2011). This chapter describes a protocol for the measurement of PLD activity in vivo. The protocol takes advantage of a unique property of PLD, i.e., its ability to substitute a primary alcohol, such as 1-butanol, for water in the hydrolytic reaction. This transphosphatidylation reaction results in the formation of phosphatidylbutanol (PBut), which is a specific and unique reporter for PLD activity. The assay is highly sensitive for detecting PLD activity in vivo, following stimulation of intact plant cells, seedlings, and tissues, being a valuable method for studying the regulation of plant PLD activity in vivo.

Key words Phospholipase D, Transphosphatidylation, Phosphatidic acid, Thin layer chromatography, Isotopic radiolabeling

1 Introduction

Phospholipase D (PLD) hydrolyzes structural phospholipids like phosphatidylcholine (PC) and phosphatidylethanolamine (PE) to produce phosphatidic acid (PA). The latter, is emerging as an important plant lipid second messenger, which is produced in response to various biotic and abiotic stresses (1, 2). In vitro, PLD activity can easily be measured by following the hydrolysis of a labeled substrate or by following the production of PA, as described in detail by Papan and Wang in Chapter 20 (3). In vivo, this is more complex as PA is also synthesized de novo as precursor for glycerolipids (phospholipids and galactolipids) and through phosphorylation of PLC-generated diacylglycerol (DAG) by diacylglycerol kinase (DGK) (2, 4). Thus, PLD activity cannot simply be measured by determining total PA levels.

Earlier studies on its in vitro activity established that PLD has a unique ability to transfer the phosphatidyl moiety of a phospholipid to a primary alcohol rather than water, producing phosphatidylalcohol

Teun Munnik and Ingo Heilmann (eds.), *Plant Lipid Signaling Protocols*, Methods in Molecular Biology, vol. 1009,
DOI 10.1007/978-1-62703-401-2_20, © Springer Science+Business Media, LLC 2013

rather than PA (5, 6). Later, this transphosphatidylation reaction was also found to occur in vivo and has subsequently been used to develop an assay measuring PLD activity in vivo, in mammalian cells (7) and in plants (8–13).

Here, a protocol for measuring in vivo PLD activity in suspension-cultured tobacco BY-2 cells, Arabidopsis seedlings, Arabidopsis leaf disks and *Vicia faba* epidermal peels is described, which can easily be adopted for other plant species or tissues. Plant material is first metabolically labeled with $^{32}P_i$ so that PLD's substrate, the structural phospholipids become ^{32}P-labeled. Then material is incubated and treated in the presence of 0.5 % 1-butanol. Subsequently, ^{32}P-phosphatidylbutanol (^{32}P-PBut) is formed in a time-dependent manner. If a PLD is activated in response to the treatment, more ^{32}P-PBut is formed. Lipids are extracted and analyzed by EtAc TLC which separates PBut and PA from the rest of the phospholipids. Quantification is done by phosphoimaging, quantifying PBut and PA as percentage of the total ^{32}P-labeled phospholipids in the extract.

2 Materials

2.1 Plant Media

1. Medium for cell suspensions: For 1 L, 4.4 g Murashige–Skoog (MS) salts with vitamins, 30 g sucrose, 100 μl BAP (6-Benzylaminopurine, 10 mM), 100 μl NAA (naphthalene-acetic acid, 54 mM in 70 % EtOH). Adjust pH to 5.8 with KOH. Autoclave 20 min at 120 °C.
2. Medium for *Arabidopsis* seedlings: 0.5× MS medium, supplemented with 1 % (w/v) sucrose and 1 % (w/v) agar.
3. Pot soil to grow mature plants.
4. Opening buffer for epidermal peels: 10 mM MES-KOH, pH 6.1, 10 mM KCl.

2.2 Plant Material and Cultivation

1. Suspension-cultured plant cells (tobacco BY-2; 4–5 days old (*see* **Note 1**).
2. Rotary shaker, 125 rpm, 24 °C, in the dark.
3. Seedlings (*Arabidopsis thaliana*, Col-0); ~5 days old.
4. Growth chamber, 21 °C, 16 h light, 8 h dark.
5. Mature *Arabidopsis* plants (~3weeks) for leaf disks.
6. Leaf disk cutter: cork borer (0.5 cm diameter).
7. Mature Broad bean (*Vicia faba*) plants (2–4 weeks) for epidermal peels.

2.3 Labeling Components

1. ^{32}P-inorganic phosphate ($^{32}P_i$; carrier-free, 10 μCi/μL).
2. 2 mL "Safe-Lock" Eppendorf tubes.
3. 3 sterile 15 mL tubes.

4. 0.45 μm and 0.22 μm pore size filters for 10 mL syringes.
5. Labeling buffer for seedlings and leaf disks: 2.5 mM MES-KOH (pH 5.7), 1 mM KCl.
6. Labeling buffer for cell suspensions: CFM (cell-free medium, 0.22 μm filtered).
7. 1-Butanol (sec- and *tert*-butanol as controls).
8. Fume hood.
9. Safety glasses.
10. Safety screen, 1 cm Perspex.
11. Perspex reaction tube holders.
12. Gloves.

2.4 Lipid Extraction Components

Keep solvents cold (*see* **Note 2**).

1. Perchloric acid (PCA): 50 % and 5 % (w/v) in H_2O.
2. CMH: chloroform/methanol/HCl (50:100:1, by vol.).
3. Chloroform.
4. 0.9 % (w/v) NaCl.
5. Theoretical upper phase (TUP): chloroform/methanol/1 M HCl (3:48:47, by vol.).
6. Microcentrifuge capable of 13,000 × *g*.
7. Vortex shaker.
8. Glass Pasteur pipets.
9. *Iso*-propanol.
10. Vacuum centrifuge with cold-trap.
11. Nitrogen (gas).

2.5 Thin Layer Chromatograph

1. TLC plates: Merck silica 60 TLC plates, 20 × 20 × 0.2 cm.
2. Oven at 110 °C.
3. Graduated cylinders (250, 100, 25 mL).
4. Ethyl acetate (EtAc) TLC solvent: Upper organic phase of a mixture of ethyl acetate/iso-octane/formic acid/water (12:2:3:10, by vol.). Shake well for 15 s and let phases settle for ~10 min. Use the organic upper phase as the TLC solvent.
5. Filter paper (21 × 21 cm).
6. Plastic wrap.
7. TLC Tank.
8. Hair dryer.
9. Autoradiography film (Kodak, X-Omat S).
10. PhosphorImager and screen.
11. Light cassette.

3 Methods

Described below are four procedures to measure PLD activity in (1) suspension-cultured plant cells, (2) *Arabidopsis* seedlings, (3) *Arabidopsis* leaf disks, and (4) *Vicia faba* epidermal peels, respectively. Once labeling is stopped through PCA, a similar lipid extraction procedure and TLC analysis is performed. Obviously, other species and tissues can be analyzed in the same way.

3.1 PLD Activity in Plant Cell Suspensions

Typically, tobacco BY-2 cells are used but we have also good experience with other cell suspensions, including tomato, potato, *Arabidopsis*, *Medicago*, coffee (14–18), or unicellular algae like *Chlamydomonas* (19). The protocol also works for pollen tubes and fungal microspores (20) (M. Parzer & T. Munnik, unpublished).

1. Cell suspensions are grown on a rotary shaker (125 rpm) at 24 °C in the dark and subcultured weekly.
2. Pour ~10 mL of 4–5 days old cell suspensions into a sterile 15 mL test tube.
3. Let the big cell clumps sink for 30 s and pipet 1,200 μL cells (for 12 samples) from the top into a 2 mL reaction tube (*see* **Note 3**). Use rest of the cells to prepare *Cell-Free Medium (CFM) if required.
4. Preparation of CFM: Let the cells sink to the bottom of the tube or centrifuge for 1 min at 1,000 ×*g*. Pipet out the liquid and filter through 0.45 μm and subsequently trough 0.22 μm syringe filters. This CFM is used for cell treatments as 1:1 dilutions.
5. To the 1,200 μL cells add 120 μCi $^{32}PO_4^{3-}$ (~10 μCi/sample) in the fume hood of an isotope lab.
6. Aliquot 12 samples of 80 μL into 2 mL "Safe-Lock" reaction tubes in a Perspex rack using large orifice tips (*see* **Note 4**). Label for ~3 h, overnight is also possible; close tubes to prevent water loss and keep the rack in the fume hood to reduce your exposure to radiation.
7. Apply stresses and controls by adding 1:1 volumes (i.e., 80 μL) using CFM containing 1 % 1-butanol at $t = 0$ with 15″ intervals (*see* **Note 5**).
8. Incubate at room temperature for the time desired.
9. Stop reactions by adding 20 μL (=1/10 vol.) 50 % PCA (5 % final concentration) with 15 s intervals; mix immediately.
10. Shake for 5 min.
11. Spin tubes for 5 s (*see* **Note 6**).

12. Add 750 μL of CMH ($CHCl_3$/MeOH/HCl (50:100:1) (*see* **Note 7**).
13. Shake for 5 min.
14. Spin 5 s (*see* **Note 6**).
15. Add 750 μL of $CHCl_3$ and then 200 μL 0.9 % NaCl.
16. Mix for 10 s; Spin 1 min.
17. Remove the aqueous upper phase as much as possible and transfer the organic lower phase to a new 2 mL Safe-Lock Eppendorf tube containing 750 μL TUP (*see* **Note 8**).
18. Mix for 10 s individually or 5 min on a shaker; Spin 1 min.
19. Discard the upper phase (*see* **Note 9**).
20. Add 20 μL iso-propanol (*see* **Note 10**).
21. Dry the lipid extract by vacuum centrifugation (45 min, heating 54 °C).
22. Dissolve into 50 μL $CHCl_3$ and store under N_2 at −20 °C or use immediately for TLC analysis.

3.2 PLD Activity Arabidopsis Seedlings

The protocol described below is for *Arabidopsis* seedling but can be used for other plant species or for hypocotyls of cucumber explants (21).

1. Grow seedlings on 0.5× MS plates (1 % agar), supplemented with 1 % sucrose, for 5–6 days in a climate room with 16 h light/8 h dark period at 21 °C.
2. Transfer (*see* **Note 11**) three seedlings per tube in 190 μL labeling buffer (2.5 mM Mes-KOH (pH 5.7), 1 mM KCl). Use 2 mL "Safe-Lock" reaction tubes (*see* **Note 4**). Alternatively, 48 wells microtiter plates can be used.
3. Add 10 μL $^{32}PO_4^{3-}$ containing 5–10 μCi per sample (*see* **Note 12**).
4. Incubate with radiolabel overnight. Leave the lights of the fume hood on to enable photosynthesis and close the tubes to prevent water loss (O_2 will be produced!).
5. Next day, treatments can be applied. If agonists are to be added, e.g., salt stress, then add them in relatively large volumes (e.g., 1:1, i.e., 200 μL) so that one does not have to mix rigorously. Include 1 % 1-butanol (0.5 % final concentration) for all treatments and controls. Start with 15 s intervals.
6. Stop reactions by 1/10 vol. 50 % perchloric acid (PCA, ~5 % final) with 15 s intervals. We use an Eppendorf repetition pipet for this.
7. Shake samples for 5–10 min (green tissue turns brownish) on a Vortex shaker.
8. Spin 5 s (*see* **Note 6**).

9. Open tubes and remove all liquid around the seedlings with a Pasteur pipet.
10. Add 400 μL of "CMH" ($CHCl_3$/MeOH/HCl (50:100:1)).
11. Shake on the Vortex shaker for 5 min (until leaf material is not colored anymore and the extracts are yellowish).
12. Spin 5 s (*see* **Note 6**).
13. Add first 400 μL of $CHCl_3$, followed by 200 μL of 0.9 % NaCl (*see* **Note 13**).
14. Mix for 10 s/tube or shake 5 min on a Vortex shaker.
15. Spin 1 min.
16. Remove upper phase (aqueous, free label) as much as possible and transfer the lower phase with a Pasteur pipet to a clean tube containing 400 μL TUP (do not transfer the seedlings!) (theoretical upper phase; $CHCl_3$/MeOH/1 M HCl, 3:48:47 v/v). Use the two-phases in the Pasteur pipet for a clean transfer (*see* **Note 8**).
17. Mix for 10 s or shake 5 min.
18. Spin 1 min.
19. Remove and discard the upper phase completely (*see* **Note 9**).
20. Add 20 μL of iso-propanol (*see* **Note 10**).
21. Dry the lipid extract by vacuum centrifugation (~45 min, heating 54 °C).
22. Dissolve into 100 μL $CHCl_3$ and store at −20 °C (under N_2 gas) or use immediately for TLC analysis (load 20 μL only!).

3.3 PLD Activity in Arabidopsis Leaf Disks

The protocol described below is for Arabidopsis leaves but can be used for any other plant species, e.g., *Craterostigma plantagineum* (9), rice (13), or tissues, e.g., flower petals of carnation (22).

1. Punch leaf disks (0.5 cm in diameter) from 2 to 3 weeks old *Arabidopsis* plants.
2. Transfer disks to 2 mL Safe-Lock reaction tubes containing 190 μL labeling buffer. Use 1 leaf disk per tube and use a yellow tip, with a drop of buffer, for the transfer and guidance into the tube. Make sure all leaf disk orientations are the same (e.g., adaxial side up).
3. Add 10 μL $^{32}PO_4^{3-}$ (5–10 μCi/sample), diluted in buffer to each sample.
4. Incubate with radiolabel overnight. Leave the lights of the fume hood on to enable photosynthesis and close the tubes to prevent water loss (O_2 will be produced!).
5. Next day, treatments can be applied. Apply (e.g., salt stress) in relatively large volumes (1:1) so that one does not have to mix

rigorously (i.e., 200 μL). Include 1 % 1-butanol for a final concentration of 0.5 %. Start treatments with 15 s intervals.

6. Stop reactions by adding 1 vol. (in this case 400 μL) 10 % PCA (~5 % final) with 15 s intervals (*see* **Note 14**).
7. Shake samples for 5–15 min (until green tissue is brownish).
8. Continue as in Subheading 3.2, step 8.

3.4 PLD Activity in Guard Cells

The protocol described below is for *Vicia faba* epidermal peels (23).

1. Prepare epidermal peels peeled from the abaxial surface of fully expanded leaves of 2- to 4-week-old plants.
2. Place two epidermal peels of 0.25 cm^2 (containing an average of approx. 1,000 stomata each) per treatment in a 24-well plate. It is important to have the same surface of epidermal peels per treatment. Only guard cells remain alive, epidermal cells die during the peeling procedure (23).
3. Incubate for 5 h in 400 μL of opening buffer containing 0.1 mCi $^{32}PO_4^{3-}$ mL^{-1} (carrier free) under white light to promote stomatal opening. Longer incubations are not recommended since the stomata lose their functionality.
4. Treat the peels (e.g., ABA) in relatively large volumes (1:1) so that one does not have to mix rigorously (i.e., 400 μL). Include 1 % 1-butanol for a final concentration of 0.5 %. Start treatments with 1 min intervals.
5. Stop reactions by transferring the peels to a 2 mL Safe-Lock reaction tube containing 170 μL opening buffer and 20 μL 50 % PCA (v/v).
6. Continue as in Subheading 3.1, step 8.

3.5 Thin Layer Chromatography

1. Use glass silica-60 (Merck) TLC plates which are stored at 110 °C for at least 2 days (*see* **Note 15**).
2. Before loading, mark the TLC plate with a soft pencil where the samples are to be spotted: draw a line, 2 cm from the bottom. Mark the samples with a dot, 0.5–1.5 cm apart. The solvent front tends to "smile" so distribute the samples equally and stay at least 2 cm away from the sides of the TLC plate (20 × 20 cm).
3. Return the TLC plate in the oven to keep it heat-activated (*see* **Note 15**).
4. Prepare the EtAc TLC solvent. To separate PA and PBut from the rest of the phospholipids, use the upper organic phase of a mixture of ethyl acetate/iso-octane/formic acid/water (12:2:3:10, by vol.). Keep 80 mL separate to run the TLC plate and use the rest to wash the tank and filter paper at the back of the tank which keeps the chamber saturated.

5. If all phospholipids are to be analyzed, run an additional TLC with the Alkaline solvent system (*see* **Note 16**).
6. Wash two to three times and close the tank quickly. The tank is ready now and can be used two to three times within 1 week.
7. Get your lipid samples out of the −20 °C and spin them for 2–4 min in a microcentrifuge to get the samples quickly to room temperature (*see* **Note 17**).
8. Meanwhile, take the TLC plate from the oven and put it with the silica-side onto a sheet of plastic wrap, just above the 2 cm line marked with a pencil, and wrap it. The wrapping protects the rest of the plate when loading.
9. Place the TLC plate horizontally on the table, behind a 1-cm thick Perspex screen. Put another plate (glass or Perspex) on top, but keep the 2-cm line clear to load.
10. Place a hair dryer ~30 cm away from the TLC plate, so that a gentle stream of air is streaming over the plate. This speeds up the evaporation of the chloroform and reduces the size of the loading spots.
11. Place the ^{32}P-lipid samples in a block behind the TLC.
12. Use a reaction tube filled with chloroform to wash your pipet tip in between the samples.
13. Equilibrate the tip of a 20 μL pipet with chloroform by pipetting it up and down for a number of times (*see* **Note 18**).
14. Load the pipet tip with 20 μL lipid extract and spot gently onto the TLC (*see* **Note 19**).
15. Rinse the tip with chloroform to spot the next sample.
16. Repeat until all samples are loaded.
17. After the last sample, wait approx. 1 min to dry the last spot.
18. Remove the plastic wrap and start the TLC by placing the plate into the solvent. Let the plate stand almost vertically and quickly close the tank to prevent the loss of solvent vapor.
19. Run the EtAc TLC for ~1 h; Alkaline TLC for 1.5 h (*see* **Note 20**).
20. Take the plate out and let it dry in the fume hood for 1 h to eliminate solvent fumes which will affect the PhosphorImager screen or autoradiography film (*see* **Note 21**).
21. Wrap the plate in plastic wrap and expose a PhosphorImager screen for 1 h. Quantify PBut and PA as percentage of total ^{32}P-lipids.
22. For higher resolution pictures, expose the TLC plate to autoradiography film for 1 h and overnight for a short and long exposure, respectively.

23. A typical EtAc- and Alkaline-TLC profile of *Arabidopsis* seedlings treated with salt or mannitol is shown in Fig. 1 (*see* **Note 22**).

4 Notes

1. The protocol works for any cell culture, including *Arabidopsis*, coffee, tomato, and *Medicago,* green algae like *Chlamydomonas*, and pollen tubes and fungal microspores (14–18).
2. Use cold solvents. They pipet better (less gas expansion within the pipet) and decrease chemical breakdown.
3. As most cell suspensions tend to clump a bit, the use of large orifice tips are preferred or simply cut-off the pipet tip to increase the opening. The latter is less accurate though.
4. The use of Safe-Lock reaction tubes are preferred as normal tubes may open due to the gas pressure of the solvent.
5. The CFM is a conditioned medium to which the cells are not, or hardly, responding. This is in huge contrast to using plane water or a buffer, which activates lipid signaling instantly, likely via osmotic stress.
6. A short spin in the microcentrifuge to remove the radioactive liquid from the lid to prevent contamination when opening.
7. It is very important that the CMH completely dissolves the aqueous fraction. By using a 3.75 vol. ratio, this will always occur. The MeOH functions to dissolve the water and to be dissolved by the chloroform. By increasing either chloroform or water fraction, the solution will split into two phases, an aqueous upper phase and an organic lower phase. This is used later (Subheading 3.1, step 15) to separate the lipids from the proteins, sugars, and DNA.
8. Use the two phases in the Pasteur pipet to separate the lipid fraction from the upper aqueous fraction to transfer it to a clean 2 mL Safe-Lock Eppendorf tube, already containing 750 μL TUP.
9. It is very important to remove all of the aqueous upper phase. Water will cause chemical hydrolysis of the lipid extract. First remove as much as possible the upper phase. Then suck up all extract and use the organic phase to "wash" the wall of the tube two to three times. Every time, suck up the complete extract again and let the phases split (Note that it is important to keep enough "balloon air." Use the two-phase separation in the Pasteur pipet to remove aqueous fraction. Due to a faster evaporation of the chloroform than the water, two phases may reappear and this cause chemical hydrolysis (breakdown) of the lipids.

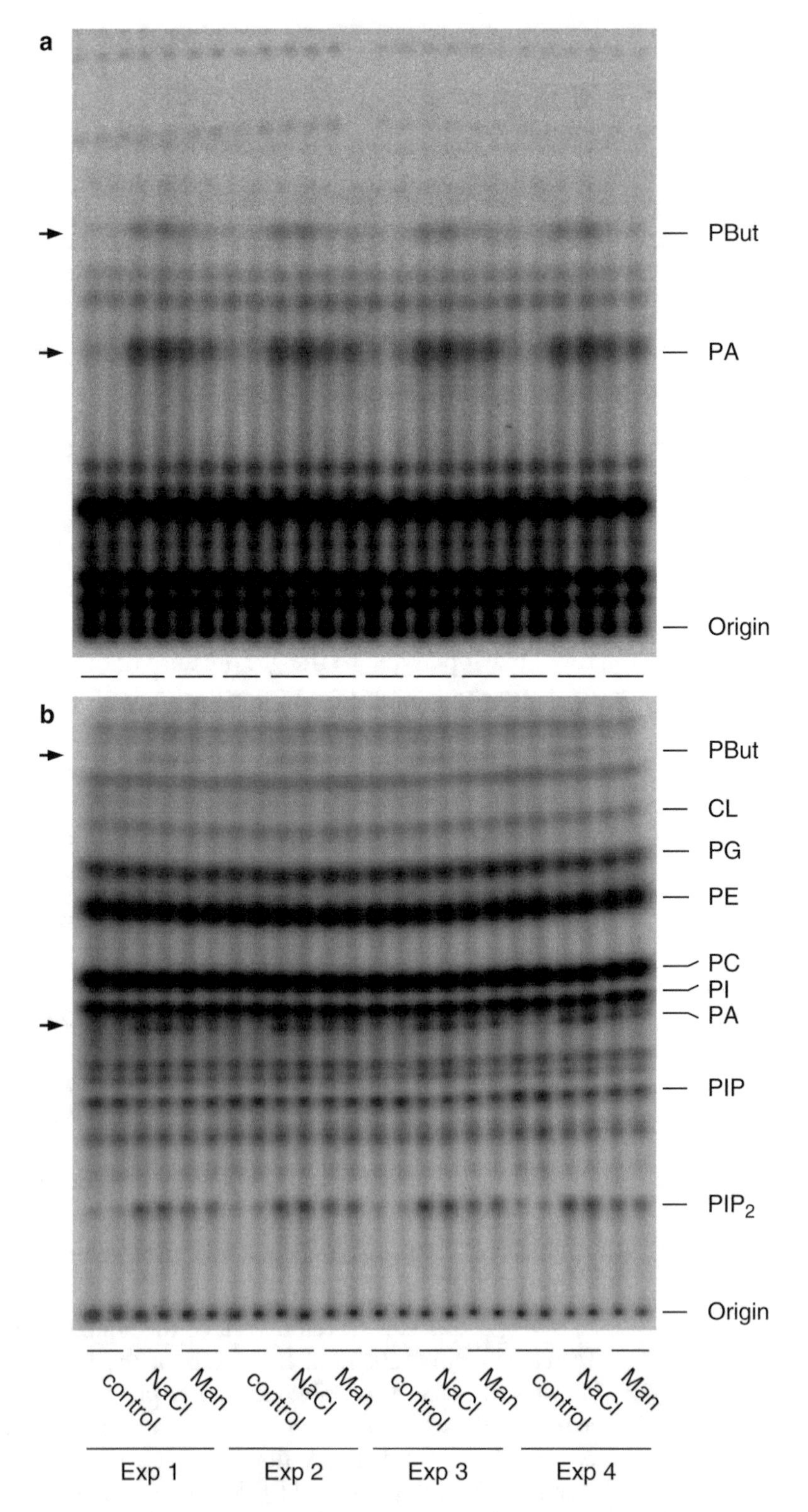

Fig. 1 Typical EtAc- (**a**) and Alkaline- (**b**) TLC profile of *Arabidopsis* seedlings treated with salt or mannitol. *Arabidopsis* seedlings were labeled overnight with $^{32}P_i$ and the next day treated with buffer containing 0.5 % 1-Butanol (control) or supplemented with 500 mM NaCl or 1 M mannitol for 30 min. Lipids were extracted and separated by EtAc TLC (**a**) for PA and PBut analysis and Alkaline TLC (**b**) too see the rest of the phospholipids. Treatments were applied in duplicates from four different seed batches. The salt- and mannitol-induced PA and PBut formation are indicated by *arrows*. A typical PIP_2 increase is also evident (**b**)

10. The iso-propanol keeps the solution into one phase, keeping the water dissolved into the chloroform.
11. First transfer the seedlings to a drop of buffer (2.5 mM Mes-KOH, pH 5.7, 1 mM KCl) in an empty petri dish. This discharges the static electricity of the roots and allows a "clean" transfer of all three seedlings in one go into the tube (do not touch the wall!). Use a yellow 20–200 μL tip for the transfers, not a tweezer (damage). Alternatively, use a 24-well microtiter plate containing 400 μL buffer with two or three seedlings per well.
12. Add the ^{32}P-label diluted, i.e., as 10 μL in buffer to lower the label variability. Usually, the label is 10 μCi/μL, carrier free.
13. The extract now splits into two phases with an "aqueous" upper phase (free label) and an organic lower phase (lipids). At the interphase, solid material (seedling remainings, protein precipitates) settles.
14. We use an Eppendorf repetition pipet for this.
15. By storing the plates above 100 °C, it prevents the silica from binding the water that is in the air. The latter affects the composition of the TLC solvent and lowers the silica's capacity to bind lipids.
16. If all phospholipids are to be analyzed, an alkaline TLC system should be used consisting of $CHCl_3$/MeOH/25 % (w/v) ammonia/H_2O (90:70:4:16, by vol.). Use potassium-oxalate impregnated plates for this type of TLC and pre-equilibrate plates 30 min prior to developing as described in detail in Chapter 2 (24).
17. When the samples are cold, the water in the air can condensate on top of the extracts causing hydrolysis and smearing of the lipid spots.
18. Chloroform evaporates easily and the gas that is building up inside the pipet tends to push out the liquid within the tip. By equilibrating first the pipet and tip with chloroform, this is strongly reduced and increases the handling and loading of the lipid samples.
19. Because of the rapid evaporation of chloroform, the extract tends to be pushed out. Use that drop to touch the pencil-marked spot on the TLC plate. Keep the pipet vertical at all times. It is also no problem to gently touch the silica with the tip. Prevent the spot from becoming too big, by loading small drops of extract at the time.
20. After 1.5 h, the front will still be a few centimeters from the top. It has, however, no use to run it completely until the end. The speed of the migration front slows down exponentially while spots diffuse in all directions, so that there is no gain in resolving power by running the TLC longer.

21. If you do not need to use the lipids any further, one can speed up the solvent evaporation by placing the plate in the 110 °C oven for 5 min.
22. To identify PLD's transphosphatidylation product PBut, reactions can be performed ±1-butanol. Alternatively, other primary alcohols can be used which run differently on EtAc TLC (25).

Acknowledgments

This work was supported by grants from the Netherlands Organization for Scientific Research (NWO; VIDI 864.05.001), the EU (COST FA0605), the Universidad Nacional de Mar del Plata (UNMdP), Consejo Nacional de Investigaciones Científicas y Técnicas (CONICET), and Agencia Nacional de Promoción Científica y Tecnológica (ANPCyT). We thank Jiorgos Kourelis for performing the experiment described in Fig. 1.

References

1. Li M, Hong Y, Wang X (2009) Phospholipase D- and phosphatidic acid-mediated signaling in plants. Biochim Biophys Acta 1791: 927–935
2. Testerink C, Munnik T (2011) Molecular, cellular, and physiological responses to phosphatidic acid formation in plants. J Exp Bot 62:2349–2361
3. Pappan KL, Wang X (2012) Assaying different types of plant phospholipase D activities in vitro. Meth Mol Biol. 1009:205–217
4. Arisz SA, Testerink C, Munnik T (2009) Plant PA signaling via diacylglycerol kinase. Biochim Biophys Acta 1791:869–875
5. Heller M (1978) Phospholipase D. Adv Lipid Res 16:267–326
6. Quarles RH, Dawson RM (1969) The distribution of phospholipase D in developing and mature plants. Biochem J 112:787–794
7. Walker SJ, Brown HA (2004) Measurement of G protein-coupled receptor-stimulated phospholipase D activity in intact cells. Methods Mol Biol 237:89–97
8. Munnik T, Arisz SA, De Vrije T, Musgrave A (1995) G protein activation stimulates phospholipase D signaling in plants. Plant Cell 7:2197–2210
9. Frank W, Munnik T, Kerkmann K, Salamini F, Bartels D (2000) Water deficit triggers phospholipase D activity in the resurrection plant *Craterostigma plantagineum*. Plant Cell 12:111–124
10. den Hartog M, Musgrave A, Munnik T (2001) Nod factor-induced phosphatidic acid and diacylglycerol pyrophosphate formation: a role for phospholipase C and D in root hair deformation. Plant J 25:55–65
11. Bargmann BO, Laxalt AM, ter Riet B, Testerink C, Merquiol E, Mosblech A, Leon-Reyes A, Pieterse CM, Haring MA, Heilmann I, Bartels D, Munnik T (2009) Reassessing the role of phospholipase D in the Arabidopsis wounding response. Plant Cell Environ 32:837–850
12. Bargmann BO, Laxalt AM, ter Riet B, van Schooten B, Merquiol E, Testerink C, Haring MA, Bartels D, Munnik T (2009) Multiple PLDs required for high salinity and water deficit tolerance in plants. Plant Cell Physiol 50:78–89
13. Darwish E, Testerink C, Khaleil M, El-Shihy O, Munnik T (2009) Phospholipid-signaling responses in salt stressed rice leaves. Plant Cell Physiol 50:986–987
14. Van der Luit AH, Piatti T, van Doorn A, Musgrave A, Felix G, Boller T, Munnik T (2000) Elicitation of suspension-cultured tomato cells triggers the formation of phosphatidic acid and diacylglycerol pyrophosphate. Plant Physiol 123:1507–1516
15. Munnik T, de Vrije T, Irvine RF, Musgrave A (1996) Identification of diacylglycerol pyrophosphate as a novel metabolic product of phosphatidic acid during G-protein activation in plants. J Biol Chem 271:15708–15715
16. Ramos-Diaz A, Brito-Argaez L, Munnik T, Hernandez-Sotomayor SM (2007) Aluminum

inhibits phosphatidic acid formation by blocking the phospholipase C pathway. Planta 225:393–401

17. Munnik T, Meijer HJ, Ter Riet B, Hirt H, Frank W, Bartels D, Musgrave A (2000) Hyperosmotic stress stimulates phospholipase D activity and elevates the levels of phosphatidic acid and diacylglycerol pyrophosphate. Plant J 22:147–154
18. den Hartog M, Verhoef N, Munnik T (2003) Nod factor and elicitors activate different phospholipid signaling pathways in suspension-cultured alfalfa cells. Plant Physiol 132:311–317
19. Meijer HJ, ter Riet B, van Himbergen JA, Musgrave A, Munnik T (2002) KCl activates phospholipase D at two different concentration ranges: distinguishing between hyperosmotic stress and membrane depolarization. Plant J 31:51–59
20. Zonia L, Munnik T (2004) Osmotically induced cell swelling versus cell shrinking elicits specific changes in phospholipid signals in tobacco pollen tubes. Plant Physiol 134: 813–823
21. Lanteri ML, Lamattina L, Laxalt AM (2011) Mechanisms of xylanase-induced nitric oxide and phosphatidic acid production in tomato cells. Planta 234:845–855
22. De Vrije T, Munnik T (1997) Activation of phospholipase D by calmodulin antagonists and mastoparan in carnation petal tissue. J Exp Bot 48:1631–1637
23. Distefano AM, Garcia-Mata C, Lamattina L, Laxalt AM (2008) Nitric oxide-induced phosphatidic acid accumulation: a role for phospholipases C and D in stomatal closure. Plant Cell Environ 31:187–194
24. Munnik T, Zarza X (2013) Rapid analysis of plant signaling phospholipids through $32P_i$-labeling and TLC. Meth Mol Biol. 1009:3–15
25. Munnik T (2001) Phosphatidic acid: an emerging plant lipid second messenger. Trends Plant Sci 6:227–233

Chapter 21

Assaying Plant Phosphatidic Acid Phosphatase Activity

Yuki Nakamura

Abstract

Phosphatidic acid phosphatase (PAP; EC 3.1.3.4) catalyzes the dephosphorylation of phosphatidic acid (PA) to produce diacylglycerol (DAG) and inorganic phosphate. In seed plants, PA plays pivotal roles both as a precursor to membrane lipids and as a signaling molecule. As more information on the roles of PAP in plants becomes available and the importance of PAP is revealed, protocols for assaying plant PAP activity are of interest to an increasing audience. This chapter describes procedures to assay plant PAP activity that are based on recent publications.

Key words Phosphatidic acid, Phosphatidic acid phosphatase, Diacylglycerol

Abbreviations

DAG	*sn*-1,2 Diacylglycerol
DGK	Diacylglycerol kinase
LPP	Lipid phosphate phosphatase
PA	Phosphatidic acid
PAH	Phosphatidate phosphohydrolase
PAP	Phosphatidic acid phosphatase

1 Introduction

Phosphatidic acid phosphatase (PAP; EC 3.1.3.4) catalyzes the dephosphorylation of phosphatidic acid (PA) to produce diacylglycerol (DAG) and inorganic phosphate. In seed plants, PA plays pivotal roles both as a precursor to membrane lipids and as a signaling molecule. PA is synthesized from glycerol-3-phosphate via two sequential reactions of acylation named the Kennedy pathway (1). On the other hand, the reaction product DAG can be utilized for the biosynthesis of different classes of glycerolipids, such as phospholipids, galactolipids, triglycerides, and sulfolipids.

Teun Munnik and Ingo Heilmann (eds.), *Plant Lipid Signaling Protocols*, Methods in Molecular Biology, vol. 1009,
DOI 10.1007/978-1-62703-401-2_21, © Springer Science+Business Media, LLC 2013

PA is also widely recognized as an important signaling molecule that is involved in various physiological functions such as responses to biotic/abiotic stresses or phytohormone responses (2). Upon activation of a signaling cascade, PA is synthesized mainly from membrane phospholipids by phospholipase D (3) or from phosphoinositides by a two-step reaction of phosphoinositide-specific phospholipase C and diacylglycerol kinase (DGK) (4, 5). In addition, PA can be produced by nonspecific phospholipase C and DGK (6–8). PA can be dephosphorylated to DAG by phosphatidic acid phosphatase (PAP), further phosphorylated by PA kinase to diacylglycerol pyrophosphate or deacylated to lyso PA by phospholipase A. However, it is yet to be understood PA originating from which pathway is involved in lipid metabolism and which in signaling. In addition, it is unclear which pathway attenuates and terminates PA signaling.

PAP isoforms are categorized into either PAP1 or PAP2 according to their enzymatic properties (9). PAP1 is primarily a soluble enzyme and involved in glycerolipid metabolism (10, 11). The activity is Mg^{2+}-dependent. The encoded protein, phosphatidate phosphohydrolase (PAH), was first purified from the yeast *Saccharomyces cerevisiae* and was found to be a homologue of a human lipin protein (12). Plant homologues were also cloned from Arabidopsis (10) and *Brassica napus* (13). Arabidopsis has two PAHs, named PAH1 and PAH2, which redundantly function in glycerolipid metabolism (10) and phospholipids turnover (11). On the other hand, PAP2 is an integral membrane protein which dephosphorylates not only PA but also other lipid phosphates, such as lyso PA, diacylglycerol pyrophosphate, sphingosine-1-phosphate, and ceramide-1-phosphate in vitro (14). Proteins possessing PAP2 activity are therefore named lipid phosphate phosphatase (LPP). PAP2 activity is independent of Mg^{2+}. LPP is considered to be mainly involved in signaling function. There are at least 9 LPPs encoded in the Arabidopsis genome and these enzymes have been found to be involved in the mediation of stress responses, in abscisic acid signaling and in chloroplast lipid metabolism (15–18).

Due to the increased information and revealed importance of plant PAP in recent years, there is increasing demand for protocols to assay plant PAP activity. This chapter describes procedures to assay plant PAP activity that are based on recent publications.

2 Materials

2.1 Preparation of Sample from Total Plant Tissue

1. Mortar and pestle.
2. Liquid N_2.
3. Suspension buffer: 50 mM Tris–HCl, pH 7.3, 50 mM NaCl, 5 % (v/v) glycerol.

2.2 Preparation of Material from Yeast PAP Mutants Complemented with a Plant PAP Sequence

1. Yeast *Δdpp1Δlpp1Δpah1*, which has significantly reduced PAP activity (12).
2. Yeast binary vector, such as pDO105 (19).
3. Bead-beater.
4. Suspension buffer as in Subheading 2.1.

2.3 Preparation of Purified Recombinant PAP Proteins from E. coli

1. *E. coli* expression vector such as pET28.
2. *E. coli* strain for protein expression such as ArcticExpressRIL (Stratagene).
3. Induction solution: 1 mM isopropyl β-D-thiogalactopyranoside.
4. Wash buffer: 20 mM Tris–HCl, pH 8.0.
5. Buffer A: 20 mM sodium phosphate, pH 7.7, 0.5 M NaCl, 20 mM imidazole, 7 mM 2-mercaptoethanol, and 1:20 (v/v) Protein Inhibitor Cocktail (Sigma-Aldrich).
6. Glass homogenizer.
7. French press.
8. Purification resin such as Ni-NTA (Invitrogen).
9. Linear gradient of Buffer A containing up to 500 mM imidazole.
10. Dialysis buffer: 20 mM Tris–HCl, pH 7.0, containing 10 % (v/v) glycerol and 7 mM 2-mercaptoethanol.

2.4 PAP Enzyme Assay In Vitro

1. Radiolabeled PA such as [^{14}C]-dipalmitoyl PA.
2. Assay buffer: 50 mM Tris–HCl, pH 6.5, containing 0.1 % (w/v) TritonX-100.
3. 0.18 mM PA dispersed in assay buffer.
4. $MgCl_2$.
5. $CHCl_3$:methanol (1:1).
6. Salt solution: 1 M KCl, 0.2 M H_3PO_4.
7. Glass pipette.
8. $CHCl_3$:methanol (2:1).
9. Heating block.

2.5 Analysis and Quantification of Reaction Product by TLC and Image Analyzer

1. Development solvent: petroleum ether–ethyl ether–acetic acid (50:50:1, v/v/v).
2. Glass tank with lid for TLC.
3. TLC plates (Silica gel 20 cm × 20 cm, Merck).
4. Commercial DAG standard.
5. Glass cutter.
6. Imaging plate and cassette.
7. Heat block.

2.6 Data Analyses

Image analyzer, such as Storm or Typhoon (GE healthcare).

3 Methods

3.1 Preparation of Sample from Total Plant Tissue

1. Grind frozen tissue into powder with mortar and pestle in the presence of liquid N (*see* **Note 1**). If organelle fractionation is needed (e.g., an assay with intact chloroplasts), appropriate fractionation procedures must be applied. An example of PAP assay with intact chloroplasts is described in ref. (17).
2. Suspend the tissue in suspension buffer and keep on ice until the reaction is performed (*see* **Note 2**).

3.2 Preparation of Material from Yeast PAP Mutants Complemented with a Plant PAP Sequence

This method is to assay activity of recombinant plant PAP expressed in the yeast triple mutant *Δdpp1Δlpp1Δpah1*, which has significantly reduced endogenous PAP activity (12). This method requires less biochemical skills compared to the *E. coli* method described in Subheading 3.3. Furthermore, this method is powerful, especially when assaying membrane-integrated PAP2 (LPP) activity, as these proteins are very difficult to express recombinantly in a soluble form. Although this method is not designed for detailed enzymological analyses, a method optimized well for further biochemical study is reported in ref. (13).

1. Clone plant PAP protein coding sequence of your interest into a yeast binary vector, such as pDO105 (19), and transform into *Δdpp1Δlpp1Δpah1* mutant background.
2. Grow the obtained transformants in liquid culture and harvest the cells during the exponential phase.
3. Wash cells twice with suspension buffer at room temperature. Suspend the pellet in suspension buffer, keep on ice for 15 min, add beads and disrupt cells with bead-beater.
4. Transfer samples to new tubes without including the beads. When assaying PAP2 activity, separation of soluble fraction by centrifugation for 10 min at 15,000 × *g* will reduce background PAP activity derived from soluble PAP activity present in the *Δdpp1Δlpp1Δpah1* mutant. Resuspend the isolated membrane fraction in suspension buffer and keep on ice until performing the assay reaction.

3.3 Preparation of Purified Recombinant PAP Proteins from E. coli

This method is suitable for detailed enzymological characterization of PAP. However, applying this method to study PAP2 is not practical since this membrane-spanning type of PAP is difficult to express recombinantly in a soluble form (*see* **Note 3**).

1. Clone coding sequence for PAP of your interest into an *E. coli* expression vector pET28 and introduce into *E. coli* strain ArcticExpressRIL (Stratagene) (*see* **Note 4**).

2. Grow the *E. coli* cells over night in LB media, dilute it (1:50) and incubate at 30 °C for 3 h, followed by temperature equilibration at 13 °C for up to 30 min.
3. Add 1 mM Induction solution and incubate for 20 h to induce the expression of protein (*see* **Note 5**).
4. After induction, recover the cells by centrifugation, wash cells once with 20 mM Tris–HCl, pH 8.0, and suspend in 40 ml of Buffer A. All procedures hereafter must be performed at 4 °C or on ice.
5. Disrupt cells by resuspending with a glass homogenizer, followed by three passes through a French press at 25,000 p.s.i. (*see* **Note 6**).
6. Remove cell debris and unbroken cells by centrifugation at 40,000 × *g* for 30 min.
7. Filtrate the supernatant through a 20-mm filter and load onto a 5 ml Ni-NTA column (Qiagen), followed by a 5 ml wash with Buffer A.
8. Elute the tagged protein from the column in 1 ml fractions with a total of 50 ml of a linear gradient of Buffer A containing up to 500 mM imidazole (*see* **Note 7**).
9. Find the protein-enriched fraction by SDS–PAGE, then dialyze it twice against 20 mM Tris–HCl, pH 7.0, containing 10 % (v/v) glycerol and 7 mM 2-mercaptoethanol (*see* **Note 8**).

3.4 PAP Enzyme Assay In Vitro

The following sections include the use of radioisotopes. The experiments are therefore performed in a secure manner according to the regulation/guideline for working with radioisotopes, as outlined for each individual research institution.

1. Add 45 nCi of radiolabeled PA to a 15 ml centrifuge tube and incubate at 37 °C in a heat block until the solvent is dried up completely.
2. Add 50 ml of 0.18 mM PA dispersed in assay buffer and vortex for 30 s. The pH might have to be adjusted depending on the type of PAP activity assayed. While extraplastidic PAP has optimal pH at neutral or slightly acidic condition, a plastidic PAP shows maximum activity at pH 9 (20).
3. Add 100 ml of $MgCl_2$ solution at a given concentration, or assay buffer in case of Mg^{2+}-independent assay. Whereas plant PAP1 (or PAH) shows maximum activity in the presence of a final concentration of 1 mM Mg^{2+}, many PAP2 show maximum activity without Mg^{2+} (10, 11, 17).
4. Incubate for 15 min at 30 °C to equilibrate the temperature of assay mix.
5. Add 50 ml of sample to the mix and incubate for 1 h at 30 °C.
6. Stop the reaction by adding 720 ml of $CHCl_3$:Methanol (1:1) and 280 ml of salt solution, vortex and centrifuge at 1,500 × *g* for 5 min (*see* **Note 9**).

7. Discard aqueous phase (upper layer) with glass pipette, add another 280 μl of salt solution, vortex and centrifuge at 1,500 × *g* for 5 min.
8. Isolate fixed amount of organic phase (e.g., 100 ml) from each sample, transfer to new microfuge tube and dry at 75 °C in a heating block.
9. Dissolve dried lipid sediment in 50 ml of $CHCl_3$:methanol (2:1). This lipid mixture will be analyzed by TLC. Samples can be stored at −80 °C for a few months if [^{14}C]-PA or [^{3}H]-PA is used for the assay.

3.5 Analysis and Quantification of Reaction Products by TLC and Image Analyzer

PAP activity is measured by quantifying radioactivity incorporated into the reaction product, DAG. Quantification methods differ depending on which radionuclide is used; TLC and image analyzer for [^{14}C] as described below, or scintillation counter for [^{32}P] or [^{3}H]. If crude extract (e.g., plant tissues or yeast total membranes) is used for the assay, the [^{14}C] method is highly recommended as it can separate radioactive DAG from other labeled secondary products enriched in the organic phase. The [^{32}P] method provides good in sensitivity and a choice of combinations of fatty acyl moieties are available, as labeled substrates are usually prepared by labeling commercial nonradioactive PA with [^{32}P]-ATP by DGK (11). Here, we describe the [^{14}C] method.

1. Pour development solvent into a glass tank with lid for TLC and wait for 30 min until the tank is fully saturated with vapor. The solvent should be measured such that the bottom of the tank is covered with ca. 1 cm of solvent.
2. Spot lipid extract onto a silica gel TLC plate, leave the plate to dry for 5 min and set in the TLC box. Make sure the solvent level is shallow enough not to submerge the spotted lipids. A small amount of commercial DAG should be spotted as an indicator of DAG migration. Once the development is started, do not open the TLC box as solvent vapor may escape and affect migration of the lipids.
3. Take out the TLC plate when top line comes to approximately 1 cm of the edge.
4. Leave the TLC plate for 15 min in a chemical hood to dry up the solvent.
5. Separate the lane for DAG standard by cutting TLC plate with a glass cutter.
6. Wrap the TLC plate with saran wrap and expose it to an imaging plate for 3 days. The exposure time may differ depending on the strength of radioactivity. If absolute value of enzyme activity is needed, spot a dilution series of radioactive substrate with known specific activity onto the TLC plate and expose together to draw a standard curve.

3.6 Data Analyses

1. Take out the imaging plate from the cassette and analyze it with an image analyzer (such as Storm or Typhoon). If radioactive spots are unclear, either (a) expose the plate again for a longer time, (b) increase the proportion of radioactive PA in the substrate mix, or (c) increase the amount of enzyme used for the assay.
2. Quantify the spots corresponding to DAG. If absolute activity needs to be determined, create a standard curve with the dilution series, calculate radioactivity of each DAG spot and divide it by the amount of protein in the sample. The specific activity can be reported as the molar amount of DAG produced per 1 min of reaction per 1 mg of protein (pmol/min/mg).

4 Notes

1. Endogenous PA levels are very sensitive to stress which can be affected in a few seconds of sampling time. Therefore, tissue sampling must be done as quickly as possible.
2. It is recommended to perform the enzyme activity assay as soon as the sample is prepared. The author has found significant reduction of specific activity in the samples if kept over night on ice.
3. The author also tested in vitro transcription and translation system for assaying Arabidopsis PAP2; however, these experiments were not successful.
4. Other expression vectors, such as pET, pQE, or pGEX, could be used to express tagged proteins. Similarly, other *E. coli* strains could also be used, but may need to be optimized with regard to expression conditions.
5. Avoid expressing protein at higher temperature, as protein will form inclusion bodies and is not amenable to be purified in a soluble form.
6. If a French press is not available, vigorous sonication might be used as an alternative for cell disruption at a smaller scale.
7. If small amounts of protein need to be purified, a batch purification method could be used, although some optimization might be required.
8. The proteins purified with this method can be stored at −80 °C, as has previously been reported (11).
9. Phase separation occurs when the ratio of $CHCl_3$ to methanol to inorganic mixture reaches around 2.0:2.0:0.7–1.0 (21). Equal volumes of ethyl acetate and 0.45 % (w/v) NaCl solution were used initially for the separation (6); however, reduced recovery of PA was reported later in this method (22).

Acknowledgments

This research was supported by Japan Science and Technology Agency (JST), PRESTO, and Core research grant of Academia Sinica, Taiwan ROC.

References

1. Kennedy EP (1957) Biosynthesis of phospholipids. Fed Proc Am Soc Exp Biol 16: 853–855
2. Testerink C, Munnik T (2011) Molecular, cellular, and physiological responses to phosphatidic acid formation in plants. J Exp Bot 62:2349–2361
3. Li M, Hong Y, Wang X (2009) Phospholipase D- and phosphatidic acid-mediated signaling in plants. Biochim Biophys Acta 1791:927–935
4. Arisz SA, Testerink C, Munnik T (2009) Plant PA signaling via diacylglycerol kinase. Biochim Biophys Acta 1791:869–875
5. Heilmann I (2009) Using genetic tools to understand plant phosphoinositide signaling. Trends Plant Sci 14:171–179
6. Nakamura Y, Awai K, Masuda T, Yoshioka Y, Takamiya K, Ohta H (2005) A novel phosphatidylcholine-hydrolyzing phospholipase C induced by phosphate starvation in Arabidopsis. J Biol Chem 280:7469–7476
7. Gaude N, Nakamura Y, Scheible WR, Ohta H, Dörmann P (2008) Phospholipase C5 (NPC5) is involved in galactolipid accumulation during phosphate limitation in leaves of Arabidopsis. Plant J 56:28–39
8. Peters C, Li M, Narasimhan R, Roth M, Welti R, Wang X (2010) Nonspecific phospholipase C NPC4 promotes responses to abscisic acid and tolerance to hyperosmotic stress in Arabidopsis. Plant Cell 22:2642–2659
9. Carman GM, Han GS (2006) Roles of phosphatidate phosphatase enzymes in lipid metabolism. Trends Biochem Sci 31:694–699
10. Nakamura Y, Koizumi R, Shui G, Shimojima M, Wenk MR, Ito T, Ohta H (2009) Arabidopsis lipins mediate eukaryotic pathway of lipid metabolism and cope critically with phosphate starvation. Proc Natl Acad Sci USA 106:20978–20983
11. Eastmond PJ, Quettier AL, Kroon JT, Craddock C, Adams N, Slabas AR (2010) Phosphatidic acid phosphohydrolase 1 and 2 regulate phospholipid synthesis at the endoplasmic reticulum in Arabidopsis. Plant Cell 22:2796–2811
12. Han GS, Wu WI, Carman GM (2006) The *Saccharomyces cerevisiae* lipin homolog is a Mg^{2+}-dependent phosphatidate phosphatase enzyme. J Biol Chem 281:9210–9218
13. Mietkiewska E, Siloto RM, Dewald J, Shah S, Brindley DN, Weselake RJ (2011) Lipins from plants are phosphatidate phosphatases that restore lipid synthesis in a pah1Δ mutant strain of *Saccharomyces cerevisiae*. FEBS J 278:764–775
14. Brindley DN, Pilquil C (2009) Lipid phosphate phosphatases and signaling. J Lipid Res 50(Suppl):S225–S230
15. Pierrugues O, Brutesco C, Oshiro J, Gouy M, Deveaux Y, Carman GM, Thuriaux P, Kazmaier M (2001) Lipid phosphate phosphatases in Arabidopsis. Regulation of the AtLPP1 gene in response to stress. J Biol Chem 276: 20300–20308
16. Katagiri T, Ishiyama K, Kato T, Tabata S, Kobayashi M, Shinozaki K (2005) An important role of phosphatidic acid in ABA signaling during germination in *Arabidopsis thaliana*. Plant J 43:107–117
17. Nakamura Y, Tsuchiya M, Ohta H (2007) Plastidic phosphatidic acid phosphatases identified in a distinct subfamily of lipid phosphate phosphatases with prokaryotic origin. J Biol Chem 282:29013–29021
18. Paradis S, Villasuso AL, Aguayo SS, Maldiney R, Habricot Y, Zalejski C, Machado E, Sotta B, Miginiac E, Jeannette E (2011) *Arabidopsis thaliana* lipid phosphate phosphatase 2 is involved in abscisic acid signalling in leaves. Plant Physiol Biochem 49:357–362
19. Ostrander DB, O'Brien DJ, Gorman JA, Carman GM (1998) Effect of CTP synthetase regulation by CTP on phospholipid synthesis in *Saccharomyces cerevisiae*. J Biol Chem 273:18992–19001
20. Joyard J, Douce R (1979) Characterization of phosphatidate phosphohydrolase activity associated with chloroplast envelope membranes. FEBS Lett 102:147–150
21. Bligh EG, Dyer WJ (1959) A rapid method of total lipid extraction and purification. Can J Biochem Physiol 37:911–917
22. Andersson MX, Larsson KE, Tjellström H, Liljenberg C, Sandelius AS (2005) Phosphate-limited oat. J Biol Chem 280:27578–27586

Chapter 22

Assay of Phospholipase A Activity

Michael Heinze and Werner Roos

Abstract

Phospholipases of the A type constitute a large family of esterases that catalyze the hydrolysis of the fatty acid ester bonds in phospholipids and thus generate lysophospholipids and fatty acids. Both products or their metabolites are important signal molecules in the cellular adaptation to stress, developmental processes and several diseases in plants and animals. The assay of PLA activity has been much promoted by the availability of phospholipid substrates with fluorophores at one or two fatty acids. The double labeled compounds display an increase of fluorescence due to the escape from intramolecular quenching or FRET. They thus allow the sensitive monitoring of PLA activity even without a separation of the hydrolysis products. This chapter is focused on the proper use of fluorescent (BODIPY) labelled substrates for assays of PLA activity in cells and subcellular fractions by fluorimetric analysis and classical or confocal microscopy.

Key words Phospholipase A_1, Phospholipase A_2, PLA activity assays, Fluorescent phospholipase substrates, Phospholipids, Lysophospholipids, Plasma membrane phospholipase, Microscopic phospholipase assay

1 Introduction

Phospholipases A catalyze the hydrolysis of ester phospholipids to yield a fatty acid plus a lysophospholipid. By far the most known PLA enzymes are PLA_2 (EC 3.1.1.4), i.e., they split phospholipids preferentially at the sn2 position (1, 2), but enzymes with a preference for the sn1 position (PLA_1, EC 3.1.1.32) are also documented (3). The classical assays of enzyme activity measure the liberation from an unlabelled substrate of fatty acids that are quantified in a subsequent enzymic or binding analysis, e.g., via conversion into acyl-CoA and subsequent H_2O_2 production by acyl-CoA oxidase (4, 5) or by using fluorescent fatty acid binding proteins (6, 7). More recently, substrates with radio- or fluorescent-labelled fatty acid(s) became routinely available, enabling the quantification of the liberated fatty acid after separation via TLC or HPLC in combination with or even solely by MS-based methods (8–10).

Teun Munnik and Ingo Heilmann (eds.), *Plant Lipid Signaling Protocols*, Methods in Molecular Biology, vol. 1009, DOI 10.1007/978-1-62703-401-2_22, © Springer Science+Business Media, LLC 2013

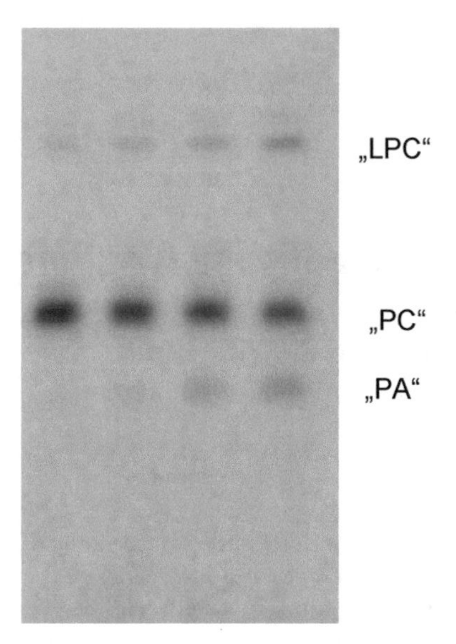

Fig. 1 Activity assay of PLA_2 in a plasma membrane preparation. According to Subheadings 3.1.1 and 3.2, a plasma membrane fraction prepared from cultured cells of *Eschscholzia* was incubated with an sn1-ether phospholipid labelled with BODIPY at the sn1 alkyl chain (sn1B in Table 1). Samples were withdrawn at 1, 2, 4 and 8 min (from left to right), extracted and separated by TLC, running top to bottom. Fluorescence was imaged with the STORM imager. Annotations are derived from TLC separations with authentic compounds: the substrate phospholipid ("PC"), the corresponding lysophospholipid ("LPC") generated by PLA_2 activity, and the corresponding phosphatidic acid ("PA") generated by PLD activity. The increase of "LPC" (1–8 min) is a measure of PLA_2 activity

Actually, assays using fluorescent phospholipids are the method of choice as they do not require expensive instrumentation and allow sufficient sensitivity and selectivity to assay PLA activities in most biological samples (11–14). Phospholipids with one labelled fatty can be used as substrates to determine the substrate specificity, i.e., to distinguish between PLA_2 and PLA_1. For this purpose, supplemental information can be obtained by using substrates that have one alkyl chain bound via an ether bond instead of the native fatty acid ester and thus cannot be liberated by phospholipase activity. Figure 1 provides a compilation of commonly used fluorescence-labelled PLA substrates.

Substrates with a fluorescent label at either fatty acid allow activity assays of high sensitivity: the most popular bis-BODIPY® FL C11-PC (sn1,2A in Table 1), after incorporation into cell membranes or micelles (15), is subject to intramolecular self quenching caused by the proximity of the BODIPY® FL fluorophores on adjacent phospholipid acyl chains. Separation of the fluorophores upon hydrolytic cleavage of one acyl chain by either PLA_1 or PLA_2 results in increased fluorescence. This increase can also be used to monitor the enzymatic conversion without separating the products, at least

Table 1
Compilation of fluorescent (BODIPY) labelled substrates for PLA activity assays

Abbreviation	Name, remarks	Formula
sn1A	*N*-((6-(2,4-DNP)amino) hexanoyl)-1-(BODIPY®FLC5)-2-hexyl-sn-glycero-3-phosphoethanolamine; contains quencher, one fluorescent product	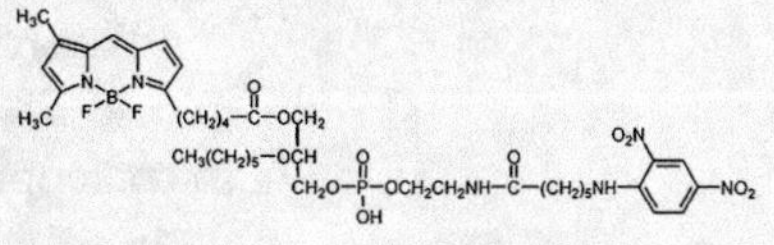
sn1B	2-decanoyl-1-(O-(11-(BODIPY®FLC11)-sn-glycero-3-phosphocholine; ether bond alkyl at sn-1; one fluorescent product	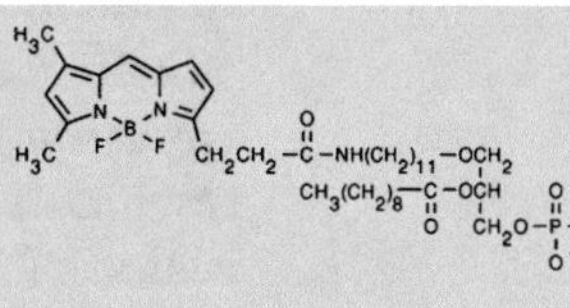
sn2B	2-(BODIPY®FLC5)-1-hexadecanoyl-sn-glycero-3-phosphocholine; one fluorescent product	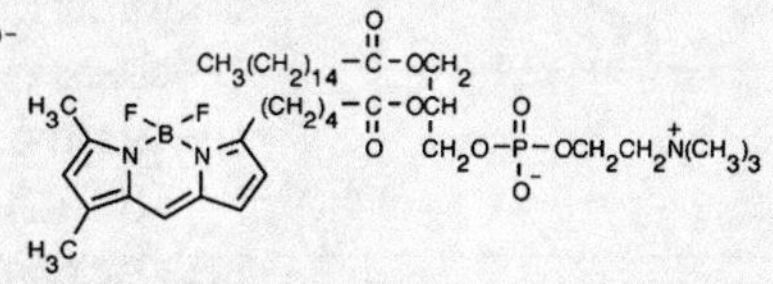
sn2A	N-((6-(2,4-dinitrophenyl)amino)hexanoyl)-2-(BODIPY®FLC5)-1-hexadecanoyl-sn-glycero-3-phosphoethanolamine; contains quencher, one fluorescent product	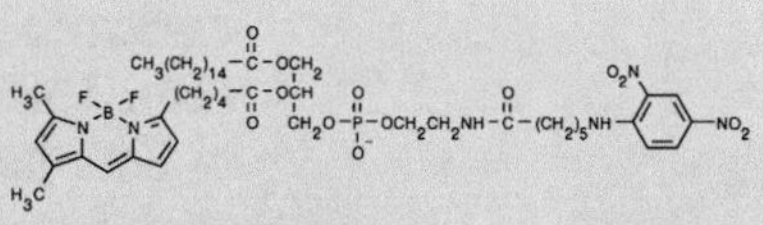
sn1,2A	1,2-bis-(BODIPY®FL C11)-sn-glycero-3-phosphocholine; intramolecular self-quenching, two fluorescent products	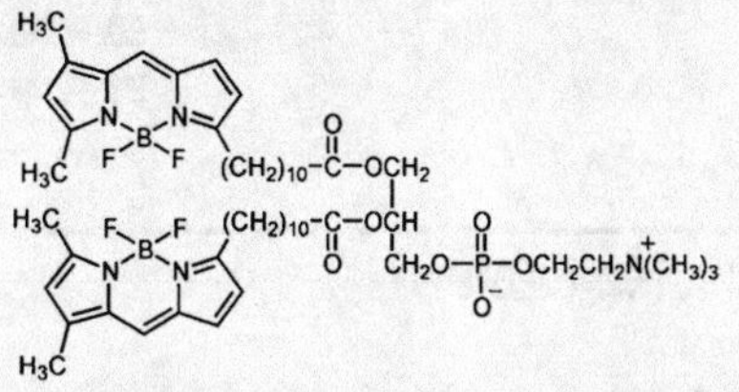
sn1,2B	1-O-(6-BODIPY®558/568-aminohexyl)-2-BODIPY®FL C5-sn-glycero-3-phosphocholine; intramolecular self-quenching by FRET; two fluorescent products	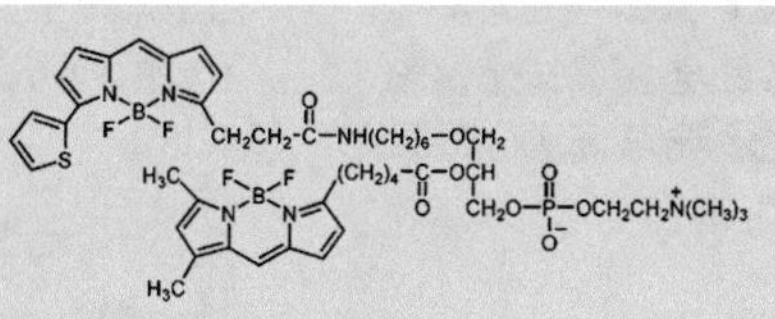

BODIPYFL C11: 4,4-difluoro-5,7-dimethyl-4-bora-3a,4a-diaza-s-indacene-3-undecanoyl
BODIPYFL C5: 4,4-difluoro-5,7-dimethyl-4-bora-3a,4a-diaza-s-indacene-3-pentanoyl
For more information of fluorescent PLA substrates, the reader is referred to the Molecular Probes Handbook (www.probes.invitrogen.com)

during an early phase of incubation when excess substrate is present. Another substrate-type which yields increased fluorescence due to the uncoupling of the quenched state, bears the BODIPY fluorophore at one fatty acid and the fluorescence quencher dinitrophenol at the phospholipid head group (sn1A and sn2A in Table 1).

A specific difficulty of PLA assays is presented by the amphiphilic properties of the substrate. In the useful concentration range, phospholipids do not form solutions of single molecules but rather organize in micelles of different complexity, or similar hyperstructures resembling their native state within cellular bilayers. In order to allow sufficient supply of substrate in a form accessible to the enzyme, mixed micelles of phospholipids are prepared, usually via sonication of the labelled substrate together with a compatible non-labelled phospholipid (15).

Fluorogenic substrates that generate fluorescence emission via the release from intramolecular quenching (cf. above) can be used for microscopic assays of PLA activity in cells and tissues. Digitized fluorescence images are obtained that allow both the localization and quantitation of cellular PLA activity (cf. Subheading 3.3). As the BODIPY fluorophore is a dye that can be imaged with the popular argon ion laser (Ex = 488 nm, Em = 530 nm), confocal images of high resolution can be obtained and used for a more accurate, in vivo localization of active PLA in cellular membranes (8, 16–18). Microscopic assays are especially useful when employed in connection with the biochemical detection of PLA in isolated membrane fractions, e.g., the plasma membrane prepared by two phase separation (19).

If PLA activities are to be quantified and compared between intact cells or tissues, it must be taken into account that cells may have a high capacity to reacylate the lysophospholipids formed by PLA-hydrolysis (16). This reacylation may lead to an underestimation of PLA activity or limit its detection, if high local product concentrations are generated. Substrates with an ether bond alkyl chain at the sn1-positon give rise to a lysolipid that is less or not reacylated and thus are more suited for PLA assays under such conditions (8, 16).

2 Materials

2.1 Fluorescent-Labelled Substrates for the Assay of PLA Activity

Table 1 provides a compilation of the most useful BODIPY-labeled fluorogenic substrates.

2.2 Buffers and Solutions

1. Incubation buffer: 20 mM Tris/HCl, 1 mM EDTA, pH 7.5, 50 mM NaCl, 0.5 % (w/v) sodium cholate.
2. Substrate solution: Generally, 1 mM stock solutions in ethanol are useful that can be diluted to give substrate solutions in the most useful range of about 0.05–2 μM for a given substrate lipid (cf. Table 1) in an appropriate buffer, e.g., 20 mM Tris/HCl, pH 7.5, that contains a low concentration of a compatible detergent, e.g., 0.02 % (w/v) CHAPS (3((3-cholamidopropyl) dimethylammonio)-propanesulfonic acid). If required, an unlabelled phospholipid, e.g., phosphatidylcholine or phosphatidylethanolamine is added to reach the final substrate concentration. This mixture is immediately sonicated (e.g., with the sonicator Bandelin Sonoplus HD70-2 with microtip, Power 70 %, Cycle 30 %, two pulses of 5 s separated by 5 s) in order to obtain sufficient unilamellar micelles, and stored at 4 °C in the dark (*see* **Notes 2** and **3**).

3 Methods

3.1 Monitoring Fluorescence Development in Microtiter Plates and in Single Probes

The following protocols refer to typical assays of PLA_2 activity in a microplate reader using subcellular samples or intact cells (*see* **Note 1**). For the quantification of high PLA_2 activities, the increase in fluorescence resulting from cleavage of bis-BODIPY FL C11 PC (sn1, 2A in Table 1) was monitored using an FLX 800 fluorescence microplate reader (Bio-Tek Instruments, http://www.biotek.com) over maximally 10 min and the initial rates (estimated either after 1 min of reaction or by non-linear curve fitting) were taken as a measure of enzyme activity. The increase in fluorescence reflects the extent of substrate hydrolysis as shown earlier (11). For the quantification of low PLA_2 activities, which do not substantially decrease the substrate concentration, longer incubation times (30–120 min) may be required, and in this case the products should be quantified after chromatographic separation (cf. Subheading 3.2).

3.1.1 Detecting Activity in Subcellular Fractions

In a black microtiter plate designed for fluorescence assays each well receives:

- 20 μl Test sample (soluble proteins, immunoprecipitates, microsomal membranes, plasma membranes).
- 30 μl Incubation buffer, mix by shaking horizontally.
- Add 180 μl substrate solution, mix by shaking in the fluorimeter and monitor the fluorescence over 10–60 s, using the top mode of fluorescence detection (*see* **Note 4**).

3.1.2 Detecting Activity in Cell Suspensions

In a black microtiter plate with transparent bottom compatible with fluorescence detection each well receives:

- 50 μl Substrate solution (*see* **Note 5**), equilibrate for 2 min.
- Add 20 μl cell suspension (about 1 mg fresh weight in 100 mM sorbitol), shake thoroughly for 2 s and monitor the fluorescence at Ex 485 nm, Em 528 nm over 60–120 s, using the bottom mode of fluorescence detection.

3.1.3 Monitoring Fluorescence Development in a Stirred Cell Suspension

A simple but less accurate version of PLA_2 assays is the monitoring of fluorescence development from intramolecular quenched substrates in a suspension of intact cells or membrane vesicles that is agitated in a spectrofluorimeter cuvette by a magnetic stirrer. A typical assay mixture contains 1 ml cell suspension (about 5 mg dry weight in an appropriate nutrient solution) or a plasma membrane preparation (about 25 μg protein, in 5 mM K-phosphate buffer, pH 7.8, 330 mM betain, 5 mM KCl, 0.1 mM EDTA).

Bis-BODIPY FL C11 PC is added from a stock solution in DMSO to give a final concentration of 10 μM (DMSO≤1 %) and the mixture gently stirred in a quartz cuvette. In our hands, fluorescence was recorded using a Shimadzu RF 5000 fluorimeter at Ex 485±3 nm and Em 515±5 nm.

3.2 Assays Requiring Separation of Products

The conversion of substrates with one labelled fatty acid requires the separation of hydrolysis products (lysophospholipid and/or fatty acids). For this purpose, assays according to Subheading 3.1.1 or 3.1.2 are performed in polypropylene reaction tubes in the dark (12, 13). At fixed time points, 100 μl of the samples are withdrawn (cell suspension assays are diluted before to an appropriate volume with buffer containing 100 mM sorbitol) and immediately transferred into 300 μl $CHCl_3$–methanol (1:2, v/v) and the mixture shaken horizontally at 400 rpm for 30 min. After addition of another 100 μl $CHCl_3$ and 100 μl 150 mM NaCl, and 5 min incubation by shaking at room temperature, two phases separate, which can be facilitated by mild centrifugation. The bottom phase containing the PLA_2 substrates and products is analyzed by thin layer chromatography (TLC).

For TLC analysis, 4 μl of the lower phase are spotted onto silica 60 TLC glass plates (Merck) and the chromatogram is developed with a mobile phase consisting of $CHCl_3$–methanol–water (65/25/4, v/v/v). After drying on air the fluorescence or radioactivity of the peaks is quantified on a densitometer (e.g., STORM 860, Molecular Dynamics) or by using a digital camera and an appropriate densitometric software. A typical chromatogram is shown in Fig. 1.

Alternatively, lipid products of the PLA-assay can be analyzed by HPLC (10) and/or by mass spectrometry-based procedures. For instance, phospholipids and their hydrolysis products can easily be analyzed by MALDI-TOF MS (8, 20). A typical readout of MS-based lipid analysis of PLA-products is shown in Fig. 2.

For either method applied, the spot or peak intensity is correlated to concentrations of PLA_2 products by calibration graphs obtained from known substrate concentrations after complete hydrolysis by bee venom PLA_2.

3.3 Cellular PLA Activity Assayed by Fluorescence Microscopy

Fluorescence images of PLA products distributed in intact cells and tissues can be used both for quantitation and for localization of the enzyme activity in living cells. The bis-BODIPY FL-labelled phospholipids can be used for this purpose with classical fluorescence microscopy or with confocal imaging.

The following is a typical protocol tested with cultured cells of *Eschscholzia californica*.

Cell suspensions are harvested by filtering through a nylon mesh of 50 μm square pore size, washed with fresh, phosphate-free culture liquid and resuspended in the same medium (about 4 mg dry weight/ml). The fluorogenic substrate, e.g., bis-BODIPY FL C11 PC (sn1,2B in Table 1) is added from a 1 mM stock solution in DMSO to give a final concentration of 10 μM. 5–10 μl of this suspension are spotted onto a microscopic slide

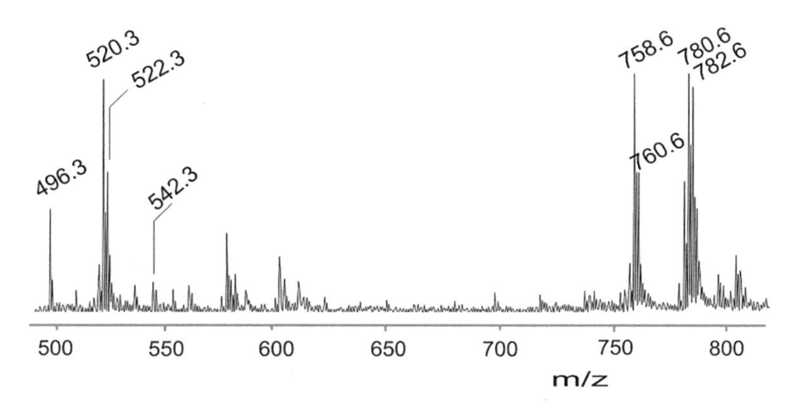

Fig. 2 Detection of substrates and products of PLA2 by MALDI-TOF MS. Cell suspensions of *Eschscholzia californica* were extracted with chloroform–methanol as described in Subheading 3.2 and the phospholipid fraction subjected to MALDI-TOF MS. Mass peaks of phosphatidylcholines and lysophosphatidylcholines are as follows

496.3	LPC 16 : 0 + H^+	758.6	PC 16 : 0 + 18 : 2 + H^+
520.3	LPC 18 : 2 + H^+	760.6	PC 16 : 0 + 18 : 1 + H^+
522.3	LPC 18 : 1 + H^+	780.6	PC 16 : 0 + 18 : 2 + Na^+
542.3	LPC 18 : 2 + Na^+	782.6	PC 16 : 0 + 18 : 1 + Na^+

Taken with permission from (8)

and covered with gas permeable biofoil (Heraeus, Osterode, Germany). Optical filters are selected for Ex 490 ± 10 nm and Em < 520 nm. Fluorescence images are captured and digitized with a video camera (e.g., SONY 3CCD) connected to a computerized image analysis system which allows to quantify fluorescence intensity at distinct areas.

In order to follow PLA activity over extended periods of time (e.g., >5 min) the use of a flowthrough cell is advisable. In this case, cells can be kept in place by a non-fluorescing nylon mesh of 150 μm square pore size.

A more precise localization of PLA-derived fluorescence requires confocal microscopy. In this case, 5 μl samples of the cell suspension to be tested are spotted onto an Isopore® membrane (Millipore, www.millipore.com) that is placed on top of an agarose (3 %) gel disk of 0.5 mm thickness made from the same liquid, sealed by a cover glass and mounted at a confocal microscope. In our hands, experiments were successfully performed using a Leica TCS-SP confocal microscope. The substrate solution (2 μM bis-BODIPY FL C11 PC in diluted culture liquid, containing effectors as required), is injected with a microsyringe into the agarose disk. Fluorescence images are scanned at 63-fold optical magnification at Ex = 488 nm (Argon laser line) and Em = 510–530 nm. Typical images are shown in Fig. 3.

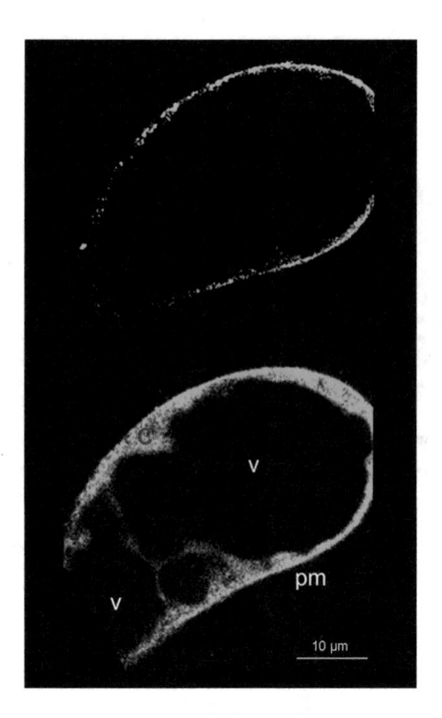

Fig. 3 Confocal fluorescence images of PLA activity in a plant cell. According to the protocol in Subheading 3.3, 5 μl of a cell suspension of *Eschscholzia californica* were incubated with 400 nM of the fluorogenic substrate bis-BODIPY FL C11 PC (sn1,2A in Table 1) and the fluorescence emission monitored at a Leica TCS-SP confocal microscope. After 5 min of equilibration with the substrate, confocal fluorescence images were scanned with the argon ion laser (Ex 488 nm, Em 510–530 nm) and the detector sensitivity adjusted to cut off the (low) fluorescence emission of control cells. Yeast glycoprotein elicitor was then added, which activates PLA_2. Further confocal scans were performed in 30 s distances. Images are shown as obtained after 1.5 min (top) and 4 min (bottom) after elicitor contact. Note that fluorescence arises first at the plasma membrane (pm) and rapidly spreads over the cytoplasm (c) but not the vacuole (v). The spread of fluorescence does not reflect the travelling of the substrate (which equilibrated prior to the elicitor contact) but rather of all BODIPY-labelled hydrolysis products, as LPC, fatty acid and LPC acylated by non-fluorescent genuine fatty acids. Taken with permission from ref. 16

4 Notes

1. Experiments can be performed at ambient temperature, but an accurate temperature control is advisable if the comparison of activity data is critical.
2. To ensure good reproducibility, the substrate solutions need to be prepared daily. Radiolabelled phospholipid substrates are handled in the same way.
3. Due to the formation of unilamellar liposomes, the working solution should be turbid but not precipitate. If precipitate occurs, the solution can be cleared by mild centrifugation.

4. Blanks are required to eliminate the initial and non-enzymatic generation of fluorescence. In these wells, the substrate or the enzyme sample, respectively, is replaced by the appropriate buffers.
5. For activity tests in suspension cultured cells, prepare substrate solution in buffer containing 100 mM sorbitol and 0.01 % (w/v) CHAPS.

References

1. Burke JE, Dennis EA (2009) Phospholipase A2 structure/function, mechanism and signaling. J Lipid Res 50:S237–S242
2. Wang X (2001) Plant Phospholipases. Annu Rev Plant Physiol Plant Mol Biol 52:211–231
3. Seo YS, Kim EY, Mang HG et al (2008) Heterologous expression, and biochemical and cellular characterization of CaPLA1 encoding a hot pepper phospholipase A1 homolog. Plant J 53:895–908
4. Hoffmann GE, Schmidt D, Bastian B (1986) Photometric determination of phospholipase A. J Clin Chem Clin Biochem 24:871–875
5. Mansfeld J, Ulbrich-Hofmann R (2007) Secretory phospholipase A2-alpha from *Arabidopsis thaliana*: functional parameters and substrate preference. Chem Phys Lipids 150:156–166
6. Kinkaid AR, Wilton DC (1993) A continuous fluorescence displacement assay for PLA2 using albumin and medium chain phospholipid substrates. Anal Biochem 212:65–70
7. Muzzio IA, Gandhi CC, Manyam U et al (2001) Receptor-stimulated phospholipase A2 liberates arachidonic acid regulates neuronal excitability through protein kinase C. J Neurophysiol 85:1639–1645
8. Viehweger K, Dordschbal B, Roos W (2002) Elicitor-activated phospholipase A2 generates lysophosphatidylcholines that mobilize the vacuolar H^+ pool for pH signaling via the activation of Na^+-dependent proton fluxes. Plant Cell 14:1509–1525
9. Fuchs B, Süß R, Schiller J (2010) An update of MALDI-TOF mass spectrometry in lipid research. Prog Lipid Res 49:450–475
10. Karkabounas A, Kitsiouli EI, Nakos G et al (2011) HPLC-fluorimetric assay of phospholipase A(2). Application to biological samples with high protein content and various reaction conditions. J Chromatogr B Analyt Technol Biomed Life Sci 879:1557–1564
11. Roos W, Dorschbal B, Steighard J, Hieke M, Weiß D, Saalbach G (1999) A redox-dependent, G-protein-coupled phospholipase A of the plasma membrane is involved in the elicitation of alkaloid biosynthesis in *Eschscholzia californica*. Biochem Biophys Acta 1448:390–402
12. Viehweger K, Schwartze W, Schumann B, Lein W, Roos W (2006) The G protein controls a pH-dependent signal path to the induction of phytoalexin biosynthesis in *Eschscholzia californica*. Plant Cell 18:1510–1523
13. Heinze M, Steighardt J, Gesell A et al (2007) Regulatory interaction of the G protein with phospholipase A2 in the plasma membrane of *Eschscholzia californica*. Plant J 52: 1041–1051
14. Hendrickson HH, Hendrickson EK, Johnson ID et al (1999) Intramolecularly quenched BODIPY-labeled phospholipid analogs in phospholipase A2 and platelet-activating factor acetylhydrolase assays and *in vivo* fluorescence imaging. Anal Biochem 276:27–35
15. Meshulam T, Herscovitz H, Casavant D et al (1992) Flow cytometric kinetic measurements of neutrophil phospholipase A activation. J Biol Chem 267:21465–21470
16. Schwartze W, Roos W (2008) The signal molecule lysophosphatidylcholine in *Eschscholzia californica* is rapidly metabolized by reacylation. Planta 229:183–191
17. Balestrieri B, Hsu VW, Gilbert H et al (2006) Group V secretory phospholipase A2 translocates to the phagosome after zymosan stimulation of mouse peritoneal macrophages and regulates phagocytosis. J Biol Chem 281: 6691–6698
18. Balsinde J, Perez R, Balboa MA (2006) Calcium-independent phospholipase A2 and apoptosis. Biochim Biophys Acta 1761:1344–1350
19. Larsson C, Sommarin M, Widell S (1994) Isolation of highly purified plant plasma membranes and separation of inside-out and right-side-out vesicles. Methods Enzymol 228:451–469
20. Eibisch M, Schiller J (2011) Sphingomyelin is more sensitively detectable as a negative ion than phosphatidylcholine: a matrix-assisted laser desorption/ionization time-of-flight mass spectrometric study using 9-aminoacridine (9-AA) as matrix. Rapid Commun Mass Spectrom 25:1100–1106

Part III

Protein–Lipid Interactions

Chapter 23

Lipid-Binding Analysis Using a Fat Blot Assay

Teun Munnik and Magdalena Wierzchowiecka

Abstract

Protein–lipid interactions play an important role in lipid metabolism, membrane trafficking and cell signaling by regulating protein localization, activation, and function. The Fat Blot assay is a relatively simple and inexpensive method to examine these interactions using nitrocellulose membrane-immobilized lipids. The assay is adapted from the method by Dowler et al. (Sci STKE 129:pl6, 2002) and provides qualitative and quantitative information on the relative affinity with which a protein binds to a particular lipid. To perform a Fat Blot assay, serial dilutions of different phospholipids are spotted onto a nitrocellulose membrane. These membranes are then incubated with a lipid-binding protein possessing a GST (or other epitope) tag. The membranes are washed and the protein, which is bound to the membrane by virtue of its interaction with the lipid's head group, is detected by immunoblotting with an antibody against GST (or other epitope). The procedure only requires a few micrograms of protein and is quick, simple and cheap to perform.

Key words Polyphosphoinositides, Phosphatidic acid, Diacylglycerol, Inositol lipids, Lipid binding, Lipid targets, Lipid-binding domains, Lipid binding overlay, Protein–lipid interactions, FYVE, FAPP1, PLCδ1, Pleckstrin homology domain

1 Introduction

Eukaryotic cells contain a wide range of lipids, including sphingolipids, neutral lipids, glycolipids, and phospholipids, all exhibiting unique biophysical properties. While the majority of these lipids have a role in membrane structure and fat storage, a minor fraction is involved in signal transduction and membrane trafficking. Hallmarks of such "signaling lipids" are their low abundance, rapid turnover, and transient increases in response to environmental cues or endogenous signals, so that downstream signaling pathways are activated, and specific cellular events and physiological responses triggered. In plants, typical signaling lipids include polyphosphoinositides (PPIs) and phosphatidic acid (PA) (2–8). PA can be generated via phospholipase D (PLD) hydrolysis of structural phospholipids or via phosphorylation of diacylglycerol (DAG) by

Teun Munnik and Ingo Heilmann (eds.), *Plant Lipid Signaling Protocols*, Methods in Molecular Biology, vol. 1009,
DOI 10.1007/978-1-62703-401-2_23,

DAG kinase (DGK) (4, 8, 9). PPIs are derivatives of phosphatidylinositol (PI), synthesized by reversible phosphorylation of the inositol head group at the 3-, 4-, and/or 5-position (1). In plants, five different PPIs can be distinguished, i.e., PI3P, PI4P, PI5P, $PI(3,5)P_2$, and $PI(4,5)P_2$ (2, 3, 5).

Signaling lipids can specifically interact with protein targets via selective lipid-binding domains which recognize the lipid's head group. As a consequence, the intracellular localization and/or enzymatic activity of target proteins are affected. Through crystallography, several of these lipid-binding domains have been characterized over the years, including FYVE, Phox homology (PX), and pleckstrin homology (PH) domains (10). FYVE and PX domains predominantly bind PI3P, for PH domains the specificity varies per protein. For instance, $PH_{PLC\delta1}$ binds to $PI(4,5)P_2$ while PH_{FAPP1} binds specifically toPI4P. There are also PH domains that bind PPIs nonspecifically or not at all (10).

Genomes of *Arabidopsis* and other plant species are predict to encode many proteins exhibiting one or more lipid-binding domains but only a few have been characterized (5, 11, 12). Here, a relatively quick and easy Fat blot assay is described to test whether such domains bind lipids, and to which lipids in particular. Alternative lipid-binding assays are described in Chapter 24 (13), using raffinose loaded-phospholipid vesicles, and Chapter 25, using lipid-affinity beads (14).

For a Fat Blot assay, serial dilutions of different phospholipids are spotted onto a nitrocellulose membrane. These membranes are then incubated with a lipid-binding protein fused to a GST (or other epitope) tag. Membranes are then washed and the protein, bound to the membrane by virtue of its interaction with lipid(s), is detected by immunoblotting using an antibody against the epitope tag. The procedure requires a few micrograms of protein and is relatively quick, simple and cheap to perform (1).

2 Materials

2.1 Fat Blot Preparation

1. Chloroform.
2. Resuspension Buffer for PPIs: $CHCl_3/MeOH/H_2O=20:9:1$ (*see* **Note 1**). Resuspension Buffer for other lipids: $CHCl_3$.
3. Ponceau S dye: 0.1 % (w/v) in 5 % HAc.
4. Spotting buffer: $CHCl_3$/MeOH/50 mM HCl/Ponceau S = 250:500:200:2 (by. vol.).
5. Nitrocellulose membrane: Amersham Hybond-C Extra (RPN203E). Cut into the appropriate size and notch the top left corner for orientation. Normal size: 4.75 cm × 2.75 cm ($l \times b$) for 3 × 8 spots (see Fig. 1). Bigger blots can be prepared accordingly.

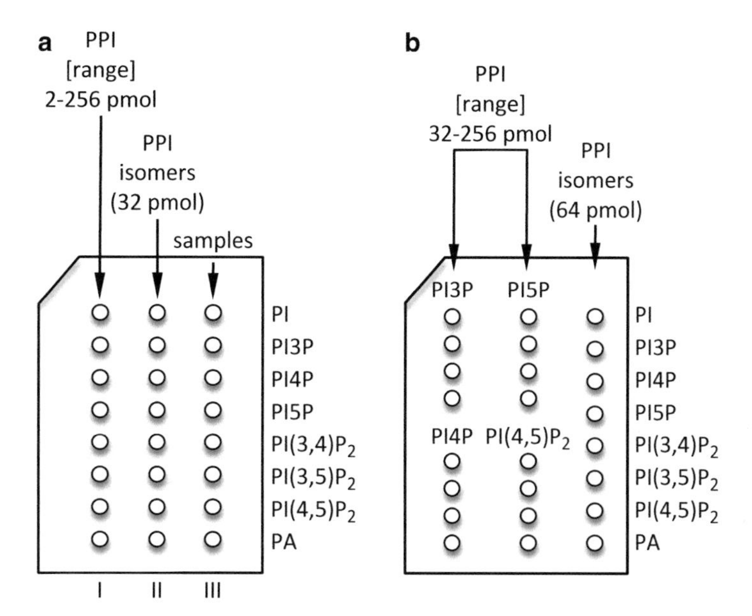

Fig. 1 Schematic drawing of Fat Blot-spotting examples to determine protein binding to a PPI. In principle, any lipid can be spotted in any concentration or order. Here two examples are given. (**a**) From *top to bottom* and from *left to right*, eight spots for *(I)* a PPI concentration range (dilution series), *(II)* eight spots for a series of PPI isomers in one concentration, and *(III)* eight spots for a series of samples to be applied later (e.g., lipid extracts, or other phospholipids to be tested). (**b**) Alternative blot, in this case designed to focus on PI4P binding

6. Scissors.
7. Soft pencil.
8. 0.5 ml reaction tubes.
9. Cold room.
10. Glass vials, 1 and 4 ml, with screw caps. Preferably brown glass.
11. 1 mM phospholipid stocks: PA, PC, PE, PG, PI, PI3P, PI4P, PI5P, PI(3,4)P_2, PI(3,5)P_2, PI(4,5)P_2, PI(3,4,5)P_3, PS. Use long-chain lipids, synthetic or natural (*see* **Note 2**). Prepare 1 mM stock solutions by dissolving normal phospholipids into chloroform; Dissolve the polar PPIs into chloroform:methanol:water (20:9:1, by vol.) (*see* **Note 1**). Store at −20 °C in glass vials under N_2(g) or Ar(g).
12. N_2 or Ar gas.
13. Gilson pipets P1000, P200, P20, P2, and their tips, or equivalent pipets.
14. Aluminum foil.

2.2 Fat Blot Protein Binding

1. Fat Blot buffer: 50 mM Tris–HCl, pH 7.5, 150 mM NaCl.
2. Fat Blot buffer-0.06 % Tween: 50 mM Tris–HCl, pH 7.5, 150 mM NaCl, 0.06 % Tween-20.

3. Fat Blot buffer-0.1 % Tween: 50 mM Tris–HCl, pH 7.5, 150 mM NaCl, 0.1 % Tween-20.
4. Blocking buffer: 3 % bovine serum albumin (BSA; fatty acid free; *see* **Note 3**) in Fat Blot buffer.
5. Blocking buffer-0.06 % Tween: 3 % BSA in Fat Blot buffer-0.06 % Tween.
6. Blocking buffer-0.1 % Tween: 3 % BSA in Fat Blot buffer-0.1 % Tween.
7. GST-fusion protein. For example, 2xFYVE (PI3P), PH_{FAPP1} (PI4P), $PH_{PLC\delta 1}$ [for PI(4,5)P_2] or any other phospholipid-binding domain in blocking buffer-0.1 % Tween (0.5–1.0 μg/ml).
8. Graduated cylinders (100, 25, and 5 ml).
9. Microtiter plates with four compartments.
10. 50 ml Polypropylene tubes.
11. Plastic wrap.
12. ECL kit (General Electric).
13. Anti-GST (or other tag) antibodies.
14. Secondary antibodies.
15. Autoradiography film (Kodak, X-Omat S).
16. PhosphorImager and screen.
17. Light cassette.

3 Methods

3.1 Fat Blot Spotting

PPIs are relatively expensive and usually bought as dried ammonium salts in very small quantities. Therefore, when stock and dilution series have to be prepared, we usually spot 6 or 12 blots. Stored in the cold room (4 °C) in the dark (aluminum foil), these blots stay remarkably stable, for at least 6 months.

1. Cut 6 or 12 pieces of nitrocellulose membrane into the appropriate size and notch the top left corner for orientation. Size: ~4.75 cm × 2.75 cm ($l \times b$) for 3 × 8 spots (see Fig. 1).
2. For each blot, mark with a soft pencil, from top to bottom and from left to right, eight spots for *(I)* a PPI concentration range (dilution series), *(II)* eight spots for a series of PPI isomers in one concentration, and *(III)* eight spots for a series of samples to be applied later (e.g., lipid extracts, or other phospholipids to be tested), as indicated in Fig. 1a. Alternative blots can be designed, focusing on the domain's specificity, e.g., see Fig. 1b.
3. Set-up a cold room for the blot spotting. Label 0.5 ml reaction tubes for the dilution range of a lipid species. Also label tubes

for the lipids to be spotted in a single concentration. Conduct all further steps in the cold.

4. Resuspend the dried PPIs in Resuspension Buffer to make a 1 mM stock solution. Work in small, 1 ml glass vials and make sure to wash the walls well to ensure that all lipid is resuspended. For example: wash the walls of a 0.1 mg vial of PI4P with 102.5 μl Resuspension Buffer. Make 1 mM stock for all lipids.
5. Dilute the lipids in the labeled 0.5 ml Eppendorf tubes. Use Spot Buffer as a diluent.
   (a) To make a 128 pmol spot, 1 μl of a 128 μM solution is required.
   (b) To make the 128 μM sub-stock, add 12.8 μl 1 mM stock to 87.2 μl Spot Buffer.
   (c) Serial dilute this 128 μM sub-stock 1:1 to obtain the 64 μM sub-stock by adding 50 μl of 128 μM lipid to 50 μl Spot Buffer.
   (d) Dilute further until the 32, 16, 8, 4, 2, and 1 μM sub-stocks are obtained.
   (e) For the PPI isomers and other lipids (e.g., PA), dilute directly from the 1 mM Stock in Spot Buffer to the appropriate concentration (II).
6. Align all Fat Blots (e.g., 12) in 3×4 rows (*see* **Note 4**).
7. Spot 1 μl of all dilutions using Spot Buffer-equilibrated pipet tips, from top to bottom, from left to right.
8. Dry the blots at room temperature for 1 h in a dark place such as an empty drawer.
9. Transfer all 1 mM lipid stocks back into the −20 °C freezer under N_2(g).
10. Store the blots in Alu foil at 4 °C until further use.

3.2 Fat Blot Procedure

1. Take one of the Fat Blots spotted according to Subheading 3.1 from its 4 °C storage and place in a small plastic container. We prefer microtiter plates with four compartments for normal blots, or 50 ml Falcon tubes for big blots.
2. Block the Fat Blot with 5 ml blocking buffer w/o Tween (*see* **Note 5**). Gently agitate for 1 h at room temperature.
3. Prepare 2 ml lipid-binding domain protein solution (0.5 μg/ml) in blocking buffer-0.1 % Tween (4 ml for big blots in 50 ml polypropylene tubes).
4. Discard blocking solution and add the lipid-binding domain solution to the Fat Blot.
5. Gently agitate O/N at 4 °C.

6. Discard the solution and wash the Fat Blot, three times with gentle agitation in 5–6 ml Fat Blot buffer-0.1 % Tween (5 min/wash) in microtiter plates (or 50 ml polypropylene tubes for bigger blots).
7. Prepare primary antibody solution. For anti-GST, we add 3 μl a 1:2,000 stock to 5 ml of blocking buffer-0.06 % Tween.
8. Add primary antibody solution to the Fat Blot and gently agitate for 1–2 h at room temperature.
9. Discard GST antibody and wash blot three times with gentle agitation in 5 ml Fat Blot buffer-0.1 % Tween for 5 min/wash.
10. Prepare 5 ml secondary antibody solution in blocking buffer-0.06 % Tween, i.e., the HRP-conjugated anti-mouse 1:3,300 (1.5–2 μl).
11. Add secondary antibody solution to Fat Blot. Gently agitate for 1–1.5 h at RT.
12. Discard secondary antibody solution and wash blot four times with gentle agitation in 8 ml Fat Blot buffer-0.1 % Tween for 5 min/wash.
13. Detect with chemiluminescent developing solution according to supplier's instructions and visualize by exposing to an autoradiography film (1–30 min).
14. An example of a Fat Blot using a GST-fusion of the 2xFYVE domain is shown in Fig. 2.

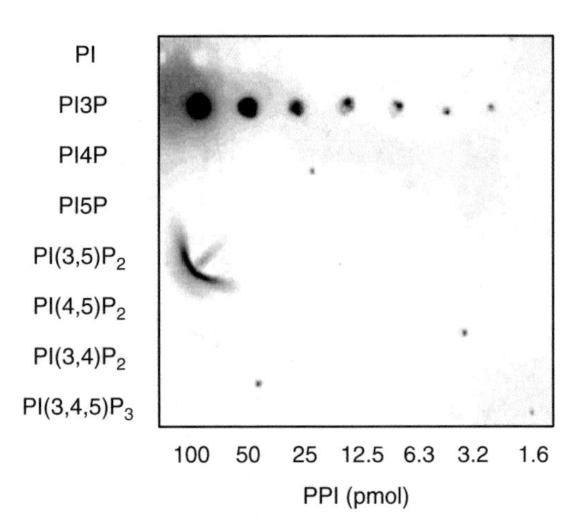

Fig. 2 Fat Blot showing specific binding of GST-2xFYVE to PI3P. Fat Blot containing serial dilutions of all PPI isomers was incubated with 1 μg/ml GST-2xFYVE, washed, incubated with an anti-GST antibody, and visualized using ECL. Adapted from Vermeer et al. (12)

4 Notes

1. PPIs are very polar and do not dissolve in chloroform. Water is required to increase the polarity of the solvent and MeOH is required to dissolve the water into the chloroform.
2. Avanti polar lipids are preferred for their quality. Use unsaturated, e.g., C18:1 lipids as they are easier to solubilize than saturated lipids. Use chain lengths of C16 or longer; the shorter the fatty acids, the less they bind to the nitrocellulose.
3. Make sure to use fatty-acid free BSA since "normal" BSA has lots of lipids bound to it, including phospholipids, which may introduce increased background levels and reduce the level of "detector" protein for the spotted lipids.
4. To align all Fat Blots (e.g., 12) and to keep them flat for spotting, microscopic glass slides are used.
5. Blocking works better without Tween-20. For all antibody incubations we prefer 0.06 % Tween, and for the lipid binding-protein incubations and washings, 0.1 % Tween works better.

References

1. Dowler S, Kular G, Alessi DR (2002) Protein lipid overlay assay. Sci STKE 129:pl6
2. Xue HW, Chen X, Mei Y (2009) Function and regulation of phospholipid signalling in plants. Biochem J 421:145–156
3. Munnik T, Vermeer JE (2010) Osmotic stress-induced phosphoinositide and inositol phosphate signalling in plants. Plant Cell Environ 33:655–669
4. Testerink C, Munnik T (2011) Molecular, cellular, and physiological responses to phosphatidic acid formation in plants. J Exp Bot 62:2349–2361
5. Munnik T, Nielsen E (2011) Green light for polyphosphoinositide signals in plants. Curr Opin Plant Biol 14:489–497
6. Ischebeck T, Seiler S, Heilmann I (2010) At the poles across kingdoms: phosphoinositides and polar tip growth. Protoplasma 240:13–31
7. Wang X, Devaiah SP, Zhang W, Welti R (2006) Signaling functions of phosphatidic acid. Prog Lipid Res 45:250–278
8. Arisz SA, Testerink C, Munnik T (2009) Plant PA signaling via diacylglycerol kinase. Biochim Biophys Acta 1791:869–875
9. Li M, Hong Y, Wang X (2009) Phospholipase D- and phosphatidic acid-mediated signaling in plants. Biochim Biophys Acta 1791: 927–935
10. Lemmon MA (2008) Membrane recognition by phospholipid-binding domains. Nat Rev Mol Cell Biol 9:99–111
11. van Leeuwen W, Okresz L, Bogre L, Munnik T (2004) Learning the lipid language of plant signalling. Trends Plant Sci 9:378–384
12. Vermeer JE, van Leeuwen W, Tobena-Santamaria R, Laxalt AM, Jones DR, Divecha N, Gadella TW Jr, Munnik T (2006) Visualization of PtdIns3P dynamics in living plant cells. Plant J 47:687–700
13. Julkowska MM, Rankenberg JM, Testerink C (2013) Liposome-binding assays to assess specificity and affinity of phospholipid–protein interactions. Methods Mol Biol. 1009: 261–271
14. McLoughlin F, Testerink C (2013) Lipid affinity beads; from identifying new lipid binding proteins to assessing their binding properties. Methods Mol Biol. 1009: 273–280

Chapter 24

Liposome-Binding Assays to Assess Specificity and Affinity of Phospholipid–Protein Interactions

Magdalena M. Julkowska, Johanna M. Rankenberg, and Christa Testerink

Abstract

Protein–lipid interactions play an important role in cellular protein relocation, activation and signal transduction. The liposome-binding assay is a simple and inexpensive method to examine protein–lipid binding in vitro. The phospholipids used for liposome production are dried and hydrated. Subsequent extrusion of the phospholipid mixture ensures the production of large unilamellar vesicles (LUV) filled with raffinose. Those LUVs can be easily separated from the aqueous solution by centrifugation. By incubating a protein of interest with the LUVs and subsequent centrifugation steps, the bound protein fraction can be determined using Western Blot or Coomassie staining. This technique enables analysis of protein–lipid binding affinity and specificity.

Key words Large unilamellar vesicle (LUV), Liposome assay, Phospholipid binding, Binding affinity, Phosphatidic acid, Phospholipid signaling

1 Introduction

Protein–lipid interactions are important for eukaryotic cellular signaling pathways (1). If a protein has a predicted lipid-binding domain, or when there is other evidence that it might bind phospholipids, it is important to establish its in vitro lipid-binding characteristics. Several methods to do this have been described. These include precipitation using lipid affinity beads (2–5) lipid overlay assays (6), vesicle (liposome) binding assays (7–9), and SPR-based methods (10, 11).

All of these techniques have their own advantages and purposes. The vesicle-binding technique has the advantage over fat blots and beads that the lipids are presented in a more natural way, i.e., in a lipid bilayer. The surface plasmon resonance (SPR) method is especially suitable for getting quantitative binding data on affinity, but is rather complicated to perform and requires expensive equipment

Teun Munnik and Ingo Heilmann (eds.), *Plant Lipid Signaling Protocols*, Methods in Molecular Biology, vol. 1009,
DOI 10.1007/978-1-62703-401-2_24,

as well as specific expertise (11). Liposome-binding assays can be carried out using a simple procedure and standard, inexpensive tools on any recombinant, tagged protein. Thus, it is the method of choice for a quick validation of candidate phospholipid-binding proteins.

The protocol described here was successfully used in our lab to assess lipid-binding affinity and specificity of both phosphatidic acid (PA)- and polyphosphoinositide (PPI)-binding proteins (8, 12) (MMJ, JMR, CT; unpublished data).

2 Materials

2.1 Stock Solutions of Phospholipids for Liposome Production

Long-chain synthetic or natural phospholipids can be used for preparing liposomes (*see* **Note 1**). To prepare stock solutions, the phospholipids are dissolved in chloroform, or chloroform–methanol–water (20:9:1) in the case of polyphosphoinositides, and stored at −20 °C.

2.2 Stock Solutions for Liposome Binding Assay

EDTA, KCl, and Tris buffers can be made beforehand and stored at room temperature. Laemmli sample buffer 4× and DTT can be made beforehand and stored at −20 °C. Extrusion and binding buffers are to be prepared fresh and kept at room temperature.

1. 1 M DTT: Dissolve 1.54 g dithiothreitol in 10 ml distilled water. Store in 200 μl aliquots at −20 °C.
2. 0.5 M EDTA: Dissolve 73 g ethylenediaminetetraacetic acid in 250 ml distilled water. Adjust the pH to 8.0 with concentrated NaOH otherwise EDTA will not dissolve. Store at room temperature.
3. 1 M Tris–HCl pH 7.5: Dissolve 121.14 g Tris in 500 ml distilled water. Adjust the pH to 7.5 with concentrated HCl. Bring the total volume to 1 l with distilled water. Store at room temperature.
4. 1 M Tris–HCl pH 6.8: Dissolve 121.14 g Tris in 500 ml distilled water. Adjust the pH to 6.8 with concentrated HCl. Bring the total volume to 1 l with distilled water. Store at room temperature.
5. 1 M KCl: Dissolve 74.55 g potassium chloride in 1 l distilled water. Store at room temperature.
6. Extrusion buffer: 250 mM Raffinose pentahydrate, 25 mM Tris–HCl pH 7.5, 1 mM DTT. Add 3.72 g of raffinose pentahydrate to 20 ml distilled water. Microwave briefly to dissolve all raffinose (*see* **Note 2**). Let the solution cool down and add 25 μl of 1 M DTT and 625 μl of 1 M Tris–HCl pH 7.5. Bring the total volume to 25 ml. Prepare fresh. Keep at room temperature to prevent raffinose from precipitating.

7. Binding buffer 6×: 750 mM KCl, 150 mM Tris–HCl pH 7.5, 6 mM DTT, 3 mM EDTA. Combine 7.5 ml of 1 M KCl, 1.5 ml of 1 M Tris–HCl pH 7.5, 60 μl 1 M DTT and 60 μl of 0.5 M EDTA. Bring the total volume to 10 ml. Prepare fresh. Keep at room temperature.
8. Binding buffer 1×: 150 mM KCl, 25 mM Tris–HCl pH 7.5, 1 mM DTT, 0.5 mM EDTA. Dilute binding buffer 6× in distilled water in 1:5 ratio.
9. Laemmli sample buffer 4×: Dissolve 0.8 g of SDS in 4.6 ml 87 % glycerol, 2 ml β-mercaptoethanol, 2.4 ml 1 M Tris–HCl pH 6.8 and 1 ml distilled water. Add 8 mg bromophenol blue. Store in 1 ml aliquots at −20 °C.
10. Laemmli sample buffer 1×: Dilute the sample buffer 4× in distilled water in 1:4 ratio.

2.3 Materials for Liposome Binding Assay

1. 2 ml Safe-Lock Eppendorfs with round bottom.
2. 1.5 ml Regular Eppendorf tubes.
3. Vacuum centrifuge (Savant).
4. Probe sonicator.
5. Mini-Extruder (Avanti).
6. Two 1.0 ml Hamilton Syringes with short needles (Avanti).
7. Filter Supports (Avanti).
8. Polycarbonate membrane, pore size 0.2 μm (Avanti) (*see* **Note 3**).
9. 48 Well cell culture plate (Greiner Bio-One).
10. Beckman Coulter J-E centrifuge and JA-20.1 rotor with adaptors for 2 ml Eppendorf tubes.
11. Orbital shaker.
12. Table top Eppendorf centrifuge.
13. Heat block, 95 °C.

3 Methods

It is essential to carry out all procedures at or above room temperature. At lower temperatures raffinose pentahydrate in the extrusion buffer will precipitate.

1. Mix phospholipids in the required molar ratios (*see* **Note 4**) in 2 ml Safe-Lock Eppendorf tubes with round bottom. Use 400 nmol of total lipid mix for every protein sample. If polyphosphoinositides are added to phospholipids dissolved in chloroform, add appropriate amounts of MeOH (*see* **Note 5**).

2. Dry the lipid mixtures in a vacuum centrifuge for at least 30 min at medium heating. Lipid pellets can be stored at −20 °C under N_2(g).
3. Add 500 μl extrusion buffer to the dried lipid mixture. Let the lipids rehydrate for 1–1.5 h at room temperature on the orbital shaker with occasional vortexing to promote multilamellar vesicle (MLV) formation.
4. Using a probe sonicator, sonicate lipid mixtures briefly for 30 s with an amplitude of 4 μm (*see* **Notes 6** and 7).
5. Assemble the extruder (see Fig. 1). Prepare petri dishes containing distilled water, extrusion buffer and a third for waste. Place the Teflon Bearing in the Extruder Outer Casing and put one of the Internal Membrane Supports on top of it with the O-ring facing up. Put another Internal Membrane Support in the Retainer Nut. Soak four Filter Supports in extrusion buffer. Place two of the Filter Supports over the orifice inside the O-ring on the Internal Membrane Support in the Extruder Outer Casing. Place the polycarbonate membrane over the Filter Supports and O-ring. Polycarbonate membrane can be either soaked in extrusion buffer or left dry. Place the two remaining Filter Supports on the Internal Membrane Support in the Retainer Nut. Carefully attach the Retainer Nut (with O-ring and Filter Supports facing the membrane) to the threaded end of Extruder Outer Casing and tighten by hand until it is finger tight. Wash the assembled Mini-Extruder 2× with distilled water and 3× with extrusion buffer (*see* **Note 8**).
6. Extrude lipid suspension 13× over a polycarbonate membrane (*see* **Note 9**) to produce an optically clear suspension of large unilamellar vesicles (LUV). The Mini-Extruder does not have to be reassembled when extruding multiple compatible lipid mixtures, provided the Mini-Extruder is rinsed in between (*see* **Note 8**). Alternatively, if any cross-contamination between mixtures has to be avoided, disassemble the extruder, rinse thoroughly with distilled water and assemble again with new Filter Supports and polycarbonate membrane, starting from Subheading 3, step 5 for every new lipid mixture.
7. Dilute the liposome suspension in three volumes (1.5 ml) of binding buffer (1×). Pellet the liposomes by centrifugation at 50,000 × *g* for 15 min at 22 °C (*see* **Note 10**). Promptly and carefully remove the supernatant (*see* **Note 11**).
8. Resuspend liposome pellets in 25 μl binding buffer (1×) per sample (*see* **Note 12**).
9. Prepare protein mix: Use 500 ng of protein (*see* **Note 13**) per sample. Dilute the protein in distilled water to a volume of 20.8 μl, then add 4.2 μl of binding buffer (6×) (*see* **Note 14**) to get a final volume of 25 μl per sample, containing the protein in 1× binding buffer.

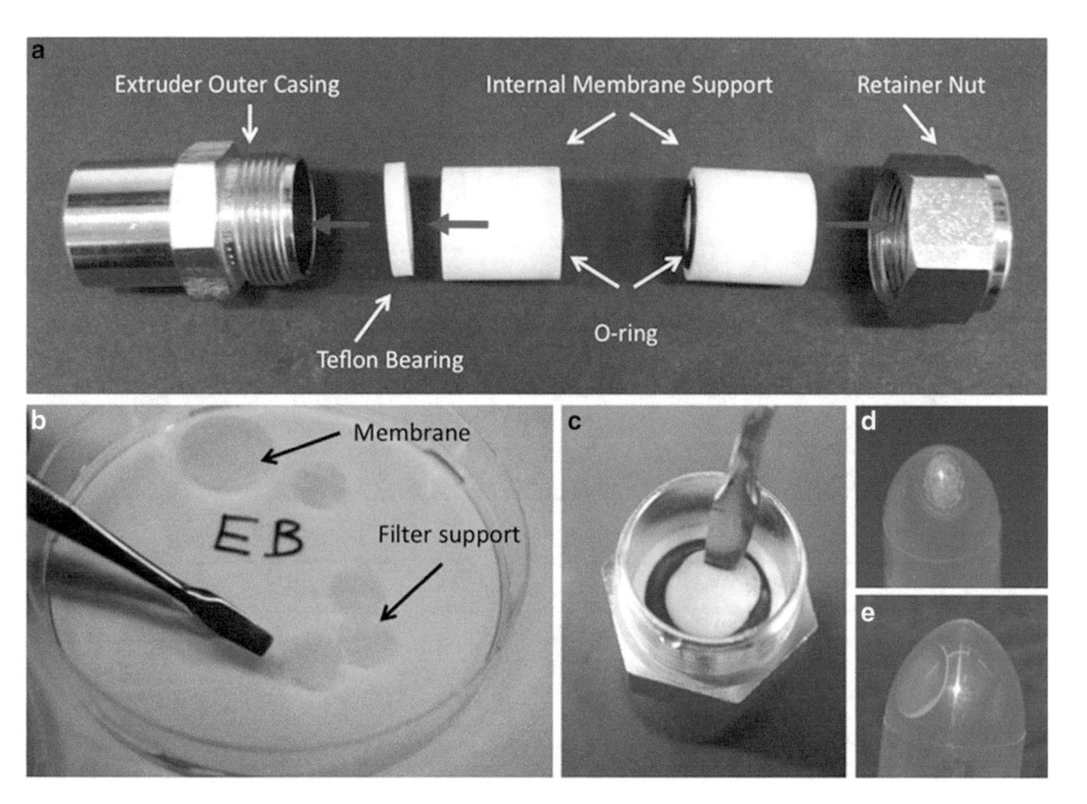

Fig. 1 Preparation of the liposomes. (**a**–**c**) Assembly of the Mini-Extruder. (**a**) Teflon Bearing and one Internal Membrane Support are placed in the Extruder Outer Casing with the O-ring facing up. The other Internal Membrane Support is placed in the Retainer Nut also with the O-ring facing up. (**b**) Four Filter Supports and one 0.2 μm polycarbonate membrane are soaked in the extrusion buffer. (**c**) Two Filter Supports are placed within the O-ring on each Internal Membrane Support and polycarbonate membrane is placed over the Filter Supports in the Extruder Outer Casing Internal Membrane Support (membrane placing not shown). (**d**, **e**) Appearance of lipid pellets. (**d**) Dried lipid pellet prepared at Subheading 3, step 2. containing 1,320 nmol of PC/PE (1:1) lipid mix, which in a typical experiment is a mix for testing six protein samples with one lipid concentration each. (**e**) Liposome pellet after extrusion and centrifugation obtained at step 7 containing 1,320 nmol of PC/PE (1:1) Note that slight variations in pellet appearance may occur as an effect of lipids used

10. Centrifuge protein mix for 5 min at 16,000 × *g* to get rid of any possible precipitates.
11. Combine 25 μl of protein mix with 25 μl liposome suspension mix prepared in Subheading 3, steps 1–8. and incubate on an orbital shaker platform for 30–45 min at room temperature.
12. Centrifuge samples at 16,000 × *g* for 30 min at room temperature (*see* **Note 15**).
13. Harvest supernatant (*see* **Note 15**). Resuspend the liposome pellet (*see* **Note 16**) in 300–500 μl 1× binding buffer and transfer to a new tube (*see* **Note 17**). Centrifuge samples at 16,000 × *g* for 30 min at room temperature and remove supernatant (*see* **Note 15**).

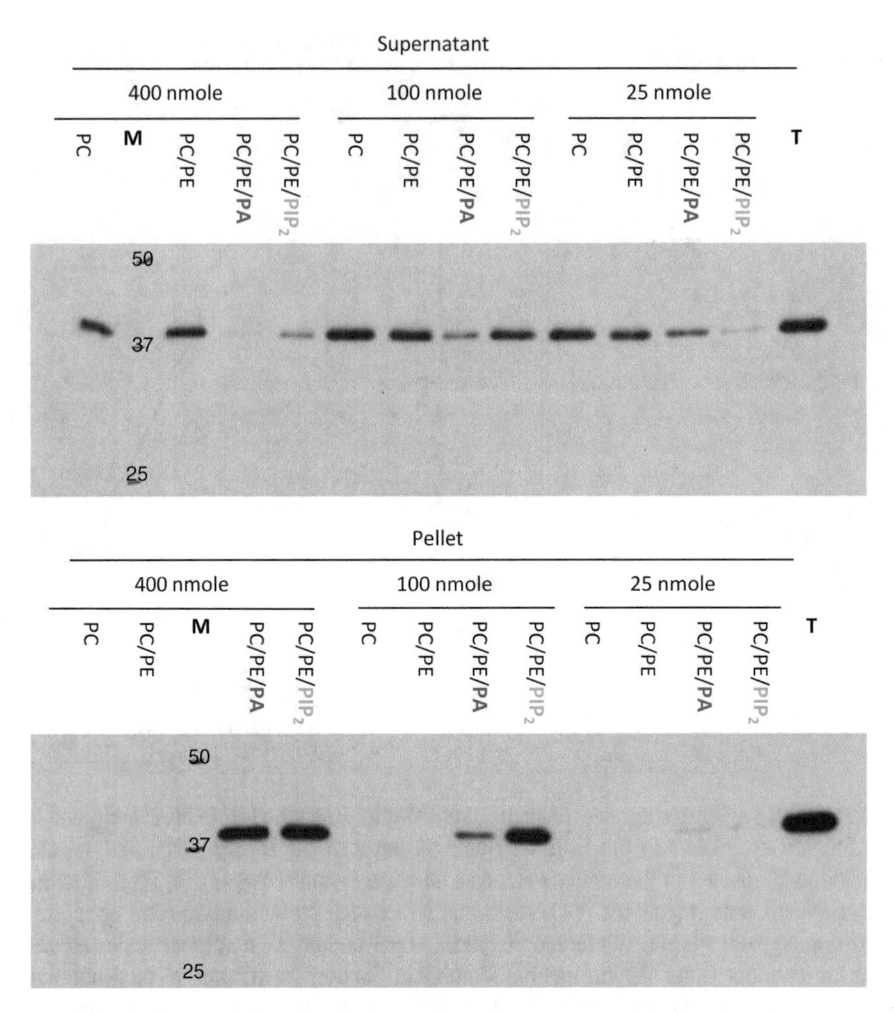

Fig. 2 Liposome binding assay of GST-tagged PDK1-PH domain. GST tagged PDK1-PH2 protein (500 ng) was incubated with liposomes composed of structural lipids (PC or PC/PE, 1:1), or liposomes containing PA or PIP_2 (PC/PE/PA or PC/PE/PIP_2, 9:9:2) in decreasing concentrations (400, 100 and 25 nmol total lipid). (*M*) Precision plus marker was used to estimate the size of the protein; MW indicated in kDa, and (*T*) total protein is a loading control from protein mix prepared at Subheading 3, step 9. For each sample 10 μl was loaded on the gel. Total protein (*T*) amount is equal to 75 ng

14. Resuspend the liposome pellet in 33 μl of 1× Laemmli sample buffer. Add 16.7 μl of 4× Laemmli sample buffer to supernatant fraction. By doing so the protein amount present in the pellet faction is twice as concentrated as protein present in the supernatant fraction.
15. Incubate all samples at 95 °C for 5 min (*see* **Note 18**) and load 10 μl on SDS-Page for Western blot analysis. An example blot of lipid binding of GST-tagged PDK1 protein is shown in Fig. 2. Incorporate a loading control of the protein used on each blot.

4 Notes

1. Phospholipids with various lipid head groups and fatty acid chains can be used for liposome formation. Usually mixtures of lipids are used. In our experiments we have used liposomes composed of structural lipids with diC18:1 acyl chains with or without additional lipids of interest (LOI) present. Note that phospholipids with a gel-liquid crystalline phase transition temperature (T_m) above room temperature have to be heated up during rehydration and liposome extrusions (Subheading 3, step 6). The latter can be achieved by using a heating block provided by Avanti.
2. The extrusion buffer is a saturated solution of raffinose pentahydrate and it is thus necessary to heat up the solution in order to entirely dissolve all raffinose pentahydrate.
3. Extrusion of MLV suspensions through membranes with a pore size >0.2 μm will produce a polydisperse suspension of MLV, rather than unilamellar vesicles (LUV). For the purpose of liposome sedimentation it is optimal to use largest homogeneously sized LUV (namely 0.2 μm) possible.
4. In our lab we use a mix of structural cell membrane (lipid bilayer forming) phospholipids, such as 1,2-dioleolyl-*sn*-glycero-3-phosphatidylcholine (PC) and 1,2-dioleolyl-*sn*-glycero-3-phosphatidylethanolamine (PE) in a 1:1 ratio, as a negative control for lipid binding. It is also possible to choose for one background phospholipid if your protein of interest might bind to one of the background lipids. The LOI can be added typically in a range of 1–50 % of total lipid, depending on the expected affinity of the protein tested. In each case, the ratio of PC:PE is kept constant. Some proteins are known to bind specific lipids present in small local patches, while others require high local lipid concentrations. Therefore strength of protein affinity towards LOI can be tested either by altering the percentage of lipid of interest in the vesicles (usually in a range of 1–50 % of total lipid) or by making dilution series of a fixed lipid mix. The dilution series used in our lab go down to 25 nmol total lipid per protein (note that only lipids in the outer monolayer of the LUV are accessible for protein binding). Depending on the physiochemical characteristics of the LOI and its molar percentage in the lipid mixture, the physical properties of the liposome vesicles and thus also protein binding to the liposomes may vary.

   One lipid mix is usually composed of lipid quantities needed for more than one protein sample (for example for various proteins of interest, positive and negative control samples and various lipid concentrations). Furthermore it is advised to prepare an additional 10 % to ensure having enough lipid suspension. When doing dilution series the lipids may be dried for the whole series as one lipid mix and diluted from the original lipid mix after liposome centrifugation (Subheading 3, step 8).

5. Polyphosphoinositides (PPIs) are bought and stored in chloroform–methanol–water ($CHCl_3$–MeOH–H_2O) 20:9:1 (v/v), and may precipitate when the $CHCl_3$–MeOH ratio increases. Therefore combining PPIs with phospholipids dissolved in pure chloroform requires a correction of the $CHCl_3$–MeOH ratio. Before adding PPIs, add the appropriate amount of MeOH, so that the $CHCl_3$–MeOH ratio of the chloroform-dissolved lipids will be equal to 20:9. After this methanol addition the desired amount of PPI's can be added.
6. Brief sonication of the lipid suspension will decrease the size of MLV to enhance formation of the liposomes. Make sure to clean the sonicator probe between different lipid mixtures with distilled water. Make sure not to sonicate the samples excessively to limit the formation of small unilamellar vesicles.
7. Often MLV formation efficiency is increased via subjecting the hydrated lipid suspension to 3–5 freeze–thaw cycles. However, for this protocol freeze–thaw cycles are not recommended, since they might result in precipitation of the raffinose pentahydrate from the extrusion buffer.
8. Take two 1.0 ml Hamilton Syringes and make sure that the needles are tightened in order to prevent leakage. Insert Hamilton Syringes at both ends of the extruder and mark one side of the extruder as IN site and the other as OUT site. Do not switch the syringes from now on as it might give a risk of contamination of the LUV sample with MLVs. Take approximately 1 ml of distilled water into the IN syringe and flush it through the extruder once into the OUT syringe. Discard the water from OUT syringe into petri dish marked "waste." Repeat this step once more with distilled water and three times with the extrusion buffer. It is important to not twist the syringes when pushing them into or pulling them out of the Mini-Extruder and during extrusion, because the needle might unscrew from the syringe, resulting in sample loss. If the Hamilton Syringe cannot be pushed in or out, you may carefully turn them, but only in the direction in which the needle is tightened to the syringe, so that it will be not be unscrewed from the syringe.
9. It is impossible to take up the lipid suspension from 2 ml Eppendorf tubes using a Hamilton Syringe with extrusion needles. Therefore, the lipid suspension has to be transferred first to a well in 48 wells plate and drawn up from there with the Hamilton Syringe while tilting the plate to minimize loss of sample. Make sure that the needles are tightened to the syringes in order to prevent leakage. When lipid suspension is sucked

up into the IN Hamilton Syringe, air bubbles should be avoided and removed if they occur. At this step it is normal that the liposome suspension is milky. This opaqueness is caused by the fact that liposomes are still large and multilamellar. Put the IN Hamilton Syringe into the IN end of extruder (*see* **Note 8**) and push the solution through the extruder to the OUT Hamilton Syringe. Then push the solution back through the extruder to IN Hamilton Syringe. Lipid suspension has to be pushed through extruder for 13 times (or if longer extrusion is desired; at least an odd number of times). After the 13th time the solution will be in OUT Hamilton Syringe. It should be more translucent than before extrusion. The increase in transparency of liposome suspension is an indication that liposomes are now smaller and unilamellar.

10. In our lab, a Beckman Coulter J-E centrifuge and JA-20.1 rotor with adaptors for Eppendorf tubes is used for centrifugation of the liposomes, but any centrifuge that will go up to 50,000 ×*g* at 22 °C will be suitable for this purpose. With this centrifuge the liposomes are pelleted at 50,000 ×*g* for 15 min at 22 °C.

11. Liposome pellets are usually transparent and it is difficult to see them before removal of supernatant (Fig. 1e). Therefore it is advisable to put the tubes in the centrifuge in a fixed position, so that the position of the pellet could be easily predicted. Depending on the final lipid amount and the type of lipids used in the lipid mix, the size and transparency of the pellet may differ.

12. The amount of 1× binding buffer added will determine the concentration of the liposomes used per protein sample. When 400 nmol lipid is used per protein sample, the total lipid concentration for liposome preparation will be 8 mM when dissolved in 1× binding buffer. In this step you may also dilute the liposome suspension for the dilution series. When doing so, liposomes should be diluted to desired concentration using 1× binding buffer.

13. In our lab, proteins induced in and purified from *E. coli* containing GST-tags are used in liposome binding assays. In our experience, GST elution buffer (composed of 20 mM reduced glutathione and 50 mM Tris–HCl pH 8.0) or the GST-tag itself do not interfere with the lipid binding. Other tags (e.g., 6×His, YFP) can be used as well in the liposome binding assay. Yet, additional control experiments, testing lipid binding of the protein tag and possible interference of elution/storage buffer need to be performed.

    Purified protein concentrations are quantified by comparing protein concentration to dilution series of BSA of known concentration on SDS-gel stained with Coomassie Colloidal

Blue. Alternatively, one may use Coomassie staining to determine the fraction of the target protein in conjunction with a spectroscopic protein assay to determine the overall protein concentration and quantify protein concentrations. Before preparing a master mix of purified protein for the liposome assay, the tubes with purified protein elutions need to be briefly centrifuged (for 30 s at 16,000 × *g*) in order to get rid of potential protein aggregates or Sepharose beads.

It is advisable to prepare a master mix of the protein samples for all lipid mixes plus two extra samples that can be used as protein loading control on SDS gel. For novel binding assays it is advisable to include one negative control sample in the binding assay by testing protein without addition of any lipids.

14. Proteins are usually eluted in quite low concentrations. In order to avoid having to dilute the purified proteins, concentrated binding buffer (6×) is added to protein fraction. Minimum concentration of purified eluted protein is 24 ng/μl (500 ng/20.1 μl).

15. At this step a cooled centrifuge should be used for degradation-sensitive proteins, with the temperature set at 20–22 °C, to avoid heating up of the sample during spinning. It is important to harvest supernatant as quickly as possible in order to avoid pellet resuspension into the supernatant fraction and/or protein bound to the liposomes to dissociate from the pellet. It is best to harvest supernatant from only two or four samples at a time while still spinning the rest of the samples in the centrifuge. Harvesting of supernatant can be performed using a pipette, take care to remove the entire supernatant fraction in one step placing the pipette tip as not to disturb the pellet. Make sure to use a fresh pipette tip for each sample.

16. At this point the liposome pellets are much smaller than at Subheading 3, step 7 and 8. If you are making dilution series you might not see the pellet at all at the lowest lipid concentrations. Therefore at this step it is crucial to place the tubes in the centrifuge in a fixed position, so that the position of the pellet can be easily predicted. By harvesting supernatant quickly (*see* **Note 15**) disruption of the pellet fraction will be minimized.

17. Transfer of the resuspended pellet fraction into the new tube will reduce the nonspecific binding substantially. The protein from the supernatant fraction bound to the walls of the tube but not to the liposomes can increase the background binding signal.

18. It is important to heat the samples up just before loading on gel to prevent precipitation of KCl from the binding buffer with the SDS present in Laemmli sample buffer. The samples need to be placed in a 95 °C heat block for 5 min just before

loading on gel. Place a heavy object on the top over the tubes in order to prevent lids to open due to pressure build up inside the tube. The samples containing sample buffer can be stored at −20 °C and need to be heated up again before running them on gel.

Acknowledgments

The authors thank Carlos Galvan-Ampudia for his help in cloning the PH-domain of AtPDK1 in the GST expression vector and Edgar Kooijman for support and advice on lipid physiochemical properties. This work was supported by the Netherlands Organisation for Scientific Research (NWO) grants Vidi 700.56.429 and ALW 820.02.017, STW Perspectief 10987, NGI Horizon project 93511011, NSF Grants DMR-0844115 and CHE-0922848, Kent State, and an ICAM fellowship to JMR (ICAM Branches Cost Sharing Fund). Financial support from EU-COST Action FA0605 is also gratefully acknowledged.

References

1. Stahelin RV (2009) Lipid binding domains: more than simple lipid effectors. J Lipid Res 50(Suppl):S299–S304
2. Catimel B, Schieber C, Condron M, Patsiouras H, Connolly L, Catimel J, Nice EC, Burgess AW, Holmes AB (2008) The PI(3,5)P2 and PI(4,5)P2 interactomes. J Proteome Res 7:5295–5313
3. Monreal JA, McLoughlin F, Echevarria C, Garcia-Maurino S, Testerink C (2010) Phosphoenolpyruvate carboxylase from C4 leaves is selectively targeted for inhibition by anionic phospholipids. Plant Physiol 152: 634–638
4. Testerink C, Dekker HL, Lim ZY, Johns MK, Holmes AB, Koster CG, Ktistakis NT, Munnik T (2004) Isolation and identification of phosphatidic acid targets from plants. Plant J 39: 527–536
5. Krugmann S, Anderson KE, Ridley SH, Risso N, McGregor A, Coadwell J, Davidson K, Eguinoa A, Ellson CD, Lipp P, Manifava M, Ktistakis N, Painter G, Thuring JW, Cooper MA, Lim ZY, Holmes AB, Dove SK, Michell RH, Grewal A, Nazarian A, Erdjument-Bromage H, Tempst P, Stephens LR, Hawkins PT (2002) Identification of ARAP3, a novel PI3K effector regulating both Arf and Rho GTPases, by selective capture on phosphoinositide affinity matrices. Mol Cell 9:95–108
6. Dowler S, Kular G, Alessi DR (2002) Protein lipid overlay assay. Sci STKE 129:pl6
7. Loewen CJR, Gaspar ML, Jesch SA, Delon C, Ktistakis NT, Henry SA, Levine TP (2004) Phospholipid metabolism regulated by a transcription factor sensing phosphatidic acid. Science 304:1644–1647
8. Testerink C, Larsen PB, van der Does D, van Himbergen JAJ, Munnik T (2007) Phosphatidic acid binds to and inhibits the activity of Arabidopsis CTR1. J Exp Bot 58: 3905–3914
9. Levine TP, Munro S (2002) Targeting of Golgi-specific pleckstrin homology domains involves both PtdIns 4-kinase-dependent and -independent components. Curr Biol 12:695–704
10. Guo L, Mishra G, Taylor K, Wang X (2011) Phosphatidic acid binds and stimulates Arabidopsis sphingosine kinases. J Biol Chem 286:13336–13345
11. Besenicar M, Macek P, Lakey JH, Anderluh G (2006) Surface plasmon resonance in protein–membrane interactions. Chem Phys Lipids 141:169–178
12. Kooijman EE, Tieleman DP, Testerink C, Munnik T, Rijkers DT, Burger KN, de Kruijff B (2007) An electrostatic/hydrogen bond switch as the basis for the specific interaction of phosphatidic acid with proteins. J Biol Chem 282:11356–11364

Chapter 25

Lipid Affinity Beads: From Identifying New Lipid Binding Proteins to Assessing Their Binding Properties

Fionn McLoughlin and Christa Testerink

Abstract

Lipid affinity beads can be used to identify novel proteins with lipid binding capacity or to determine binding prerequisites of known lipid-binding proteins. Here we describe several applications for which this tool can be used and which considerations have to be taken into account. In addition to a precise protocol, several suggestions are made for experimental setups to facilitate identification of in vivo lipid binding targets.

Key words Phosphatidic acid, Phosphatidic acid target proteins, Phosphatidic acid beads, Lipid binding, Phospholipid signaling

1 Introduction

Cellular membranes primarily consist of phospholipids. Only few of these, including phosphatidic acid (PA) and polyphosphoinositols (PPIs), have a function as signaling phospholipids (1). Several proteins implicated in signaling cascades upon abiotic and biotic stress, have been described to bind PA in plants. In addition, PA also binds proteins that function in growth and general metabolism (as reviewed in ref. 2). Different lipid affinity beads can be used to isolate proteins with affinity for various signaling lipids. The protocol described here has been optimized for PA.

PA beads can be used to isolate and identify proteins from complex cellular extracts. Strategies to identify less abundant PA-binding proteins generally include a pre-purification step based on protein characteristics (e.g., size, charge) (3–5) to avoid saturation of the PA beads by very abundant PA binding proteins. Alternatively, pre-fractionation based on cellular localization (i.e., presence on the membrane, subcellular localization) can be performed (6); (Fig. 2).

Teun Munnik and Ingo Heilmann (eds.), *Plant Lipid Signaling Protocols*, Methods in Molecular Biology, vol. 1009, DOI 10.1007/978-1-62703-401-2_25, © Springer Science+Business Media, LLC 2013

Not all the proteins that bind to PA beads will be bona fide in vivo PA targets since PA present on the beads is not in its natural form. To exclude false positives or to get a better idea which proteins are genuine PA targets, several approaches are possible. One option is to determine if the protein of interest is competed off the beads by a more naturally occurring form of PA, through the addition of liposomes containing PA or soluble PA to the PA-binding assay. Tomato (*Solanum lycopersicum*) phosphoenolpyruvate carboxylase (PEPC) was identified in this way, as it was competed off the beads by the presence of soluble PA (Fig. 1a (4)). In addition, it was shown that PEPC has a higher PA binding affinity after a hypo-osmotic stress treatment, which is another indication that PEPC is an in vivo target of PA (Fig. 1b (4)). Later, it was shown that Sorghum C_4 PEPC activity was inhibited in the presence of anionic lipids and that the PEPC fraction targeted to the membrane of sorghum (*Sorghum bicolor*) leaf cells is largely modified (7).

Depending on the nature of the protein of interest, the nucleotide binding or phosphorylation state or several other posttranslational variations can be varied in vitro in order to determine if this has an effect on the PA binding affinity. It has been shown that the nucleotide binding state of NSF, a protein involved in intracellular trafficking in sheep brain, is important for PA binding. The binding to PA occurs only in its ADP-bound state (3). Additionally, the PA binding specificity was determined using beads coupled to another anionic lipid, phosphatidyl inositol 4,5-bisphosphate ($PI(4,5)P_2$) as a negative control. The specificity of a protein binding to a certain anionic lipid and not to others indicates specific binding capacities. Although beads can be used for this application, a liposome-binding assay is a more suitable technique to determine this, since it is easier to vary between different lipid compositions (as described in ref. (8)).

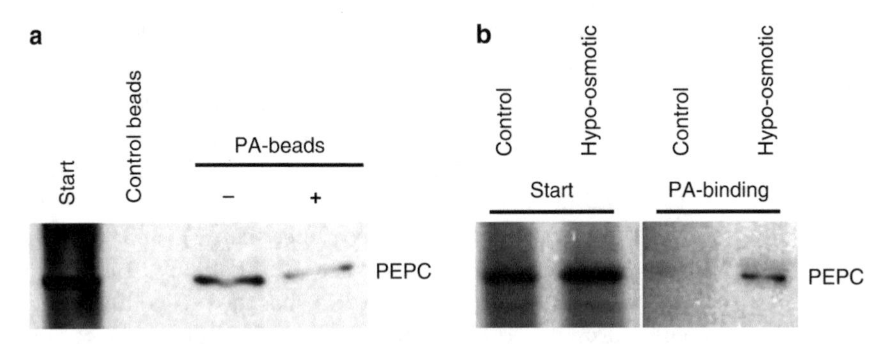

Fig. 1 PA-binding characteristics of tomato PEPC (modified from ref. 4 with permission). (**a**) PEPC is competed off the beads in the presence of soluble PA. Soluble protein extract of tomato suspension-cultured cells was mixed with PA in the absence (–) or presence (+) of soluble PA. (**b**) PEPC PA-binding affinity increases after hypo-osmotic stress treatment. Tomato cell suspension-cultured cells were treated for 15 min with one volume of control (cell-free medium) or hypo-osmotic medium (water). PEPC was detected using Western analysis using anti-PEPC antiserum

Another way to determine if a protein is targeted to PA is to determine if proteins are relocated in vivo upon a PA inducing stimulus. An *Arabidopsis thaliana* cell suspension was exposed to control and PA-inducing conditions (250 mM NaCl, 10 min). Subsequently the peripheral membrane protein pool was isolated from the control (C) and the salt induced (T) sample and a PA binding assay was conducted on these samples (Fig. 2). Proteins

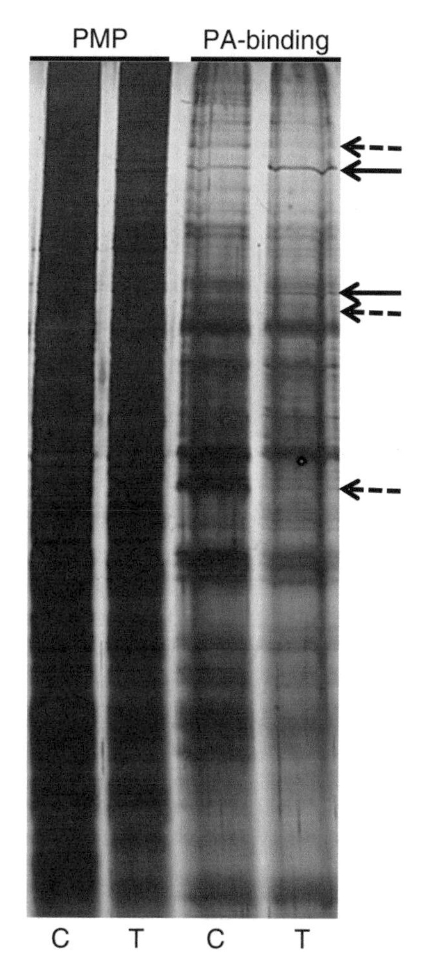

Fig. 2 Several proteins with PA binding affinity are either enriched or decreased in the peripheral membrane protein pool after salt stress treatment. *Arabidopsis thaliana* cell culture was either control or salt (250 mM NaCl for 10 min) treated. The peripheral membrane protein pool (PMP) was isolated by isolating and washing cellular membranes and subsequently eluting the peripheral bound proteins with 100 mM Na_2CO_3. These protein pools were used as input for the PA binding assay. *Solid arrows* indicate proteins that are enriched in the PA-binding peripheral membrane protein pool when salt stimulated and the *dashed arrows* indicate proteins that are decreased in this sample compared to the control. This approach allows identification of proteins that are re-localized upon a PA-inducing stimulus. The proteins were visualized using silver staining

that were altered in their location upon a PA inducing stimulus, in particular proteins that are enriched on the membrane upon salt treatment, are very likely to be targeted to PA in the membrane in vivo (6).

2 Materials

2.1 Sepharose Beads Coupled to Different Kind of Lipids

Synthesis of phosphatidic acid and phosphatidyl inositol beads has been described in ref. 9. Beads can be stored at 4 °C in distilled water containing 0.02 % sodium azide.

2.2 Competition Assay

In case of a competition assay, specific requirements and additional steps are performed. Essential steps, solutions and equipment are marked with #. Additional steps will be described in the notes.

2.3 Stock Solutions for PA-Binding Assay

The Tris–HCl, KCl, EDTA, NaCl, and azide stock solutions can be made beforehand and stored at room temperature.

1. 1 M Tris–HCl pH 6.8 and pH 7.5 and 8.0: Dissolve 121.14 g Tris in 500 ml of distilled water. Adjust the pH to the desired pH with concentrated HCl. Bring the total volume to 1 l with distilled water. Store at room temperature.
2. 1 M KCl: Dissolve 74.55 g potassium chloride in 1 l of distilled water. Store at room temperature.
3. 0.5 M EDTA: Dissolve 73 g ethylenediaminetetraacetic acid in 250 ml distilled water. Adjust the pH to 8.0 with concentrated NaOH otherwise EDTA will not dissolve. Store at room temperature.
4. 5 M NaCl: Dissolve 146.1 g NaCl in 500 ml of distilled water. Store at room temperature.
5. 1 M DTT: Dissolve 1.54 g dithiothreitol in 10 ml distilled water. Store in 200 μl aliquots at −20 °C.
6. Phosphatidic acid (di C18:1) dissolved in chloroform according to the instructions (Avanti). Alternatively dissolve C8:0 lipids to a 1 M concentration in 1 M Tris–HCl pH 7.5.
7. 2 % Sodium azide: Dissolve 200 mg of sodium azide in 10 ml of distilled water.

2.4 Working Solutions for PA Binding Assay

Lysis buffer (without the complete protease inhibitors) and IPP buffer can be stored at 4 °C up to 3 months. Sample buffer (4×) and B-buffer can be prepared beforehand and stored at −20 °C.

1. Lysis buffer 2×: 50 mM Tris–HCl, pH 8.0, 50 mM KCl, 10 mM EDTA, and autoclave. Add 1 % IGEPAL CA-630 (Sigma) after autoclaving. Add complete protease inhibitors (Boehringer) freshly.

2. IPP buffer 1×: 50 mM Tris–HCl, pH 7.5, 150 mM NaCl, 5 mM EDTA, and autoclave. Add 0.1 % Tween 20 after autoclaving.
3. Sample Buffer 4×: Dissolve 0.8 g of SDS in 4.6 ml 87 % glycerol, 2 ml β-mercaptoethanol, 2.4 ml 1 M Tris–HCl pH 6.8 and 1 ml distilled water. Add 8 mg bromophenol blue. Store in 1 ml aliquots.
4. B-buffer 1×: 50 mM Tris–HCl, pH 7.5, 2 % SDS, 100 mM DTT.

2.5 Materials for PA Binding Assay

1. 0.5 and 1.5 ml (preferably Safe-Lock) Eppendorf tubes.
2. Cooled table top Eppendorf centrifuge.
3. Rotator at 4 °C.
4. 95 °C heat block.
5. Probe sonicator (*see* **Note 3**).

3 Methods

It is essential to keep the samples between 0 and 4 °C throughout the whole procedure to prevent protein degradation.

1. Spin 225 μl soluble protein (ca. 1–2 mg/ml) in a tabletop centrifuge at 16,100 × *g* for 10 min (*see* **Note 1**).
2. Add the supernatant (soluble fraction) to 225 μl 2× lysis buffer (lysis buffer is diluted 1:1 in protein isolation buffer) (*see* **Note 2** and **3**).
3. Spin at 16 100 × *g* for 4 min to remove any insoluble matter, transfer the supernatant to a new Eppendorf tube and keep on ice (*see* **Note 4**).
4. Spin an appropriate amount of PA beads (*see* **Notes 5–7**), stored in water plus 0.02 % sodium azide, at 400 × *g* for 2 min (*see* **Note 8**).
5. Wash the beads twice with IPP buffer (by spinning at 400 × *g* for 2 min and replacing the IPP buffer) and make a 10 % suspension in this buffer.
6. Pipet 60 μl 10 % beads (=6 μl drained volume) in each 0.5 ml Eppendorf tube (*see* **Note 9**).
7. Add the soluble fraction of your protein sample to each Eppendorf tube containing the beads.
8. Rotate the samples at 4 °C for 1.5–2 h (*see* **Note 10**).
9. Spin at 400 × *g* for 2 min.
10. Wash the beads four times in 500 μl IPP (by spinning at 400 × *g* for 2 min and replacing the IPP buffer) and keep samples on ice during the process (*see* **Note 11**).

11. Elute the bound proteins by adding 30 μl 2× sample buffer and incubate for 10 min at 4 °C (*see* **Note 12**).
12. Transfer sup to new tube and boil for 3 min (*see* **Note 13**).
13. Store samples at −20 °C.

To reuse the beads:

1. Collect the used beads in an Eppendorf tube.
2. Add B-buffer to make a 25 % bead suspension and leave for 10 min at room temperature (*see* **Note 14**).
3. Wash the beads six times with IPP buffer (spin at 400 × *g* for 2 min) and store at 4 °C as a 10 % suspension in IPP.
4. Add sodium azide to a final concentration of 0.02 %.

4 Notes

1. For some applications (for example if the goal is to identify proteins using mass spectrometry) the volume of the protein sample could be bigger than 225 μl. In this case, transfer the sample to a 1.5 ml tube (or an even bigger volume) and keep the ratio of the lysis buffer and protein sample 1:1. Increasing the amount of beads accordingly to the increase in volume is advisable.
2. The pH of the mixture should be between 7.5 and 8.0. This is usually the case but if the pH of the protein extract is much higher or lower than the desired pH it is advisable to check the pH of your mixture by pipeting a small quantity on pH paper.
3. Dry down PA from the chloroform stock (e.g., di C18:1). Add lysis buffer and sonicate for 5 min or alternatively dilute soluble PA dissolved in 1 M Tris–HCl pH 7.5 in lysis buffer (end concentration 100 μM). Use either solution instead of regular lysis buffer at this step to add to the protein. Incubate for 20 min on ice before transferring the sample to the beads.
4. In some cases white precipitation occurs in later steps. The precipitation can be removed by using a tip with a very small diameter (for example Gilson p2 tips or tips for loading polyacrylamide gels).
5. Optimizing the ratio between the amount of protein and the amount of beads is best done in a trial experiment. The ratios given in the protocol are optimized for the PA beads we currently have. This can be used as a starting ratio for your experiment but it is advisable to test your protein sample with different amount of beads. This way the ratio at which the PA beads are the limiting factor can be determined.

6. Different protein samples require different bead-to-protein ratios. When using recombinant protein expressed in *E. coli* or purified protein, less beads are needed in comparison to a complex protein mix. In our experience, the ratio between beads and proteins can be reduced at least four times compared to the ratio described in the general protocol.
7. Calculate the total amount of drained beads you need and pipet double the volume out of the stock. This surplus is primarily needed to equalize the bead concentration between the first and the last sample. It is also possible that some beads are lost during the washing steps. The remaining beads can be stored again at 4 °C.
8. The tip of standard yellow tips are not broad enough to homogenously pipet the beads. Tips with a broader diameter of the tip are available. Alternatively the last 5 mm of the tip can be cut with a clean scalpel to broaden the diameter to facilitate the entering of the beads. This promotes equal distribution of the beads over the different samples.
9. For optimal distribution of the beads over the different samples, make sure the beads are homogenously distributed in the stock every time a sample is taken. This can be done by inversion or tapping the Eppendorf and quickly pipeting afterwards. Make sure to pipet from the mid-section of your stock. After the beads are distributed over the eppendorfs always check if the amount of beads is the same in each Eppendorf tube.
10. Make sure the beads rotate head over head instead of rocking to increase the binding efficiency.
11. If the binding assay is performed in a 1.5 ml tube, the beads should be transferred to a 0.5 ml Eppendorf at this step. This decreases the chance of losing any beads during the washing steps. Mind that the diameter of the tip is large enough for the beads to pass.
12. For an optimal elution of your protein, gently tap the eppendorfs after 2 and 4 min to optimize the exposure to the sample buffer.
13. It is not necessary to spin the samples before transferring the supernatant. A short spin is possible in case there is some sample stuck to the wall of the Eppendorf. Make sure you pipet the sample in the bottom of the new tube to avoid unnecessary loss of your sample.
14. Beads can be reused up to five times. When the beads are reused for a couple of times it is possible that a white precipitate occurs between the beads. This can be removed before you start a new PA-binding assay using a pipet tip with a small diameter.

Acknowledgments

This work was supported by the Netherlands Organization for Scientific Research (NWO) Vidi grant 700.56.429. Financial support from EU-COST Action FA0605 is also gratefully acknowledged.

References

1. Munnik T, Testerink C (2009) Plant phospholipid signaling: "in a nutshell". J Lipid Res 50(Suppl):S260–S265
2. Testerink C, Munnik T (2011) Molecular, cellular, and physiological responses to phosphatidic acid formation in plants. J Exp Bot 62:2349–2361
3. Manifava M, Thuring JW, Lim ZY, Packman L, Holmes AB, Ktistakis NT (2001) Differential binding of traffic-related proteins to phosphatidic acid- or phosphatidylinositol (4,5)-bisphosphate-coupled affinity reagents. J Biol Chem 276:8987–8994
4. Testerink C, Dekker HL, Lim ZY, Johns MK, Holmes AB, Koster CG, Ktistakis NT, Munnik T (2004) Isolation and identification of phosphatidic acid targets from plants. Plant J 39: 527–536
5. Krugmann S, Anderson KE, Ridley SH, Risso N, McGregor A, Coadwell J, Davidson K, Eguinoa A, Ellson CD, Lipp P, Manifava M, Ktistakis N, Painter G, Thuring JW, Cooper MA, Lim ZY, Holmes AB, Dove SK, Michell RH, Grewal A, Nazarian A, Erdjument-Bromage H, Tempst P, Stephens LR, Hawkins PT (2002) Identification of ARAP3, a novel PI3K effector regulating both Arf and Rho GTPases, by selective capture on phosphoinositide affinity matrices. Mol Cell 9:95–108
6. McLoughlin F, Arisz SA, Dekker HL, Kramer G, de Koster CG, Haring MA, Munnik T, Testerink C (2013) Identification of novel candidate phosphatidic acid binding proteins involved in the salt stress response of Arabidopsis thaliana roots. Biochem J 450:573–581
7. Monreal JA, McLoughlin F, Echevarria C, Garcia-Maurino S, Testerink C (2010) Phosphoenolpyruvate carboxylase from C4 leaves is selectively targeted for inhibition by anionic phospholipids. Plant Physiol 152: 634–638
8. Julkowska MM, Rankenberg JM, Testerink C (2013) Liposome-binding assays to assess specificity and affinity of phospholipid-protein interactions. Methods Mol Biol. 1009:261–271
9. Lim ZY, Thuring JW, Holmes AB, Manifava M, Ktistakis N (2002) Synthesis and biological evaluation of a PtdIns(4,5)P2 and a phosphatidic acid affinity matrix. J Chem Soc Perkin Trans 1: 1067–1075

Part IV

Imaging

Chapter 26

Using Genetically Encoded Fluorescent Reporters to Image Lipid Signalling in Living Plants

Joop E.M. Vermeer and Teun Munnik

Abstract

The discovery of the green fluorescent protein has revolutionized cell biology as it allowed researchers to visualize dynamic processes in living cells. The fusion of fluorescent protein variants with lipid binding domains that bind to specific phospholipids have been very instrumental in investigating the role of these molecules in living plants. Here, we describe the use of these reporters to image lipids in living Arabidopsis seedlings using fluorescence microscopy.

Key words Lipid binding domain, GFP, Microscopy, Imaging, Arabidopsis

1 Introduction

Phospholipids have been shown to be important players in plant growth and development. They are important for plants to cope with several biotic and abiotic stresses, such as water and heat stress and the defense against pathogens, but also in plant development and cell expansion. The use of genetically encoded fluorescent reporters to visualize lipids in living plant cells has been very instrumental for understanding the cellular roles of phospholipids, as they allow the monitoring of the dynamics of these molecules in living cells at the cellular resolution (1–6).

Typically, these reporters consist out of a lipid binding domain [e.g., Pleckstrin Homology (PH) domain] (7), having high specificity towards the lipid of interest, fused to a fluorescent protein, e.g., the green fluorescent protein (GFP). This gives yield to so-called FLAREs, which is short for Fluorescent Lipid-Associated Reporters (8). These reporters are suitable to study lipid signalling using transient expression as well as stable expression in living plants. In recent years, we and others have used FLAREs to reveal new aspects of phospholipid signalling that could not be easily revealed using established biochemical techniques (2, 5, 9, 10). Of course, to be able to

Teun Munnik and Ingo Heilmann (eds.), *Plant Lipid Signaling Protocols*, Methods in Molecular Biology, vol. 1009,
DOI 10.1007/978-1-62703-401-2_26,

use these reporters, researchers need to have access to a fluorescence microscope (epifluorescence or confocal). However, this nowadays is not usually a big problem, as many laboratories have access to fluorescence microscopy facilities. Here we describe the methodology to use FLAREs to visualize lipids in living plant cells. Using this approach we have successfully imaged the dynamics of PtdIns3*P*, PtdIns4*P*, and PtdIns(4,5)P_2 in living plants and have revealed novel roles in cell expansion and stress responses.

2 Materials

Prepare all solutions using ultrapure water (prepared by purifying deionized water to attain a conductivity of 18 MΩ cm at 25 °C) and analytical grade reagents. Prepare and store all reagents at room temperature, unless indicated otherwise. All work with seedlings prior to the imaging should be carried out in a sterile laminar flow hood. Diligently follow all waste disposal regulations when disposing waste materials and transgenic plant material.

2.1 Preparation of Medium and Plants

1. Prepare 0.5× Murashige and Skoog (MS) medium with vitamins and 2-(*N*-morpholino)ethanesulfonic acid (MES) (*see* **Note 1**) (Duchefa, NL) by dissolving the required amount in ultrapure water. Adjust the pH to 5.8 using 1 M potassium hydroxide (KOH). If desired the medium can be supplemented with sucrose (*see* **Note 2**). For solid medium add 1 % (w/v) plant agar (Duchefa, NL) to the medium. Sterilize the MS medium using an autoclave.
2. Vapor-phase sterilize Arabidopsis seeds expressing the FLARE of interest (*see* **Note 3**) by placing rack with open 2 ml polypropylene reaction tubes containing 50–100 μl of seeds in a desiccator jar (22 l) placed in a fume hood (*see* **Note 4**). Place a 250 ml glass beaker with 100 ml commercial bleach in the dessicator jar and add 3 ml of concentrated HCL (wear gloves!) and immediately close the desiccator jar. Allow sterilization to occur for at least 3 h. Next, carefully remove the rack with the polypropylene tubes and place it for at least 1 h in a sterile laminar flow hood, close tubes and remove surface-sterilized seeds for use.

2.2 Microscope

Basically any fluorescence microscope will be fine; however, a confocal microscope is preferred due to its better imaging properties. Custom made slides can be used for upright and inverted microscopes. If one wants to use chambered cover glasses (Labtek, Nunc GmbH & Co. KG, Germany, see Subheading 2.3), then only an inverted microscope is suitable. If possible, use a water immersion lens with a high numerical aperture; however, oil immersion lenses also work just fine.

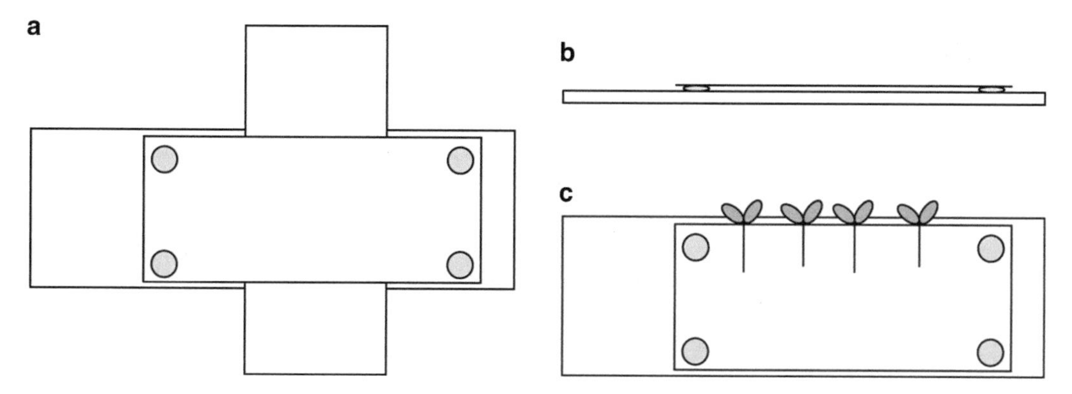

Fig. 1 Preparation of custom slides for confocal imaging of Arabidopsis seedlings. (**a**) Surface view of the imaging slide. A standard glass slide with two perpendicularly placed cover glasses functioning as spacer for the seedlings. *Gray circles* represent the four drops of silicon glue on which the third cover glass is placed. Apply pressure on the spacer slides to ensure that the third cover glass touches the silicon glue. (**b**) Side view of the slide with the two spacer slides removed revealing the chamber where the seedlings will be placed. (**c**) A complete slide with four seedlings growing inside

2.3 Preparation of Slides for Confocal Imaging

1. For imaging of seedlings on a confocal microscope we usually use two approaches. The first one is to germinate seeds on plates and subsequently transfer the seedlings to imaging slides (*see* 2). The second approach is to grow the seedlings on plates for 5–7 days and subsequently cut out an agar-block containing the seedling and to carefully place this upside down in a chambered cover glass (Labtek, Nunc GmbH & Co. KG, Germany) containing 100–200 μl 0.5× MS medium (*see* **Note 5**).
2. For the preparation of slides used for imaging (Fig. 1) of whole 5–7 days old Arabidopsis seedlings, first clean standard glass slides (75 mm × 25 mm) with 70 % ethanol and dry them with tissue paper. Place the microscope slides on a sheet of clean paper on a clean workbench in the lab (*see* **Note 6**). Next, place two coverslips (50 mm × 22 mm) in a perpendicular orientation on the middle of the object glass. This creates a spacer in which the seedlings will be placed. Subsequently, put four small drops of transparent silicone glue for glass (*see* **Note 7**) around the spacer (Fig. 1) and put a coverslip (50 mm × 22 mm) on top of the spacer. Push the cover glass down by applying gentle pressure with a finger on the spacer area. Leave the slide overnight so that the silicone glue hardens. Next day remove the spacers and wash the slides thoroughly. Place the spacers in a dedicated container, so that you can reuse them. The slides can be sterilized by autoclaving. Normally we place the slides in a glass slide container and autoclave this and store it until usage.

3 Methods

1. Put the sterilized seeds on agar plates. Seal the plates with Leukopor ventilation tape and put the plates overnight at 4 °C. Next day put the plates in a standard growth chamber (21 °C; 16 h light) and leave them to germinate and grow for 4 days (*see* **Note 8**). Just prior to placement of the seedlings in the slides, the slide container is half-filled with 0.5× MS media. Use a 1 ml pipette to fill the slides with the medium. Transfer the seedlings from the plate and place them in the spacer between the glass slide and the cover glass using a sterile toothpick. We usually do not put more than four seedlings per slide. When finished, close the container and incubate for 1–2 days in a growth chamber to recover from the treatment.
2. Gently mount the slide on the microscope. This is to minimize mechanical stress of the seedlings in the slide. Also, take care when focusing on the seedlings. When you focus to roughly the lens can squash the seedlings, which means end of experiment. So always prepare enough slides with seedlings in case the experimental work will not proceed as planned. Root hairs are especially sensitive to various kinds of stresses, so always check whether the root hairs are still alive after mounting of the slide.
3. Treatment of plants within the slide with inhibitors or osmotic agents is very straightforward. All you need is a 1 ml pipette with the solution you want to add and some paper towels. Then gently pipette the solution on one side next to the cover glass while you position the paper towels on the other side so that the solution will gently flow through the chamber with the seedlings. This can be done on or off the microscope, whatever works best for you.

4 Notes

1. Use MS-medium supplemented with MES to avoid large changes in pH that can have a negative effect on plant growth and development (e.g., root hairs).
2. One can always add the sucrose but this will increase the chance of bacterial/fungal contamination of the growth medium when using this approach.
3. The nice thing of the FLAREs is that one can design it to her/his wishes. (a) Choose the promoter of interest; we usually used a ubiquitously expressed promoter like 35S, but one can also opt

to use tissue specific regulated promoters. Take care that you can induce a phenotype by strong tissue-specific expression of FLAREs (11). This of course can be of choice and can be useful in determining in which processes different phospholipids are involved. Another note of caution is that 35S promoter is sensitive to *trans*-inactivation in several T-DNA insertion lines. Preliminary results suggest that the Arabidopsis UBQ10 is a better option (Vermeer and Munnik, unpublished observations). (b) Choose the most suitable fluorescent protein for the FLARE that you want to image. Most confocal microscopes will have a strong 488 nm laser line, which will allow imaging of GFP- and YFP-based FLAREs. RFP-based FLAREs can be very useful since there is less autofluorescence in the red. However, often the laser required for RFP excitation (e.g., 543 HeNe laser) is less strong and hence can make it more difficult to image the FLARE. (c) At the moment we have tested several FLAREs in plants: YFP-2xFYVE (PtdIns3*P*), YFP-PH_{FAPP1} (PtdIns4*P*), and YFP-$PH_{PLC}\delta_1$ (PtdIns(4,5)P_2) (Fig. 2) (4–6). FLAREs for diacylglycerol, phosphatidic acid and phosphatidyl serine are in progress (Munnik lab, in progress).

4. Chlorine gas is poisonous to humans—ensure that you work with proper ventilation.
5. Germinate and grow seedlings on 0.5× MS medium supplemented with 1 % (w/w) agar. Take care to have sufficient spacing between the seedlings (~2 cm). Five to seven days after germination (or sooner depending on the developmental stage you are interested in) use a surgical blade to cut a rectangle around the seedling and gently transfer the block of agar containing the seedling in a chambered cover glass (Nunc) containing ~150 μl 0.5× MS medium. Make sure that the seedling is facing the cover glass (to allow imaging on an inverted microscope) and close the chamber with its lid. This setup is particular useful when one needs to make long time-series, since the seedling is quite immobilized and protected from drying.
6. As these slides can be cleaned and stored for reusage, it easier to make a substantial number of them in one go.
7. Use the silicone glue for glass (available in your local do-it-yourself store), this one is the strongest and can easily stand multiple rounds of sterilization.
8. It is always best to try out which size of seedlings you are most comfortable with in transferring them in to the slide. If they are to big there will be a high chance that the root will be severely damaged during the transfer.

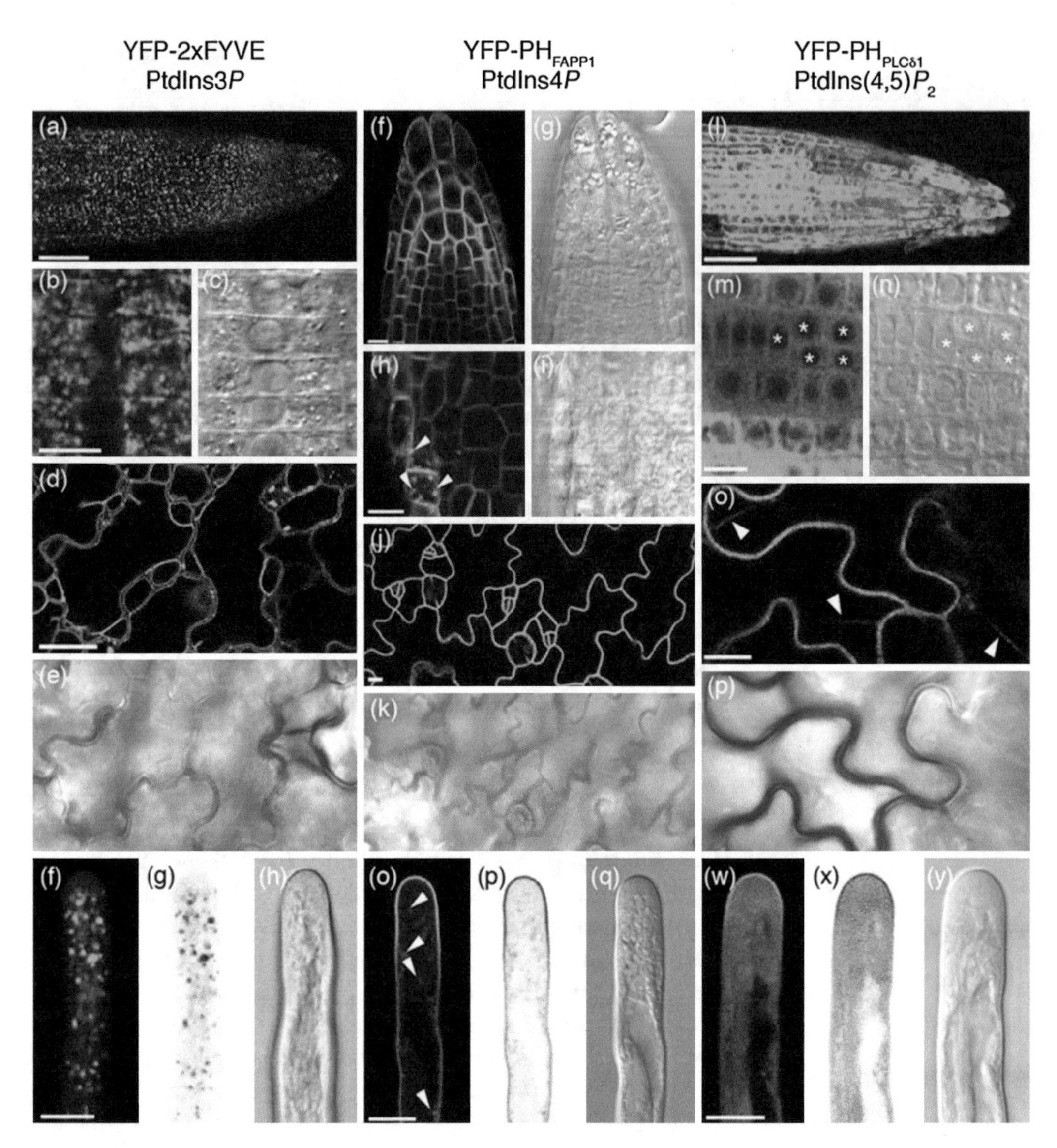

Fig. 2 FLAREs shine a new light on PtdIns3*P*, PtdIns4*P* and PtdIns(4,5)P_2 in *Arabidopsis thaliana* seedlings. Confocal images of Arabidopsis seedlings expressing the PtdIns3*P* FLARE YFP-2xFYVE (**a**–**h**), the PtdIns4*P* FLARE YFP-PH$_{FAPP1}$ (**i**–**q**), and the PtdIns(4,5)P_2 FLARE YFP-PH$_{PLC\delta1}$ (**r**–**y**). All seedlings were grown in imaging slides as depicted in Fig. 1. (**a**) Maximal image projection of a root tip, (**b**, **c**) root cortex cells showing vesicular localization of YFP-2xFYVE, (**d**, **e**) leaf epidermal cells showing strong labelling of the tonoplast membrane. (**f**, **g**) Growing root hair showing non-polarized distribution of YFP-2xFYVE-labelled vesicles. (**g**) Inverted image of (**f**) to highlight the membrane structures. Bars = 50 μm (**a**) and 10 μm (**b**, **d**, **f**). (**i**–**j**) confocal image of a root tip, (**k**–**l**) root cortex cells and (**m**–**n**) leaf epidermal cells. *Arrowheads* in (**b**) indicate YFP-PH$_{FAPP1}$-labelled punctate structures. (**o**) Growing root hair showing a tip-focussed plasma membrane gradient oh YFP-PH$_{FAPP1}$ that is only present in growing root hairs. (**p**) Inverted image of (**o**) to highlight the gradient. Bars = 50 μm (**i**), 10 μm (**k**, **o**), and 20 μm (**m**). (**r**) Maximal image projection of a root tip, (**s**, **t**) root cortex cells, (**u**, **v**) leaf epidermal cells. *Asterisks* in (**m**, **n**) indicate absence of YFP-PH$_{PLCd1}$ fluorescence in the nucleus due to the nuclear export signal present in the PH$_{PLC\delta1}$. *Arrowheads* in (**u**) highlight transvacuolar strands, indicating a cytosolic localization of the unbound probe. Note that the high signal in the cytosol reflects unbound probe due to the very low PtdIns(4,5)P_2 levels in plants. (**w**–**y**) Growing root hair revealing a small gradient of YFP-PH$_{PLC\delta1}$ in the plasma membrane at the extreme tip of the growing root hair. (**x**) Inverted image of (**w**) to highlight the gradient. Bars = 50 μm (**r**), 10 μm (**s**, **u**, **w**). Figure and parts of the legend are reproduced from ref. 12 with permission

References

1. Balla T (2007) Imaging and manipulating phosphoinositides in living cells. J Physiol 582: 927–937
2. Garrenton LS, Stefan CJ, McMurray MA, Emr SD, Thorner J (2010) Pheromone-induced anisotropy in yeast plasma membrane phosphatidylinositol-4,5-bisphosphate distribution is required for MAPK signaling. Proc Natl Acad Sci U S A 107:11805–11810
3. Gillooly DJ et al (2000) Localization of phosphatidylinositol 3-phosphate in yeast and mammalian cells. EMBO J 19:4577–4588
4. van Leeuwen W, Vermeer JE, Gadella TW Jr, Munnik T (2007) Visualization of phosphatidylinositol 4,5-bisphosphate in the plasma membrane of suspension-cultured tobacco BY-2 cells and whole Arabidopsis seedlings. Plant J 52:1014–1026
5. Vermeer JE et al (2009) Imaging phosphatidylinositol 4-phosphate dynamics in living plant cells. Plant J 57:356–372
6. Vermeer JE et al (2006) Visualization of PtdIns3P dynamics in living plant cells. Plant J 47:687–700
7. Hurley JH, Meyer T (2001) Subcellular targeting by membrane lipids. Curr Opin Cell Biol 13:146–152
8. Stefan CJ et al (2011) Osh proteins regulate phosphoinositide metabolism at ER-plasma membrane contact sites. Cell 144:389–401
9. Thole JM, Vermeer JE, Zhang Y, Gadella TW Jr, Nielsen E (2008) ROOT HAIR DEFECTIVE4 encodes a phosphatidylinositol-4-phosphate phosphatase required for proper root hair development in *Arabidopsis thaliana*. Plant Cell 20:381–395
10. Varnai P, Balla T (1998) Visualization of phosphoinositides that bind pleckstrin homology domains: calcium- and agonist-induced dynamic changes and relationship to myo-[3H]inositol-labeled phosphoinositide pools. J Cell Biol 143:501–510
11. Lee Y, Bak G, Choi Y, Chuang WI, Cho HT (2008) Roles of phosphatidylinositol 3-kinase in root hair growth. Plant Physiol 147:624–635
12. Munnik T, Nielsen E (2011) Green light for polyphosphoinositide signals in plants. Curr Opin Plant Biol 14:489–497

Chapter 27

Imaging Changes in Cytoplasmic Calcium Using the Yellow Cameleon 3.6 Biosensor and Confocal Microscopy

Sarah J. Swanson and Simon Gilroy

Abstract

Changes in the concentration of cytoplasmic calcium, $[Ca^{2+}]_{cyt}$ are central regulators in many cellular signal transduction pathways including many lipid-mediated regulatory networks. Given this central role that $[Ca^{2+}]$ has during plant growth, monitoring spatial and temporal $[Ca^{2+}]$ dynamics can reveal a critical component of cellular physiology. Here, we describe the measurement of $[Ca^{2+}]_{cyt}$ in *Arabidopsis* root cells using plants expressing Yellow Cameleon 3.6 (YC 3.6). YC3.6 is a Ca^{2+}-sensitive biosensor where the intensity of its fluorescence resonance energy transfer (FRET) signal changes as the Ca^{2+} level within the cell rises and falls. The FRET from this calcium reporter can be visualized using confocal microscopy and the resultant images converted to a quantitative map of the levels of Ca^{2+} using an approach called ratio analysis.

Key words Yellow Cameleon 3.6, Calcium, Ratio imaging, Fluorescence resonance energy transfer

1 Introduction

Plants generally maintain cytoplasmic $[Ca^{2+}]$ at very low levels (<200 nM), by sequestration of calcium into subcellular compartments or by its removal from the cell via transporters on the plasma membrane. Signaling events originating either during normal development or during a response to an environmental stimulus can trigger an increase in $[Ca^{2+}]$ that then acts as a second messenger, regulating subsequent cellular responses (1, 2). Such signal-related changes in Ca^{2+} have been inferred to play roles in a host of lipid-based signaling networks, making direct measurement of Ca^{2+} signals an important goal in understanding plant lipid-based signaling systems.

The dynamics of $[Ca^{2+}]_{cyt}$ have been successfully observed in plants using a number of small molecule, calcium-sensitive, fluorescent probes, including calcium green, fura-2, and indo-1 (3). However, use of these dyes has important limitations related

Teun Munnik and Ingo Heilmann (eds.), *Plant Lipid Signaling Protocols*, Methods in Molecular Biology, vol. 1009,
DOI 10.1007/978-1-62703-401-2_27,

to difficulties in introducing them into plant cells and ensuring that once inside, they remain in the cytosol (3). The fluorescent protein-based [Ca^{2+}] sensors of the Yellow Cameleon (YC) family obviate these problems because they can be stably expressed in the plant cells of interest and stay localized to the cytoplasm.

The Yellow Cameleons are calcium sensors based on fluorescence resonance energy transfer (FRET) between two fluorescent proteins: cyan fluorescent protein (CFP) and yellow fluorescent protein (YFP); these fluorophores are connected by a calcium sensing domain that includes calmodulin and the M13 calmodulin-binding domain of myosin light chain kinase (Fig. 1a; (4, 5)). The original YC 2.1 version of this sensor has been successfully used in plants (6, 7) but has now been superseded by YC 3.6, where the YFP is replaced by a circularly permutated form called Venus, yielding a protein with a greatly improved brightness and dynamic range (8). It is important to note that there are now many versions of Yellow Cameleon which have shifted Kd's for Ca^{2+} and a range of other Ca^{2+} reporters designed around fusing GFP to other Ca^{2+}-responsive proteins (e.g., (9, 10)); however, currently most of the GFP reporter-based Ca^{2+} imaging in plants is being made using YC 3.6.

The imaging and analysis of YC3.6 expressing plants is relatively straight-forward and readily accomplished with the confocal

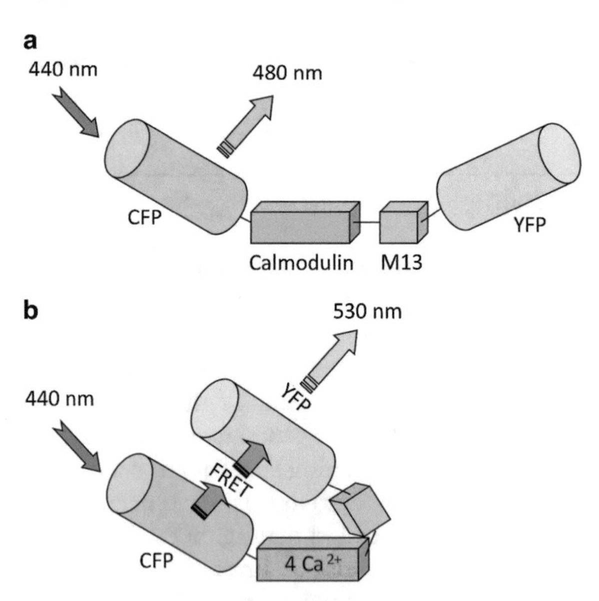

Fig. 1 Structure of the YC 3.6 calcium indicator. (**a**) Domains of YC 3.6 include cyan fluorescent protein (CFP), calmodulin (a calcium binding protein), M13 (the calmodulin-binding domain of myosin light chain kinase), and Yellow Fluorescent Protein (YFP). (**b**) Conformation of the YC 3.6 protein in the presence of high calcium. Calmodulin binds calcium, then this activated calmodulin is able to bind M13. This allows the YFP and CFP to more closely interact such that FRET is able to occur. FRET results in less CFP emission and more YFP emission from the YC 3.6 protein

microscopes available in most university imaging facilities. At low resting [Ca^{2+}], YC 3.6 shows a low FRET signal; however, upon an increase in [Ca^{2+}] the calmodulin domain becomes activated by calcium and binds to the calmodulin-binding M13 region of the sensor. Therefore, at high [Ca^{2+}] the sensor changes conformation, bringing the two fluorescent proteins into closer proximity/more optimal orientation thus facilitating an increase in the FRET signal and a concomitant decrease in CFP signal (Fig. 1b). Analysis of the fluorescence from both CFP and FRET enables accurate measurement of the [Ca^{2+}]. In practice, this is performed by calculating the ratio of FRET signal/CFP signal as the value of this ratio corrects for many potential optical artifacts in such fluorescence measurements, such as changes due to differences in reporter concentration or photobleaching (3).

In the method described below, we will focus on the measurement of [Ca^{2+}] in *Arabidopsis* roots using the confocal microscope for YC 3.6 data collection. Most confocal microscopes have: (1) an emission line from the argon ion laser suitable for CFP/FRET excitation, (2) filters able to separate the CFP and FRET fluorescence, (3) two channels capable of collecting the CFP and FRET signals simultaneously, and (4) detectors that allow quantitative analysis of the resultant fluorescent signals, making them highly suited to YC3.6 imaging.

2 Materials

2.1 Plant Material

Imaging is greatly facilitated by the use of plants stably expressing the fluorescent protein-based sensor YC 3.6 (*see* **Note 1**).

1. Choose a promoter to drive YC 3.6 expression that is appropriate for the cell type studied. For example, the CaMV35S promoter gives strong YC 3.6 expression in the epidermis, cortex, and root cap. Other regions of the plant may require a cell-specific promoter (e.g., Lat52 rather than CaMV35S for pollen, ref. (11)).
2. When developing YC 3.6 expressing lines, screen transformants for fluorescence to ensure usable signal for microscopy. In addition, have untransformed plants on hand which have been grown and handled identically as the experimental material, to provide an essential control for autofluorescence (*see* **Note 2**).

2.2 Sample Preparation

Handle plants as little as possible prior to imaging to minimize cytosolic calcium changes due to mechanical stress. Additionally, ensure that the sample will not dry out or otherwise become stressed during imaging. For cytosolic calcium studies in *Arabidopsis* seedling roots, grow plants on a cover glass coated in phytagel (an optically clear agar substitute available from Sigma-Aldrich). Suitable cover glass sizes range from 40 to 60 mm, dependent on the type of microscope stage available and/or sample chamber used during imaging.

1. Autoclave cover glasses by packing vertically into a beaker, alternating cover glass with a piece of filter paper to prevent each glass form sticking to its neighbor. Once autoclaved, work in the sterile hood and use tweezers to place each cover glass in a sterile petri dish.
2. Add 0.5 % (w/v) phytagel to growth medium and autoclave. Allow to cool until the bottle can be handled with a bare hand. Warm liquid phytagel solution can be pipetted onto the cover glass in the petri plate, forming a "pillow" shape 1–2 mm thick, then allowed to cool and solidify (*see* **Note 3**).
3. After cooling, surface-sterilized seed (2 min, 70 % ethanol) can be planted in the phytagel using a sterile spatula. Poke the seed through the gel such that it is resting on the cover glass itself. Plant 2–4 seeds per cover glass, then seal the plate with parafilm or gas-permeable tape to minimize contamination and to prevent drying (Fig. 2a).

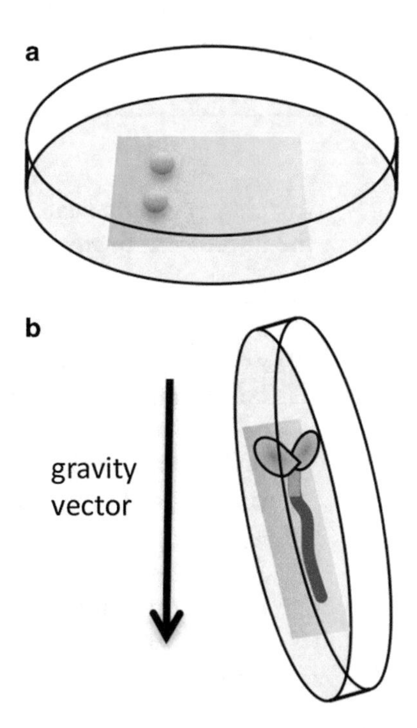

Fig. 2 Preparation of the cover glass and growth of Arabidopsis roots for microscopy. (**a**) Petri plate with sterile cover glass; seeds are planted into the "pillow" of phytagel on the cover glass. (**b**) Cartoon of a vertically grown 7-day-old Arabidopsis plant with its root growing down along the cover glass through the phytagel media

4. Germinate in growth chamber with the plate flat for one day. On day 2, set the plate on edge such that it is angled back slightly from the vertical, so that the roots grow down along the cover glass (Fig. 2b). *Arabidopsis* roots that are 4–10 days old are usable for microscopy when grown this way (*see* **Note 4**).

3 Methods

3.1 Configuration of the Confocal Microscope

1. For successful FRET, YC 3.6 requires CFP excitation. Commonly used laser lines for CFP excitation are the 458 nm line of the argon-ion laser (generally installed on most confocal microscopes) or the 405 nm emission of a diode laser.
2. Use as little excitation energy as possible to minimize photobleaching and potential damage to the sample. Excess excitation energy has the potential to alter the cellular physiology under study (*see* **Note 5**).
3. Select the correct primary dichroic mirror to separate the excitation energy from the fluorescence signal. Generally, the primary dichroic mirror for confocal microscopy matches the laser line selected (e.g., 458 or 405 nm; see Fig. 3).
4. Select the range of YC 3.6 fluorescence wavelengths collected in each channel. For example, a usable range would be 460–505 nm for the CFP channel and 525–535 nm for the FRET channel (Fig. 3; *see* **Note 6**).

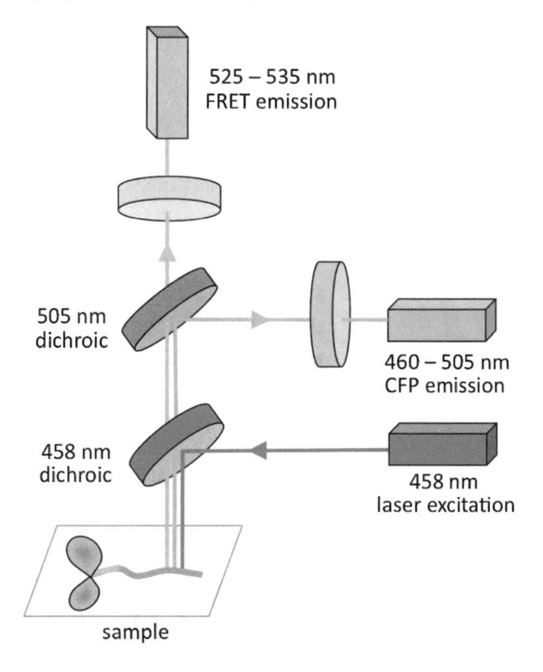

Fig. 3 Configuration of the confocal microscope for imaging the YC 3.6 calcium indicator

5. Some confocal microscopes have the option of using another channel for a transmission (bright field) image. If this is available, enable the transmission channel for the purpose of focusing during the time course.
6. Pseudocolor the image from the CFP channel to green-scale (where increasing intensity of signal is represented as an increasing green brightness), and the image from the FRET channel to red-scale (*see* **Note 7**).

3.2 Imaging Parameters of the Confocal Microscope

For ion imaging using the confocal microscope, generally some sacrifice in spatial resolution is needed to achieve the temporal resolution desired, because the higher the spatial resolution, the longer the microscope takes to acquire each image. If too much time is taken obtaining each image or too long elapses between frames of a time course, important changes in [Ca^{2+}] will be missed (*see* **Note 8**).

1. With the sample on the stage of the microscope, select the proper objective and focus on the cells of interest. Confirm that the sample is fluorescent.
2. Select a pixel resolution that is compatible with fast scan times. Smaller images will allow faster scan times and a shorter interval between each image in the time course.
3. Likewise, select a fast scan speed that will enable rapid image acquisition.
4. Reduce image averaging (usually applied to reduce noise in the image by averaging the signal in multiple frames) or remove completely; this will also allow collection of more frames in a given time period.
5. The confocal pinhole diameter for both the CFP and the FRET detection channels should be adjusted to give the same optical section thickness (*see* **Note 9**).
6. The confocal pinhole diameter should be at the physical optimum of 1 Airy unit or slightly larger (to capture a greater Z section of the sample). Reducing the pinhole below one Airy unit will reduce signal with little gain in resolution.
7. Detector gain should be carefully adjusted for the CFP and FRET detection channels such that there are no pixels that are saturated in intensity in the image (*see* **Note 10**).
8. Confirm that at resting calcium, the detector gain for the FRET channel is set such that the signal is low, allowing for increases in fluorescence from FRET during a [Ca^{2+}] increase (*see* **Note 11**).
9. Some confocal microscope detectors will have an offset that adjusts the output of the low end of fluorescence intensities (i.e., filters out low signals to improve the contrast of the image). For quantitative imaging, set the offset to 0 for both the CFP and FRET channels.

3.3 Data Collection

Once configuration and scanning parameters are established, it is important that these settings be reused for every sample. Changes in laser strength, wavelengths collected, scan time, pinhole diameter, or gain can alter the measured FRET/CFP ratio independently of [Ca^{2+}] changes. Therefore, all settings must be the same in order to compare results between samples and between data collection sessions (*see* **Note 12**).

1. With the sample on the stage of the microscope, select the proper objective and focus on the cells of interest. Confirm using epifluorescence that the sample is fluorescent.
2. Load the previously saved YC 3.6 configuration and scanning parameters.
3. Begin imaging with the scanning laser and perform a continual scan to focus on the desired optical section. Crop to zoom into the region of interest and rotate if necessary.
4. Set up the time series interval and number of frames to collect in the time course (*see* **Note 13**).
5. Start the time course. If possible, view the signal from each channel separately in addition to the overlay in order to monitor any calcium changes.
6. Collect ~30 frames for a baseline prior to addition of any treatments.
7. After ~30 frames, apply treatment (*see* **Note 14**).
8. If available, observe the transmission image during the time course to confirm that the sample stays in focus.
9. Follow any [Ca^{2+}] changes in real time by observing the overlay of the FRET and CFP channels.
10. When the [Ca^{2+}] response is finished, collect ~30 more frames at the end of your time course then save the image stack.
11. In addition to the experimental material, perform the same experiment on a number of samples that are not transformed with YC 3.6 using the saved settings to *see* the amount and distribution of autofluorescence (*see* **Note 15**).

3.4 Ratio Analysis

A number of confocal systems have software to process the CFP and FRET image pairs into a ratio image. Alternatively, it is possible to analyze the data using an image-processing program such as ImageJ (NIH) or a commercial package such as MetaFlour (Molecular Devices) or iVision (BioVision Technologies).

1. Confocal data is usually collected with pixel intensity values ranging from 0 to 255 (8 bit) or from 0 to 4,095 (12 bit). Export the images to a file type able to contain this data depth that is usable by the image analysis program, e.g., TIFF (RGB full color). Export without using compression or scaling in order to preserve the raw intensity values.

2. Export such that the CFP image and FRET image are each contained in their own color channel (red, green or blue) in the TIFF for every frame in the time course as this easily keeps the paired CFP and FRET images together (*see* **Note 16**).
3. In the image analysis program, open the TIFF and separate the CFP and FRET color channels into individual image windows containing the raw data.
4. Quantify the amount of background signal in both the CFP and FRET raw images by measuring an average intensity in an area away from the sample (*see* **Note 17**).
5. Subtract out the average CFP background signal from every pixel in the CFP image, then repeat for the FRET image with the FRET background value. Typically, image analysis programs have a "pixel arithmetic" type of command able to do this arithmetic on each pixel of the image.
6. Generate a ratio image by dividing the FRET image by the CFP image. Again there will generally be a command within the image analysis language for this purpose (*see* **Note 18**).
7. The resulting ratio image will contain extreme values generated by arithmetic performed on noise and/or areas of low signal in the raw images; defining the significant regions of the image to eliminate these extreme values will give a more robust analysis (*see* **Note 19**).
8. It is now possible to determine an average ratio of the sample region of interest. Repeat the analysis for each image in your time course to generate data for a graph showing relative changes in [Ca^{2+}] over time.
9. It is often desirable to pseudocolor a ratio image to more easily *see* differences in ratios, and therefore differences in [Ca^{2+}]. This can be performed from a "pseudocolor" command within the image analysis software or by generating a scaled image which allows a more user-controlled mapping of color to ratio value (*see* **Note 20**).

3.5 Calibration of YC 3.6

Often YC 3.6 [Ca^{2+}] data is presented as a change in ratio over time; however, it is possible to approximate an absolute [Ca^{2+}] by calibration of the YC 3.6 ratio to known [Ca^{2+}]. Two approaches exist for calibration, in vivo and in vitro. In theory, the in vivo calibration is more accurate because any impact of the cytoplasmic milieu on YC 3.6 fluorescence is included, although in practice the permeabilization/buffering approach required for in vivo calibration has proven extremely difficult to apply to intact plants (*see* **Note 21**). Therefore, most analyses calibrate to an in vitro standard. An in vitro calibration can be performed by ratio analysis of the fluorescence from purified recombinant YC 3.6 in buffer solutions of known free Ca^{2+} levels.

It is also possible to use the published K_d for YC 3.6 (250 nM, ref. (8)) to estimate $[Ca^{2+}]$ according to the equation: $[Ca^{2+}] = (K_d[R - R_{min}]/[R_{max} - R])(F_{min}/F_{max})$, where R is the measured ratio value, R_{max} and R_{min} represent the maximum and minimum ratio values, and F_{min} and F_{max} represent the maximum and minimum FRET fluorescence obtainable from the sample, respectively (12). This approach will provide a rough estimate of the cytoplasmic Ca^{2+} level.

1. To obtain R_{min}, treat the plant with 20 μM of the Ca^{2+} ionophore Br-A23187 and 5 mM EGTA and record the minimum ratio obtained.
2. Replace the EGTA solution with 5 mM Ca^{2+} and record the maximum Ca^{2+} level obtained $[R_{max}]$. Alternatively, alcohol with 1 M Ca^{2+} has been used to drive Ca^{2+} levels to R_{max} (*see* **Note 22**).

4 Notes

1. Stably transformed lines for a range of Arabidopsis mutants can be readily generated via either crossing to available CaMV35S driven YC3.6 expressing lines (e.g., ref. (13)) or generated de novo via *Agrobacterium* dipping (ref. (14)).
2. Relying only on PCR screening or screening for antibiotic resistance can yield transformed lines that are too dim or show non-uniform expression and so are unsuitable for subsequent microscopy.
3. Make medium fresh each time, do not reheat solidified phytagel solution. Keep plates in the sterile hood with the lid closed to minimize drying of the gel while cooling. Prepared plates can be stacked and stored at 4 °C in a plastic sleeve until use.
4. For experiments requiring addition of a test solution, cut a window in the gel a couple of mm ahead of the root a few hours in advance of imaging and fill the window with sterile media lacking phytagel. Allow the root to grow into the window. Add subsequent treatments to the liquid in this window.
5. For example, too much excitation energy while monitoring the tip-focussed calcium gradient in a growing root hair can slow tip growth and dissipate the $[Ca^{2+}]$ gradient. Yet the cell remains viable (i.e., cytoplasmic streaming continues) and morphologically the disruption of growth is highly cryptic.
6. When selecting the range of wavelengths, keep in mind that bleed-through of the CFP signal into the FRET channel can be a problem and can be tested for by imaging plants expressing only CFP with the imaging parameters used for YC3.6.

7. When viewed in an overlay, changes in the signal intensity from each channel (and thus changes in [Ca^{2+}]) will be easily visualized in real time, with increases in [Ca^{2+}]$_{cyt}$ appearing as increased red coloration in the image.
8. Another potential problem to avoid is scanning an individual frame so slowly (e.g., tens of seconds) that the physiological state of the region is different at the start of the scan than it is at the end of the scan. Typically for Ca^{2+} imaging, image acquisition of around 1 s and up to 5 s between images will capture most of the fast Ca^{2+} kinetics seen in plants to date.
9. This is not necessary if both channels share the same pinhole.
10. Saturating pixels do not report the true signal intensity (i.e., that area could in fact be much brighter), therefore images containing saturated pixels are not amenable to quantification and analysis.
11. For example, if intensity values range from 0 (no signal) to 255 (saturating), then adjust the gain for the FRET channel so that most pixels in the region of interest have an intensity of 100–150.
12. When collecting data for quantitative analysis, use the confocal microscope after it has been on for at least an hour to ensure that room temperature is stable and all components are warmed up. Additionally, circadian rhythm has been shown to have an effect on cytosolic [Ca^{2+}] responses in plants (15, 16), therefore, conducting YC 3.6 imaging at the same time of day may yield more reproducible results.
13. Confirm that the interval between frames is longer than the time is takes to scan a single frame.
14. For example, cold shock alters cytoplasmic [Ca^{2+}] quickly in most plant cells and is an easy treatment to administer by pipetting 100 μl ice-cold buffer onto the root during imaging. This is a useful control to test the imaging parameters to be used.
15. If there is significant autofluorescence in the plant material (not uncommon), it may not be possible to use YC 3.6 for calcium measurements. Some confocal microscopes have software to perform "linear unmixing," a mathematical calculation performed on the image based on reference spectra with the goal of separating autofluorescence from fluorescent protein signal. However, because it infers the pixel intensity values, linear unmixing cannot be used in conjunction with quantitative ratio analysis without caution.
16. Take care to note which color channel of the TIFF corresponds to each fluorescence channel of the confocal microscope data.

17. Alternatively, measure the average intensity in an image taken with identical configuration and scanning parameters in a region on the microscope slide away from your sample. Either of these background signal measurements will include detector dark noise level in addition to anything non-sample in the optical path that may generate a signal.
18. To generate a ratio image with fractional values (e.g., the calculated ratios from YC 3.6 can cover the range from 0.3 to 2.3) it may be necessary to change the data type of your raw FRET and CFP images to floating point (the raw images are generally in "byte" or "short integer" formats that cannot record fractional numbers whereas "floating point" format can).
19. To define the significant regions of your image, many image analysis programs will be able to generate a thresholded mask that excludes low-fluorescence and/or noisy areas outside the region of interest. Use the raw image that is less bright, e.g., the CFP image, to generate this mask.
20. To convert the fractional ratio image data into a form usable to display in byte-form pseudocolor, first multiply each pixel in the ratio image by a scaling factor to bring the intensity values into byte range (0–255). For example, if your ratios range from 0.3 to 2.3 then multiplying by 100 will give values from 30 to 230. Once the ratio image is in byte format, then apply the color look-up table (CLUT) from within the image processing package; a rainbow CLUT is usually effective at showing differences and the accepted convention is to map low ratio values to the blue end of the spectrum and high ratios to the red.
21. In vivo calibration involves adding Ca^{2+} buffers to the medium to set a known extracellular Ca^{2+} level and permeabilizing the plasma membrane with ionophore (such as A23187 or ionomycin) or detergent so that internal and external Ca^{2+} levels are the same.
22. It is often very difficult to obtain R_{min} and R_{max} from tissues deep within the plant, likely due to difficulties in access for the Ca^{2+} buffer and ionophore to cells below the epidermis. 1 M $CaCl_2$ plus 10 % ethanol has been used as an alternative strategy to elevate Ca^{2+} throughout the plant (17).

Acknowledgments

This work was supported by grants from the USDA (2007-35304-18327), NSF (MCB 0641288) and NASA (NNX09AK80G) to S.G.

References

1. Dodd AN, Kudla J, Sanders D (2010) The language of calcium signaling. Annu Rev Plant Biol 61:593–620
2. Kudla J, Batistic O, Hashimoto K (2010) Calcium signals: the lead currency of plant information processing. Plant Cell 22: 541–563
3. Swanson SJ, Choi WG, Chanoca A et al (2011) In vivo imaging of Ca(2+), pH, and reactive oxygen species using fluorescent probes in plants. Annu Rev Plant Biol 62:273–297
4. Miyawaki A, Llopis J, Heim R et al (1997) Fluorescent indicators for Ca2+ based on green fluorescent proteins and calmodulin. Nature 388:882–887
5. Miyawaki A, Griesbeck O, Heim R et al (1999) Dynamic and quantitative Ca2+ measurements using improved cameleons. Proc Natl Acad Sci USA 96:2135–2140
6. Miwa H, Sun J, Oldroyd GE et al (2006) Analysis of calcium spiking using a cameleon calcium sensor reveals that nodulation gene expression is regulated by calcium spike number and the developmental status of the cell. Plant J 48:883–894
7. Allen GJ, Kwak JM, Chu SP et al (1999) Cameleon calcium indicator reports cytoplasmic calcium dynamics in *Arabidopsis* guard cells. Plant J 19:735–747
8. Nagai T, Yamada S, Tominaga T et al (2004) Expanded dynamic range of fluorescent indicators for Ca(2+) by circularly permuted yellow fluorescent proteins. Proc Natl Acad Sci USA 101:10554–10559
9. Horikawa K, Yamada Y, Matsuda T et al (2010) Spontaneous network activity visualized by ultrasensitive Ca(2+) indicators, yellow Cameleon-Nano. Nat Methods 7:729–732
10. Garaschuk O, Griesbeck O, Konnerth A (2007) Troponin C-based biosensors: a new family of genetically encoded indicators for in vivo calcium imaging in the nervous system. Cell Calcium 42:351–361
11. Twell D, Klein TM, Fromm ME et al (1989) Transient expression of chimeric genes delivered into pollen by microprojectile bombardment. Plant Physiol 91:1270–1274
12. Grynkiewicz G, Poenie M, Tsien RY (1985) A new generation of Ca2+ indicators with greatly improved fluorescence properties. J Biol Chem 260:3440–3450
13. Monshausen GB, Messerli MA, Gilroy S (2008) Imaging of the Yellow Cameleon 3.6 indicator reveals that elevations in cytosolic Ca2+ follow oscillating increases in growth in root hairs of *Arabidopsis.* Plant Physiol 147:1690–1698
14. Clough SJ, Bent AF (1998) Floral dip: a simplified method for Agrobacterium-mediated transformation of *Arabidopsis thaliana.* Plant J 16:735–743
15. Dodd AN, Jakobsen MK, Baker AJ et al (2006) Time of day modulates low-temperature Ca signals in Arabidopsis. Plant J 48:962–973
16. Love J, Dodd AN, Webb AA (2004) Circadian and diurnal calcium oscillations encode photoperiodic information in *Arabidopsis.* Plant Cell 16:956–966
17. Knight H, Trewavas AJ, Knight MR (1996) Cold calcium signaling in *Arabidopsis* involves two cellular pools and a change in calcium signature after acclimation. Plant Cell 8:489–503

INDEX

Teun Munnik and Ingo Heilmann (eds.), *Plant Lipid Signaling Protocols*, Methods in Molecular Biology, vol. 1009,
DOI 10.1007/978-1-62703-401-2,

L

M

N

O

P

R

S

T

Y

Printed in the United States
by Baker & Taylor Publisher Services